Rilem Symposium 1975

FIBRE REINFORCED CEMENT AND CONCRETE

Edited by

Adam Neville

Professor and Head of Department, Department of Civil Engineering, The University of Leeds

Chairman of the RILEM Permanent Commission on Concrete

in association with

D.J. Hannant, *University of Surrey*
A.J. Majumdar, *Building Research Establishment*
C.D. Pomeroy, *Cement and Concrete Association*
R.N. Swamy, *University of Sheffield*

THE CONSTRUCTION PRESS LTD

RILEM: International Union of Testing and Research Laboratories for Materials and Structures

1975 Symposium; 14th to 17th September 1975.

ISBN 0 904406 10 5
Copyright 1975   RILEM

Published in 1975 by:

The Construction Press Ltd.,
Lunesdale House,
Hornby,
Lancaster, LA2 8NB.

All rights reserved. This book, or parts thereof, must not be reproduced in any form without the permission of the publishers.

Printed in Great Britain by
The Blackburn Times Press.

FIBRE REINFORCED CEMENT AND CONCRETE

CIMENT ET BÉTON RENFORCÉS DE FIBRES

ISBN 0 904406 10 5

Contents

This volume contains all the formal papers accepted for presentation at the 1975 RILEM Symposium. Supplementary contributions, the discussion and a combined index will be published in a separate volume which will appear as soon as possible after the Symposium is concluded.

Section 1   Uses of Fibre Concrete

1.1 **Applications of fibre concrete** — 3
D.R. Lankard

Section 2   Applications of Fibre Cement

2.1 **Applications of fibre cement** — 23
J.F. Ryder

Section 3   Theories of Fibre Concrete and Fibre Cement

3.1 **Theories of fibre cement and fibre concrete** — 39
A.S. Argon and W.J. Shack
3.2 Fibre pull-out in multiply-cracked discontinuous fibre composites — 55
D.K. Hale
3.3 Fibre spacing and specific fibre surface — 69
H. Krenchel
3.4 Mechanics of glass fibre reinforced cement — 81
N.G. Nair
3.5 Strength of concrete beams with aligned or random steel micro-reinforcement — 95
V.S. Parameswaran and K. Rajagopalan
3.6 Some results of investigations on steel fibre reinforced concrete — 105
B. Schnütgen

Section 4   Properties and Testing of Steel Fibre Concrete

4.1 **Properties and testing of steel fibre concrete** — 119
N.M. Dehousse
4.2 Qualitative testing of fibre reinforced centrifugated concrete — 137
E.F.P. Burnett
4.3 Reduction of shrinkage cracking in reinforced concrete due to the inclusion of steel fibres, R.H. Elvery and M.A. Samari — 149
4.4 Durability of steel fibre concrete — 159
D.J. Hannant and J. Edgington
4.5 Bond studies on oriented and aligned steel fibres — 171
A.E. Naaman and S.P. Shah
4.6 Fibre concrete for a folded plate structure — 179
J.P. Rammant and M. Van Laethem
4.7 Flexural behaviour of fibre concrete with conventional steel reinforcement — 187
R.N. Swamy and K.A. Al-Noori
4.8 Some properties of high workability steel fibre concrete — 197
R.N. Swamy and H. Stavrides
4.9 Full scale fibre concrete beam tests — 209
G.R. Williamson and L.I. Knab

Section 5 Properties and Testing of Concrete containing Fibres other than Steel

5.1 **Properties and testing of concrete containing fibres other than steel**
J.J. Zonsveld — 217

5.2 Contribution to the study of the mechanical behaviour of concrete reinforced with polypropylene fibres
J. Dardare — 227

5.3 Strength and deformation properties of concrete reinforced with randomly spaced steel and basalt fibres
K. Komlos — 237

5.4 The effects of fibre reinforcements on lightweight aggregate concretes
A.G.B. Ritchie and O.A. Al Kayyali — 247

Section 6 Properties and Testing of Asbestos Fibre Cement

6.1 **Properties and testing of asbestos fibre cement**
H. G. Klos — 259

6.2 Investigation of the "corrosion" of asbestos fibres in asbestos cement sheets weathered for long times
L. Opoczky and L. Pentek — 269

Section 7 Properties and Testing of Cement containing Fibres other than Asbestos

7.1 **Properties of fibre cement composites**
A.J. Majumdar — 279

7.2 Validity of flexural strength reduction as an indication of alkali attack on glass in fibre reinforced cement composites
E.B. Cohen — 315

7.3 Microstructural features in glass fibre reinforced cement composites
A.C. Jaras and K.L. Litherland — 327

7.4 Can asbestos be completely replaced one day?
H. Krenchel and O. Hejgaard — 335

7.5 Tensile stress-strain behaviour of glass fibre reinforced cement composites
D.R. Oakley and B.A. Proctor — 347

7.6 Structural properties of carbon fibre reinforced cement
S. Sarkar and M.B. Bailey — 361

Section 8 Applications

8.1 **Inflation forming of steel fibre reinforced concrete domes**
G.B. Batson — 375

8.2 Full-scale trials of a wire-fibre-reinforced concrete overlay on a motorway
J. Grejory, J.W. Galloway and K.D. Raithby — 383

8.3 The use of fibre reinforced concrete in hydraulic structures and marine environments — 395

8.4 Steel fibre reinforced concrete pavement
C.D. Johnston — 409

8.5 Investigation of fibre-reinforced material in the USSR
B.A. Krylov and V.P. Trambovetsky — 419

8.6 Effective application of steel fibre reinforced concrete
K. Nishioka, N. Kakimi, S. Yamakawa and K. Shirakawa — 425

8.7 Tailoring fibre concretes to special requirements
C.D. Pomeroy — 435

8.8 The design of glass fibre reinforced cement cladding panels
A.J.M. Soane and J.R. Williams — 445

8.9 Fort Hood fibre concrete overlay
G.R. Williamson — 453

Colloque Rilem 1975

CIMENT ET BÉTON RENFORCÉS DE FIBRES

Edité par

Adam Neville

*Professeur et Chef de Département
Département du Génie Civil
Université de Leeds*

*Président de la Commission permanente
du Béton de la RILEM*

en association avec

D.J. Hannant,   *Université de Surrey*
A.J. Majumdar, *Building Research Establishment*
C.D. Pomeroy,   *Cement and Concrete Association*
R.N. Swamy,   *Université de Sheffield*

 THE CONSTRUCTION PRESS LTD

RILEM: Réunion Internationale des Laboratoires d'Essais et de
Recherches sur les Matériaux et les Constructions

Colloque 1975; 14 à 17 septembre 1975.

ISBN 0 904406 10 5
Copyright 1975   RILEM

Publié en 1975 par:
The Construction Press Ltd.,
Lunesdale House,
Hornby,
Lancaster LA2 8NB.

Tous droits réservés. Ce livre ne peut
être reproduit, en tout ou en partie,
sous quelque forme que ce soit, sans
la permission des éditeurs.

Imprimé en Grande Bretagne par
The Blackburn Times Press.

Sommaire

Cet volume contient tous les rapports formels acceptés pour presentation à la Colloque RILEM 1975. Les contributions supplémentaires, la discussion et un index combiné apparaîtraient dans un deuxième volume qui serait publié aussitôt que possible après la fin da la Colloque.

Section 1   L'Emploi du Béton de Fibres

1.1   **L'emploi du béton de fibres** 3
       D.R. Lankard

Section 2   L'Emploi du Ciment renforcé de Fibres

2.1   **L'emploi du ciment renforcé di fibres** 23
       J.F. Ryder

Section 3   Théories du Béton et du Ciment de Fibres

3.1   **Théories du béton et du ciment de fibres** 39
       A.S. Argon et W.J. Shack
3.2   L'arrachement des fibres dans des composés à fibres discontinues, 55
       montrant des fissurations multiples
       D.K. Hale
3.3   L'esplacement des fibres et leur surfacé spécifique 69
       H. Krenchel
3.4   La mécanique du béton renforcé de fibres de verre 81
       N.G. Nair
3.5   La résistance de poutres de béton avec micro-armatures d'acier 95
       alignées ou non
       V.S. Parameswaran et K. Rajagopalan
3.6   Quelques résultats d'études sur le béton renforcé de fibres d'acier 105
       B. Schnütgen

*Section 4   Méthodes d'Essais et Caractéristiques Mécaniques des Bétons
              armés de Fibres Métalliques*

4.1   **Méthodes d'essais et caractéristiques mécaniques des bétons armés** 119
       de fibres métalliques
       N.M. Dehousse
4.2   Essais qualitatifs sur le béton de fibres centrifugé 137
       E.F.P. Burnett
4.3   Réduction de la fissuration au retrait du béton causée par l'inclusion 149
       de fibres d'acier
       R.H. Elvery et M.A. Samari
4.4   La durabilité du béton de fibres d'acier 159
       D.J. Hannant et J. Edgington
4.5   Études d'adhérence sur les fibres d'acier alignées et orientées 171
       A.E. Naaman et S.P. Shah
4.6   Béton de fibres pour une structure à plaque repliée 179
       J.P. Rammant et M. Van Laethem

4.7 Le comportement en flexion du béton de fibres avec armature traditionnelle en acier 187
R.N. Swamy et K.A. Al-Noori

4.8 Quelques propriétés des bétons de fibres d'acier à grande ouvrabilité 197
R.N. Swamy et H. Stavrides

4.9 Essais à l'échelle actuelle sur poutres de béton de fibres 209
G.R. Williamson et L.I. Knab

Section 5 Les Propriétés et les Essais des Bétons contenant des Fibres autres que les Fibres d'Acier

5.1 **Les propriétés et les essais des bétons contenant des fibres autres que les fibres d'acier** 217
J.J. Zonsveld

5.2 Contribution à l'étude du comportement mécanique des bétons renforcés avec des fibres de polypropylène 227
J. Dardare

5.3 La résistance et les caractéristiques de déformation des bétons renforcés de fibres d'acier et de basalte réparties au hasard 237
K. Komloš

5.4 Les effets des armatures de fibres sur les bétons d'agrégats légers 247
A.G.B. Ritchie et O.A. Kayyali

Section 6 Les Propriétés et les Essais du Ciment à Fibres d'Amiante

6.1 **Les propriétés et les essais du ciment à fibres d'amiante** 259
H.G. Klos

6.2 Étude de la "corrosion" des fibres d'amiante dans les plaques d'amiante-ciment après vieillissement à long terme 269
L. Opoczky et L. Pentek

Section 7 Les Propriétés et Les Essais des Cements contenant des Fibres autres que Celles d'Amiante

7.1 **Les propriétés et les essais des cements contenant des fibres autres que celles d'amiante** 279
A.J. Majumdar

7.2 Validité de la réduction de la résistance à la flexion en tant qu'indication d'une attaque des alkalis sur le verre dans les composés de ciment de fibres 315
E.B. Cohen et S. Diamond

7.3 Caractéristiques de la microstructure dans les composés de ciment de fibres de verre 327
A.C. Jaras et K.L. Litherland

7.4 Pourra-t-on jamais jour remplacer l'amiante-ciment? 335
H. Krenchel et O. Hejgaard

7.5 Comportement tension-déformation en flexion des composés de ciment renforcé de fibres de verres 347
D.R. Oakley et B.A. Proctor

7.6 Caractéristiques de stabilité du ciment armé de fibres de carbone 361
S. Sarkar et M.B. Bailey

Section 8 Applications

8.1 **Domes en béton de fibres d'acier moulé sur coffrage gonflant** 375
G.B. Batson, D.J. Naus et G.R. Williamson

8.2 Essais à l'échelle actuelle sur un revêtement d'autoroute en béton renforcé de fibres de fils d'acier 383
J. Gregory, J.W. Galloway et K.D. Raithby

8.3	L'emploi de béton de fibres dans les constructions hydrauliques et dans une atmosphère marine G.C. Hoff	395
8.4	Revêtement de route en beton de fibres C.D. Johnson	409
8.5	Étude des matériaux renforcés de fibres d'acier en URSS B.A. Krylov et V.P. Trambovetsky	419
8.6	L'application efficace du béton renforcé de fibres d'acier K. Nishioka, N. Kakimi, S. Yamakawa et K. Shirakawa	425
8.7	L'adaptation du béton de fibres à des exigences particulières C.D. Pomeroy et J.H. Brown	435
8.8	La conception de panneaux de bardage en ciment renforcé de fibres de verre A.J.M. Soane et J.R. Williams	445
8.9	Revêtement fibreux à Fort Hood G.R. Williamson	453

Section 1

Uses of Fibre Concrete

1.1 Opening Paper: Fibre concrete applications

D R Lankard
*Battelle's Columbus Laboratories,
Columbus, Ohio, USA*

Summary

Serious efforts to develop commercial applications for fibre reinforced (FR) concrete are just a few years along. Application areas in which significant field trials have taken place in the US, England, and Western Europe include overlays for bridge decks and pavements (highway and airfield), mining and tunneling applications, slope stabilization, refractory applications, concrete repair, industrial floors, and precast concrete products. Experience has been gained with steel, glass, and polypropylene fibres as reinforcement for concrete, although most of the field work to date has centered on steel fibre concrete.

The initial results of field experiences with FR concretes have been for the most part encouraging. There are, however, areas where additional work is needed if the material is to gain larger and more diverse acceptance as a construction and building material. Techniques and equipment for efficient and rapid mixing of large quantities of FR concrete need refining. Improved performance of FR concrete can be expected as ways are found to place the material with low water contents and as better fibre/matrix bonds are developed.

An increase in the use of FR concrete is expected to occur as the available information base concerning design procedures, preparation, and properties is expanded.

Résumé

Les efforts sérieux pour développer des applications commerciales du béton armé de fibres ne datent que de quelques années. Les domaines où d'importants essais sur le terrain ont été entrepris aux Etats-Unis, en Angleterre et en Europe de l'Ouest comprennent les recouvrements de tabliers de ponts et de pavements (routes et terrains d'aviation), les applications dans l'exploitation minière et le percement de tunnels, la stabilisation de pentes, les applications aux réfractaires, la réparation du béton, les farines industrielles, et les produits de béton prémoulé. Une certaine expérience a été acquise avec l'acier, le verre et les fibres de polypropylène comme armature du béton, bien que la majeure partie du travail accompli jusqu'ici sur le terrain ait été concentrée sur le béton à fibres d'acier.

La plupart des résultats initiaux d'essais sur le terrain avec le béton armé de fibres ont été encourageants. En certains domaines cependant un travail supplémentaire est nécessaire, si le matériau doit devenir plus largement répandu dans la construction. Les techniques et l'appareillage pour le mélange efficace et rapide de grande quantités de béton armé de fibres ont besoin d'être raffinés. On peut s'attendre à une amélioration du rendement de ce béton à mesure que l'on trouvera des moyens de placer le matériau faible en eau, et que de meilleures liaisons entre les fibres et la matrice seront développées.

On peut espérer une augmentation de l'usage du béton armé de fibres au fur et à mesure que s'étendront les données fondamentales disponibles concernant le mode de production, les procédés, la préparation et les propriétés.

INTRODUCTION

Fibre reinforcement of cement and concrete, the subject of the current Symposium, is not really a new idea. The use of straw in bricks and hair in mortar predates the use of conventional Portland cement concrete. Nature, too, has provided man with fibre reinforced construction materials in the form of wood and bamboo. However, excluding the use of asbestos, it has been only within the last 15 years or so that serious consideration has been given to the use of synthetic fibres to improve the performance of conventional, moldable construction materials (which include gypsum plaster, cement paste, and concrete*). Dr. Ryder has previously reviewed the situation concerning the use and benefit of fibres in cement-based materials. The use of fibres in concretes requires additional considerations both from the point of view of the mixing and placing of the concrete and that of the applications for the material in the building and construction industries.

Fibre additions to concrete offer a convenient and practical means of achieving improvements in many of the engineering properties of the material such as fracture toughness, fatigue resistance, impact resistance and flexural strength. The concept of providing the reinforcement as an integral part of the fresh concrete mass can also provide advantages in terms of the fabrication of products and components. These potential advantages and the novelty of the concept have stimulated what is now considerable worldwide interest in the use of fibre reinforcement for concrete.

The fibres that are currently being investigated as reinforcement for concrete include steel (carbon and stainless), glass (alkali-resistant) and polypropylene. Other fibres such as nylon, polyethylene, rayon, and E-glass have been investigated in the past but have been ruled out from serious consideration due either to high cost, low effectiveness, or inadequate resistance to the alkaline cement environment (in the case of Portland cement).

In general, the concretes used with fibre reinforcement differ from conventional concretes in having a higher cement content (350 to 500 kg/m^3 [600 to 800 lb/yd^3]), a lower coarse aggregate content (350 to 750 kg/m^3 [600 to 1250 lb/yd^3]), and a smaller size of coarse aggregate (10 mm [3/8] maximum size).

In most of the applications considered to date, fibre contents have ranged from about 0.3 to 2.0 volume per cent.

A review of the titles of the papers presented at the September, 1973, ACI International Symposium on Fibre Reinforced Concrete (1) provides an insight into the relative research and development interest and effort with respect to both fibre and the matrix type. Of the papers presented, 15 dealt with steel fibres, 8 with glass, 1 with carbon, and the remainder with general fibrous reinforcement concepts. With respect to the type of matrix, 22 dealt with concrete, 4 with cement, and the remainder with general considerations of brittle matrix materials. It is quite likely that the greater level of activity with steel fibre reinforcement is due in part to the fact that steel fibres have been available in reasonably large quantities for a longer time than any of the other fibre types.

The results of research work on steel FR concrete were first reported in the early 1960's (2,3). Serious efforts to study the material in commercial products and applications began in earnest in 1971 in the USA. Similar efforts in England and Western Europe shortly followed this lead. Most recently, interest in applications for steel FR concrete has become apparent in Canada, Japan, Australia, and New Zealand. Steel fibres are available in commercial quantities in the USA, England, Western Europe, Australia, and Japan.

Alkali–resistant glass fibres are now being produced in the USA, England, and possibly Japan. The primary interest in glass fibres in concrete is apparently centered in the USA at the present time, where their use has only recently been explored on a field-scale basis.

---

*In this paper, concrete refers to any mixture of hydraulic cement, aggregate, and water.

Polypropylene fibre reinforced concrete emerged on the product scene several years ago in England but the applications for this material are still somewhat limited.

The existing and potential applications for fibre reinforced (FR) concrete can be conveniently placed into two main categories,

(1) Mass concrete applications—in which the final product is produced at the construction or building site, and

(2) Precast concrete applications—in which the product is produced in a plant and shipped elsewhere for use.

In the US, it is in the mass concrete area that fibre concrete has had the greatest success in attracting the interest and active participation of potential users of the material. In England and Western Europe, the interest may be about equally divided and the overall activity is less than that in the US.

Despite the fact that FR concrete has emerged beyond laboratory status only within the last 3 to 4 years, considerable knowledge and experience has already been gained. What has happened during these few years of development work? What has been learned? And, perhaps most important, how may this learning be applied to enhance the application potential of FR concrete?

MASS CONCRETE APPLICATIONS

The mass concrete application areas studied to date for FR concrete are identified in Table 1. Table 1 also shows the type of fibres that have been used in concretes for these applications in the field and the countries where substantial field work has been undertaken.

Table 1   **Mass concrete application areas for FR concrete for which field trials have been performed**

Application area	Fibre types used in concretes evaluated in the field	Countries in which significant field work has been done
Bridge decks, overlays, and construction.	Steel	U.S.A.
Highway, street, and airfield pavement overlays and construction.	Steel, Glass	U.S.A., England, Canada
New pavement construction.	Steel	U.S.A.
Mass concrete maintenance and repairs (dams, slabs, pavements, bridges, culverts, etc.).	Steel	U.S.A., England, Western Europe
Mining and tunneling.	Steel	U.S.A., England
Rock slope stabilization.	Steel	U.S.A.
Industrial floors.	Steel, Glass	U.S.A., Italy, England
Refractory applications.	Steel	U.S.A.

In the USA, it is estimated that at least 70 per cent of the total field work effort has involved the construction, overlay, or repair of bridge decks and highway, street, and airfield pavements and the repair of other mass concrete members such as dams and culverts.

Bridge deck overlays and construction

Eight FR concrete bridge deck surfacings have been constructed in the US since 1972 (4,5). In all of the projects, the fibre used was steel* with quantities ranging from 0.75 to 1.5 volume percent.

Six of the projects were overlays of existing decks (concrete and wood decks), with overlay thickness ranging from 51 to 127 mm (2 to 5 in.), with an average of 76 mm (3 in.). Four of the overlays were fully bonded to the old deck (using epoxy and cement paste bonding agents): one of the overlays was partially bonded and one was unbonded (double thickness of polyethylene sheet between the overlay and the deck).

All of the fully and partially bonded overlays have developed some cracks over their service life which in most cases have remained very tight and have not adversely affected the riding quality of the deck. The 76 mm (3 in.) thick unbonded overlay has remained virtually crack free after 3 years of 30,000 vehicle per day trafficking. From the results obtained to date, it has not been possible to identify any significant influence of fibre size or quantity on the performance of the bridge deck overlays.

During June, 1973, the State of Virginia constructed six new bridge decks on two bridges using the two coarse, bonded construction technique. This technique involves the casting of an initial lift to a level near the top reinforcement, followed by the casting of a high quality, 51 mm (2 inch) thick wearing course. Three wearing courses were included in the project including high quality PCC, steel FR concrete, and latex modified concrete (LMC). Two test sections 16 x 12 m (52 x 38 ft) were constructed with each wearing course. The performance of the two course bonded decks will be compared with that of conventional decks (single lift construction) built at the same time on adjacent bridges.

Highway, street, and airfield pavement overlays

The interest in the US in fibre concrete as an overlay material for the rehabilitation of pavements is currently very high. This situation is due in part to ever-increasing maintenance problems and the fact that traditional overlay materials do not in many cases provide a satisfactory solution to the problem. FR concrete, by virtue of its superior strength and fatigue properties and its ability to be placed in relatively thin sections appears to offer advantages as an overlay material. A number of significant projects have been completed since 1972 involving overlays of highway pavements, residential and urban streets, airfield runways and taxiways, and parking areas (4, 6-17). Several of the experiments have been designed to comparatively study such variables as concrete mix design, fibre type and quantity, joint spacing and design, overlay thickness, and the type of bonding to the old surface (4, 11).

Highway overlays have been completed in Michigan, Iowa, and Minnesota. The overlay of a country road in Iowa in 1973 is the largest highway overlay project yet done (4.8 km [3 miles] long) (4, 10, 11).

Figure 1 shows the construction of this 6.7 m (22 ft) wide overlay which was accomplished using a slip-form paving machine.

The overlays in Iowa (4, 10, 11) and Michigan (4, 8, 9) used steel fibre concrete only (0.75 to 1.5 volume percent) while the overlay in Minnesota (in the Fall of 1974) marked the first time that steel and glass FR concrete were used on the same project. The highway overlays have varied from 51 to 102 mm (2 to 4 in.) in thickness and have included fully bonded, partially bonded, and unbonded sections. Partially bonded overlays of two residential streets with steel fibre concrete were constructed during 1972 in Iowa (4). Sixty meter (200 ft) long sections of the streets were overlayed with 64 to 102 mm (2.5 to 4 in.) of concretes containing 1.3 to 1.9 volume per cent steel fibre.

The first major experiment to assess the potential of steel FR concrete as an airfield pavement overlay material was carried out at Tampa (Florida) International Airport in February, 1972 (13). This placement followed a successful research study at the US

---

* 0.63 x 63.5 mm (0.025 x 2.5 in.)
0.25 x 0.56 x 25.4 mm (0.01 x 0.022 x 1.0 in.)
0.25 x 12.7 mm (0.010 x 0.5 in.)

Figure 1 Construction of a 4.8 km (3 mile) steel FR concrete overlay on a country highway in Iowa (U.S.A.). A slip-form paver (6.7 m [22ft] wide) was used.

Army Corps of Engineers' Waterways Experiment Station, Vicksburg, Mississippi, in which steel FR concrete outperformed plain and conventionally reinforced concrete overlays (14). The 53 m (175 ft) long overlays at Tampa were constructed on taxiway areas at thicknesses of 102 and 152 mm (4 and 6 in.). In September, 1972, a 51 to 102 mm (2 to 4 in.) thick steel fibre concrete overlay was placed on a 23 x 27 m (75 x 90 ft) area of the main taxiway at the airport in Cedar Rapids, Iowa (4). Two sizes of steel fibres were used in this project (0.41 x 25.4 mm [0.016 x 1.0 in.] and 0.63 x 63.5 mm [0.025 x 2.5 in.]). The most significant airfield overlay experiment completed to date in the US involved the construction in May, 1974, of an unbonded 127 mm (5 in.) thick steel FR concrete overlay on a 53 x 37 m (175 x 120 ft) section at the end of a main runway at JFK International Airport in New York. An adjacent 216 mm (8.5 in.) thick slab on grade (37 x 15 m [120 x 50 ft]) was constructed at the same time in the taxiway turn area. Concrete for the project contained 446 kg of cement and 105 kg of 0.63 x 63.5 mm fibre per cu m (752 lb cement and 175 lb of fibre per yd^3). Keyed and doweled construction joints were used in both the overlay and the slab on grade. The construction of the 7.6 m (25 ft) wide steel FR concrete overlay strips on a double thickness of polyethylene sheet is shown in Figure 2. The performance to date of the overlay and the slab on grade has been excellent.

Steel fibre concrete has also been used in several projects involving the overlay rehabilitation of parking areas. This work includes the largest fibre concrete project completed to date (22,572 m^2 [27 000 yd^2]) as an overlay to an Army tank parking area in Texas (15). A 102 mm (4 in.) thick steel FR concrete (1.5 volume percent of 0.25 x 12.7 mm [0.01 x 0.5 in.] fibre) was placed directly on an existing 127 mm (5 in.) thick asphalt pavement.

With a few exceptions at both extremes, the performance to date of the steel FR concrete overlays has been satisfactory and the user attitude has been one of qualified satisfaction based on the yet unknown long-term performance. It is still too early to identify the optimum type and amount of fibre required for satisfactory overlay performance. However, it does appear as if the amount of fibre and the overlay thickness required may depend on the service conditions (quality of existing pavement, traffic, etc.). It also appears that some form of debonding will be required for fibre

Figure 2. Construction of an unbonded, steel FR concrete overlay (127 mm [5 in.] thick) of a runway at Kennedy International Airport in New York in 1974.

Figure 3 Construction of a partially-bonded slip-formed glass FR concrete pavement overlay (76 mm [3 in.] thick) in St. Paul, Minnesota, in 1974.

concrete overlays to eliminate non-load related cracking expecially when overlaying existing PCC pavement.

The limited field work with glass FR concretes (16) precludes a comparison with steel FR concretes as regards the pavement overlay applications. However, recent tests on the former showed 203 mm (8 in.) thick glass FR concrete slabs outperforming a 203 mm (8 in.) thick steel bar reinforced slab under heavy loading conditions (17). The construction of a slip-formed, 76 mm (3 in.) thick glass FR concrete overlay on an urban highway in St. Paul, Minnesota, in 1974 is shown in Figure 3. The handling and placing characteristics of glass FR concrete are not unlike those of steel FR concrete.

Pavement overlay trials in England (on the M10 Motorway) will be discussed by other participants in the Symposium.

New pavement construction

Interest in fibre concrete as a material for new pavement construction has been limited. The only project in this area involved the placement, in 1971, of a 152 m (500 ft) section of steel FR concrete as the entrance to a truck weighing station on the interstate highway system in Ohio (4). One hundred and two mm (4 in.) of steel FR concrete (2 volume percent) was placed directly on a 102 mm (4 in.) asphalt concrete grading course which was placed on grade. The asphalt concrete course was tapered at each end to a feather edge and doweled expansion joints were placed at the transition between the steel FR concrete and the conventional PCC which was 203 mm (8 in.) thick.

Although 3 transverse cracks developed prior to trafficking (restrained drying shrinkage and thermal movement), the remainder of the pavement has held up extremely well under severe service conditions (rapidly decelerating trucks) over a two year traffic period.

Mass concrete maintenance and repairs

The use of FR concrete to repair deteriorated portions of concrete slabs, pavements, culverts, and other members was one of the first application areas to be investigated. Steel FR concretes have been used in a variety of repair situations in the US, England, and Western Europe since about 1969. These include

(1) Spalled areas in concrete pavements

(2) Sunken areas in tunnel decks

(3) Slab replacement in airfield taxiway, storage, and refueling areas

(4) Spalled areas along longitudinal key joints in an airfield runway

(5) Slab replacement in areas serving as access from highway to truck loading areas

(6) Deteriorated bus lane ramps

(7) Bus stopping areas on city streets

(8) Deteriorated curbing (spalled)

(9) Deteriorated sidewalks (scaled and spalled)

(10) Repair of the wearing surface on dam spillways

(11) Repair of water aqueducts.

In the case of the repairs to the runway key joints, the steel FR concrete showed performance equal to epoxy concretes over a two year observation period. Precast slabs of steel FR concrete (51 mm [2 in.] thick) have been used for the rapid repair of unserviceable areas in tunnel decks and highway pavements in the US.

In general, the performance of steel fibre concrete repairs has been excellent. The good freeze/thaw durability and spalling resistance of steel FR concrete are factors that have no doubt contributed to this situation.

Mining and tunneling

Interest in steel FR concrete for various mining and tunneling applications has been apparent in the US and England since about 1972 (18-21). The applications include,

(1) Coatings for the prevention of flame propagation on urethane foam surfaces

(2) Fan intake areas (settings)

(3) Building stopping sealings

(4) Roof and sidewall stabilization (to prevent spalling and air slacking)

(5) Fireproofing transformer and pump stations.

In these applications, the steel fibre concrete is applied by the shotcrete process. A recent estimate gave the in place cost of the material (25 mm [1 in.] thick) at $0.67 per board foot (18).

Rock slope stabilization

Initial field experiments with steel FR concrete as a material for rock slope stabilization have been very encouraging (22). Steel FR concrete (containing 1.5

Figure 4   Stabilization of an exposed rock cut using steel FR concrete applied by the shotcrete process (State of Washington, U.S.A.).

volume percent of 0.25 x 12.7 mm [0.01 x 0.5 in.] fibres) was found by the US Army Corps of Engineers to be more economical than the conventional mesh-reinforced shotcrete because the labour of pinning the mesh to the slopes could be eliminated and the steel FR concrete could be applied 76 to 127 mm (3 to 5 in.) thick versus the 178 to 228 mm (7 to 9 in.) usually required to fill behind and cover the mesh.

The application of the steel FR concrete by the shotcrete process to a basalt rock cut in the State of Washington (USA) is shown in Figure 4.

A description of the use of polypropylene fibre concrete (applied by the shotcrete technique) as a river wall stabilization material was reported in 1970 (23). However, little substantive information is available relative to its performance or anticipated advantages.

Industrial floors

Field projects using steel fibre concrete for the construction of warehouse and factory floors have been completed in the USA, England, and Italy (24, 25). The potential advantages of steel FR concrete in this application include higher allowable working stresses, reduced volume of concrete per unit floor area, and a reduction in construction and maintenance costs. The construction of a 307 m^2 (3300 sq ft) floor slab in a residential house with glass FR concrete was described recently (26). Concrete for the 102 mm (4 in.) thick slab utilized 25 mm (1 in.) long fibres in amounts of 1.5 to 2.0 volume percent. The slab (and house) was constructed as an experimental project to demonstrate materials that are beyond the experimental stage of development but not yet in common use.

Refractory applications

Considerations of the use of steel fibres as a reinforcement for refractory concretes dates to 1970 (27). Both stainless and carbon steel fibres have been used in refractory concretes to provide improved load bearing characteristics, thermal and mechanical stress resistance, and thermal shock resistance (27, 28). Applications in which the fibre reinforced concretes have shown significant improvements in performance and service life relative to conventional refractories include crucible furnace covers (non-ferrous metal industries); furnace doors, lintels, hearths, and arches (coke ovens, forge furnaces, reheat furnaces, carbon ring furnaces, aluminum reverbatory furnaces); ladle degasser covers; skid rails, piers, and guides; soaking pit covers and coping; and precast shapes used in iron desulphurizing systems (stirrers and plunging bells).

The key to the successful use of steel fibre reinforcement of refractory concretes lies in the conditions of temperature exposure. Good performance can be expected where the heating is one sided (with hot face temperatures as high as 2800°F [1538°C]) or where the refractory is subjected to rapid temperature excursions, e.g., a short duration plunge of the refractory into molten steel.

PRECAST CONCRETE APPLICATIONS

Overall, the field work with FR concrete in the precast concrete products areas has been less intensive than in the mass concrete area. This is particularly true in the US and is probably due, as much as anything, to the lack of a concerted marketing effort. Despite some familiarity with FR concrete (through publications and seminars), precasters generally have not taken the initiative in exploring how FR concrete can improve profit margins on their products. This situation may be due to the fact that precasters, relative to the mass concrete users, have fewer major problems with the performance of their products. Concrete pipe, piles, structural elements, etc. all perform adequately with conventional steel reinforcement. Improvements in performance will not necessarily mean that the salability of their product will be enhanced, especially if the initial product cost is greater. Based upon what is known of the properties of FR concretes, the potential advantages to the manufacturer appear to be,

(1) Decreased materials handling costs

(2) Increased production rate (increased number of units from the same concrete batch size assuming a reduced section thickness)

(3) Decreased breakage loss in the yard and in transit

(4) Decreased transportation costs (increased number of units per shipment assuming a reduced unit volume).

Potential advantages to the user of an FR concrete precast product include,

(1) Less weight for more economical handling

(2) More durable in service

(3) Less maintenance in service.

Admittedly, this is a simplified analysis of what is, in practice, a complex situation.

Despite the more modest development effort in the precast area, a number of FR concrete precast products have been produced and tested and in a few cases sold commercially. The most significant work is described in Table 2. The list is not complete but does identify most of the products that have been studied in more than a cursory manner.

Concrete pipe

The expected benefits of steel fibres in concrete for pipes include improved performance and thinner wall sections (29). Fibre reinforcement of pipes (and other shapes) also has the advantage of strengthening the entire volume of concrete including the extreme edges, and the possibility of accidental damage during handling is thus reduced. Concrete pipes have been produced from steel FR concrete using a number of techniques including packerhead, centrifugal casting, wet casting, and dry casting (29-31). Pipes up to 1520 mm (60 in.) diameter have been produced (30). Pipes have also been produced with glass FR concrete using the packerhead process (16).

Dolosse

Large (39 metric tonnes), steel FR concrete dolosse* were fabricated in 1972 and are currently undergoing field testing at a California site along with conventional concrete dolosse (32-34).

Utility poles

A program to investigate the suitability of steel FR concrete in centrifugally spun precast products was described recently (35). Experimental utility poles with O.D.'s up

---

*Breakwaters in the form of double ended tees whose opposite cross members are oriented at right angles to each other and are connected by the stem.

Table 2 *INTRO*   **Precast concrete product applications for FR concrete**

Product application	Type of fibre evaluated	Countries in which significant field work has been done
Car park deck slabs	Steel	England
Concrete pipe	Steel, Glass	U.S., England
Concrete piling	Polypropylene	England
Ceramic tooling	Steel	U.S.
Floating pontoon units	Steel, Glass	England
Dolosse (break waters)	Steel	U.S.
Boat hulls	Steel, Glass	U.S., England
Burial vaults	Steel	U.S.
Concrete steps	Glass	U.S.
Decorative garden units	Glass	U.S.
Utility poles	Steel	Canada
Decorative building panels	Polypropylene	England
Structural units	Steel	U.S.
Manhole assembly	Steel	England
Weight coatings for undersea gas and oil transmission line	Steel, Polypropylene	U.S., England
Pile tips	Steel, Glass	U.S.
Machine pads	Steel	U.S.
Machine frames	Steel	U.S.
Precast refractory shapes	Steel (stainless & carbon)	U.S.
Underground utility vaults	Steel	U.S.

to 152 mm (6 in.) were produced in this manner. It was concluded that the spinning technique had a favourable influence on the strength development of the concrete.

Decorative building panels

A variety of decorative cladding panels have been produced using a novel air-entrained, polypropylene FR concrete in England (36). A polypropylene FR pumice concrete was described recently by workers in New Zealand (37). Steel FR concretes have not been given serious consideration for this application due to the problem of rusting of exposed fibres.

Concrete piles

One of the most successful applications of FR concrete in the precast products area is the production of segmented piles using polypropylene FR concrete in England (23, 38). The hollow piling is made in sections about 1 m (3 ft) long with diameters ranging from 304 mm (12 in.) to 0.6 m (2 ft). The fibrillated polypropylene fibers (0.175 percent of the dry mix weight) have replaced the steel mesh reinforcement in the piles and have provided a significant improvement in impact resistance over the conventional piling. The use of the polypropylene fibres resulted in an increased production rate as well as improved performance. Installation of the polypropylene FR concrete piles in the field is shown in Figure 5.

Polypropylene fibres probably will not find widespread use in normal concrete pilings which reach lengths of 12 m (40 ft) or more. In the longer piling, steel reinforcement is needed for tensile reinforcement so that the piles can be handled and lifted prior to driving. The polypropylene fibres provide no improvement in the tensile strength of the concrete.

Structural units

One of the first applications of FR concrete in a structural situation was in the production of precast deck slabs (steel FR concrete) for a carpark in England in 1971 (29, 39). The benefits of fibre reinforcements in this application included increased crack resistance, ductility at failure, higher load capacity, and a thinner concrete section (for reduced dead load).

Figure 5 Field installation of polypropylene FR concretes piles.

Despite this early application, FR concrete has not made significant inroads into the structural precast concrete field. However, a number of significant laboratory studies have been conducted, some on full size beams, to learn how fibre reinforcement can be used to advantage in structural product applications (40-44). It is predicted that steel fibres may be used as shear reinforcement in beams and in the construction of more effective seismic joints.

Steel FR concrete has been considered as a structural element in bridge decks (53).

Other applications Several other precast FR concrete products are being produced and sold commercially on a modest basis including pile tips (steel fibre), decorative garden units (glass fibre), machine bases and frames (steel fibre), burial vaults (steel fibre), concrete steps (glass fibre), and precast refractory concrete shapes (stainless and carbon steel fibre).

Other potential applications have received cursory coverage in the literature including ceramic tooling (steel fibres) (45), floating pontoon units (steel fibre) (46), and extruded products (steel and glass fibre) (47). A hydraulically pressed, steel FR concrete manhole cover and frame recently received a favourable review by the British Agrément Board (48). Thin sections, 6 to 20 mm (0.25 to 0.75 in.) of steel FR concrete have been prepared by first casting the material on a flexible sheet metal base with longitudinal edge stiffeners and then picking up the form along the edge so as to bend the metal sheet (and the layer of steel FR concrete) into a curve that approximates to a catenary (49). A variety of special products have been prepared in this way on an experimental basis, e.g., flumes, culverts, inverted catenary pipe and roof shells.

CURRENT AND FUTURE STATUS OF FR CONCRETE

Considering the fact that serious efforts to commercially exploit FR concrete are, at most, only 3 to 4 years along, the progress that has been made is quite encouraging. Up to now, the greatest overall interest and research and development effort with regard to fibre reinforcement for concretes, has been in steel fibres. Significant field trials with glass FR concretes have been underway in the US for only a year or so. Additionally, there appears to be relatively little activity in other countries regarding the use of glass fibres in concrete for commercial applications. Most of the reported work on polypropylene FR concrete has come from England where at least one application for the material has attained a significant commercial status.

It appears that, barring any unforeseen circumstances, the majority of the development effort on FR concretes in the near future will continue to be based on steel fibres as the reinforcement material. Interest and activity in the US in FR concrete as a pavement overlay and rehabilitation material is expected to remain high. The use of steel FR refractory concretes is expected to grow in the US and experimentation with the material in other countries is likely. Another promising application area for FR concrete (principally steel) is in mining and tunneling where significant field work using shotcrete procedures will continue in both the US and England. The early successes with shotcrete stabilization of rock cuts with steel FR concrete predict a continued interest in this application area in the US and possibly the beginning of experimentation in other countries. It is likely that the future will see a greater interest in glass FR concrete applied by the shotcrete technique in mining, tunneling, and slope stabilization applications.

The use of FR concrete in the precast products field is now established, although it is difficult to predict at this point how large and diverse this market for the material will become.

The application work done to date with FR concrete (principally steel fibres) has helped to identify several factors concerned with its preparation and properties that need to be considered and improved upon if the full potential of material is to be realized.

(1) Users have expressed concern over the longer than normal mix preparation times required for FR concrete. This problem will be overcome as special equipment for handling fibres and mixing FR concrete becomes available. Already one company in the US is offering for sale a portable, self-powered machine that automatically shreds entangled masses of steel fibres and feeds the fibres at a uniform, controlled rate into the concrete mixing system.

(2) The formation of fibre balls or clumps during mixing continues to be a problem especially where high fibre contents and/or high aspect ratio fibres are used. Again, the use of special mixing equipment may be one way of overcoming this problem. Figure 6 shows a technique and equipment which has been used by Battelle researchers to minimize the fibre balling problem on several field trials. A chute which fits inside the truck mixing drum carries the fibres into the drum and drops them on the fresh concrete. Without the chute, the fibres fall into the dry feed screws at the back of the

Figure 6 Experimental fibre feed chute used to direct the flow of fibres onto the fresh concrete during mixing.

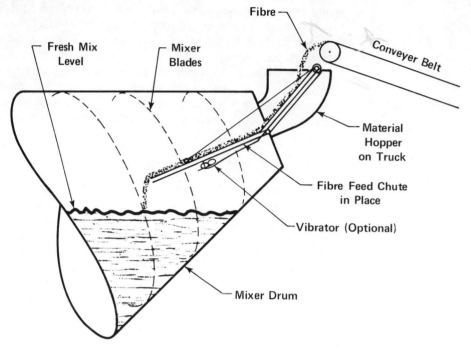

truck and must negotiate several revolutions of the drum prior to contracting the concrete. During this time, clumps of fibres can form which become hardened balls when they are mixed with the concrete. On one job, the chute was used in preparing over 300 m³ (500 yd³) of virtually ball-free steel FR concrete using fibres with an aspect ratio of 100 (0.63 x 63.5 mm [0.025 x 2.5 in.]).

Another example of a situation where special equipment development aided in overcoming FR concrete mixing and placing shortcomings is the shotcrete application area, where conventional equipment was successfully modified to accommodate the use of fibres in the mix.

(3) The properties of FR concretes prepared in the field have often been inferior to those obtained in the laboratory. This is due, in part at least, to the common practice of increasing the water content of FR concrete to satisfy the workability requirements of the workmen in the field. One solution to this problem in the mass concrete field would be the development of equipment and techniques to place low slump FR concrete. Considerable experience has already been gained in the placement of low slump conventional concretes as bridge deck overlays in the US. The use of vacuum dewatering techniques to reduce the water content of steel FR concrete slabs in-situ is currently under consideration in the US. In the precast products area, the use of hydraulic compaction to prepare FR concrete shapes (as with the manhole assembly discussed earlier) appears to offer a convenient and practical route to improved properties.

(4) Improvements in the bond between fibre and matrix would lead, it is felt, to an improvement in the beneficial effect that the fibres have on the engineering properties of the concrete. Manufacturers of both glass and steel fibres are currently considering ways to improve the fibre/matrix bond. The importance of effective fibre bond area as related to property optimization has also been recognized (50).

(5) With further regard to the performance of the material, the primary indicator of the quality of field-placed FR concrete has been flexural strength. This, in itself, is unfortunate since other engineering properties of the material doubtless influence its

performance in overlay and other applications, e.g., fracture toughness; strain at failure; E; post-cracking load-carrying behaviour; durability, etc. Looking realistically, however, truly rational design with FR concrete is probably some time away.

Despite these limitations, the work done to date has made it clear that the number of applications where the properties of fibre reinforcement in concrete can be positively used is considerable. Regular monitoring of the performance of FR concrete structures and products already in service will aid in creating confidence in the new material. Innovative uses of FR concrete such as in the planned repairs to the Dworshak Dam in Idaho (51) (using polymerized FR concrete) and in the Econocrete concept (52) will also serve to stimulate interest in its use. An increase in the use of FR concrete is expected to occur as the available information base concerning design procedures, preparation and properties is expanded and as the industries' awareness and confidence in the material grows.

ACKNOWLEDGEMENT

The author would like to express his deep appreciation to all those who helped in the preparation of this paper by providing data, photographs, and information concerned with field experience with FR concrete. In particular, thanks is due to the Battelle Development Corporation, Columbus, Ohio; the Owens-Corning Fiberglas Corporation, Granville, Ohio; and Wests Piling and Construction Company Ltd., Colnbrook, England.

REFERENCES

1. Fibre reinforced concrete. Publication SP-44, American Concrete Institute, Detroit, Michigan, 1974, p 554.
2. Romualdi, J P and Batson, G B, Mechanics of crack arrest in concrete. Journal, Proceedings of the American Ceramic Society of Civil Engineers, Vol 89 (EM3), June 1963, pp 147-168.
3. Romualdi, J P, Two phase concrete material. US Patent No. 3,429,094, assigned to the Battelle Development Corporation, Columbus, Ohio, USA, 25 February 1969.
4. Lankard, D R and Walker, A J, Pavement applications for steel fibrous concrete. Transporatation Engineering Journal, ASCE, 101, TE1, Proc. Paper 11108, February 1975, pp 137-153.
5. Sather, W R, First fibre concrete placed in Minnesota. Construction Bulletin, (USA), 3 August 1972, pp 3-5.
6. Fibrous concrete-pavement of tomorrow. American Concrete Paving Association Newsletter, Vol 8, No 10, October 1972, pp 1-6.
7. Concrete means better pavement. American Concrete Paving Association Newsletter, Vol 9, No 10, October 1973, pp 1-6.
8. Arnold, C J and Brown, M G, Experimental steel-fiber-reinforced concrete overlay. Michigan State Highway Commission, Research Report No R-852, April 1973, pp 1-17.
9. Arnold, C J, Steel fiber reinforced concrete overlay. Michigan State Highway Commission, Research Report No R-878, August 1973, pp 1-5.
10. Pioneer overlay job uses fiber reinforced concrete. Civil Engineering ASCE, January 1974, pp 38-39.
11. Iowa fibrous concrete test is largest to date. Concrete Products, Vol 77, No 1, January 1974, pp 72-74.
12. Fibrous concrete tested on highway. Engineering News Record, Vol 190, No 36, 19 October 1972, p 16.
13. Parker, F, Construction of fibrous concrete overlay-Tampa International Airport. CERL Conference Proceedings M-28, December 1972, pp 177-197.
14. Fibrous reinforced concrete performs well in airfield pavement test. Concrete Construction, Vol 18, 1974, pp 120-122.
15. Largest fibrous concrete paving project solves clumping problem. Engineering News Record, Vol 192, No 15, 11 April 1974, pp 68-73.

16 Marsh, H N and Clarke, L L Jr, Glass fibers in concrete. ACI Publication SP-44, Fiber Reinforced Concrete, 1974, pp 247-264.
17 Buckley, E L, Accelerated trials of glass fiber reinforced rigid pavements. The Construction Research Center, University of Texas at Arlington, Texas, Publication TR-3-74, 12 April 1974, p 61.
18 Chironis, N P, Sprayed fibrous concrete for mines. Coal age, Vol 20, December 1974, pp 56-59.
19 Ryan, T F, Continuous sprayed concrete tunnel lining. Tunnels and Tunneling (England), November 1973, pp 539-543.
20 Herring, K S and Kesler C E, Concrete for tunnel liners: behavior of steel fiber reinforced concrete under combined loads. Department of Civil Engineering, University of Illinois, Urbana, Illinois, Report No. FRA-ORDD 75-7 (UILU-ENG-74-2025), August 1974, p 76.
21 Poad, M E and Serbousek, M O, Engineering properties of shotcrete. US Bureau of Mines, Spokane Mining Research Center, Spokane, Washington, presented at the 1972 North American Rapid Excavation and Tunneling Conference, Chicago, Illinois, (sponsored by ASCE and AIME), 1972, pp 19.
22 Kaden, R A, Slope stabilized with steel fibrous shotcrete. Western Construction (USA), April 1974, pp 30-33.
23 Zonsveld, J J, The marriage of concrete and plastics. Plastica, No 10, October 1970, pp 1-12.
24 Sather, W R, New type of concrete placed in inside loading dock. Construction Bulletin, 15 February 1973, pp 16-17.
25 Add fiber wires to the mix for better warehouse floors. Materials Handling News (England), October 1973, pp 1-4.
26 Experimental house uses fibrous concrete floor slab. Engineering News Record, Vol 193, 16 January 1975, p 12.
27 Lankard, D R and Sheets, H D, Use of steel wire fibers in refractory castables. Bulletin American Ceramic Society, Vol 50, No 5, 1971, pp 497-500.
28 Lankard, D R, Bundy, G E and Sheets, H D, Strengthening refractory concrete. Industrial Process Heating (England), Vol 13, No 3, March 1973, pp 34-37.
29 Swamy, R N and Lankard, D R, Some practical applications of steel fiber reinforced concrete. Proc. Instn. Civil Engrs., Vol 56, August 1974, pp 235-256.
30 Henry, R L, An investigation of large diameter fiber reinforced concrete pipe. Ref No 16, pp 435-454.
31 Bortz, S A, Shipley, L E and Haughwout, L B, Centrifugal molding of ceramic tubes containing metal fibers US patent No 3,689,614, 5 September 1972.
32 Barab, S and Hanson, D, Investigation of fiber reinforced breakwater armor units. Ref No 16, pp 415-434.
33 Fiber reinforced concrete undergoes toughest test. Engineering News Record, Vol 190, 7 September 1972, p 12.
34 Coursey, G E, New shape in shore protection. Civil Engineering, ASCE, December 1973, pp 69-71.
35 Burnett, E F P, Constable, T and Cover P, Centrifugated wire fiber reinforced concrete. Ref No 16, pp 455-476.
36 Hobbs, C, Faircrete: an application of fibrous concrete. In Prospects for Fibre Reinforced Construction Materials, Proc. of an International Building Exhibition Conference sponsored by the Building Research Station, Olympia, London, November 1971, pp 59-67.
37 Garside, J H and Merwood, D L, Polypropylene fibre reinforced pumice concrete. NZ Concrete Construction, Vol 18, No 1, February 1974, pp 13-15.
38 Fairweather, A D, Use of polypropylene film fibers to increase impact resistance of concrete. Ref No 36, pp 41-44.
39 Car park at Heathrow airport. Composites, Vol 2, No 2, 1971, p 203.
40 Williamson, G R, Compression characteristics and structural beam design analysis of steel fibre reinforced concrete. US Army Corps of Engineers, Construction Engineering Research Laboratory, Technical Report M-62, December 1973, p 45.

41 Batson, G, Jenkins, E and Spatney, R, Steel fibers as shear reinforcement in beams. Journal American Concrete Institute, Vol 69, October 1972, pp 640-644.

42 Batson, G, Ball, C, Bailey, L, Landers, E and Hooks, J, Flexural fatigue strength of steel fiber reinforced concrete beams. Journal American Concrete Institute, Vol 69, November 1972, pp 673-677.

43 Henager, C H A, Steel fibrous, ductile concrete joint for seismic-resistant structures. Paper presented at the 1974 annual convention of the American Concrete Institute, 1-5 April, 1974.

44 LaFraugh, R W and Moustafa, S E, Experimental investigation of the use of steel fibers for shear reinforcement. Published by Concrete Technology Associates, Tacoma, Washington, January 1975, p 53.

45 Wire reinforcement improves performance of ceramic tooling. Iron Age, 24 May 1973, p 68.

46 Concrete paves the waves. Contract J. (England), 14 December 1972, p 44.

47 Zollo, R F, Investigation into the extrusion of a wire fiber reinforced concrete material system. National Science Foundation, Report VM-NSF-GK 37049-1, September 1973, pp 21.

48 Wexham manhole cover. Agrément board assessment Report No. 120, Cement and Concrete Association, Wexham Springs, Slough, Bucks, 26 September 1974.

49 Lankard, D R, Casting thin sections of steel fibrous concrete. Concrete Products, Vol 74, No 4, April 1971, p 52.

50 Lankard, D R, Improved flexural strength in fibre reinforced concrete. West German DOS P 2232665.4, Assignee, Battelle Development Corporation, Columbus, Ohio, USA.

51 Polymerized fibrous concrete to fix dam spillway. Engineering News Record, Vol 192, 9 January 1975, p 10.

52 Halm, H J and Eisenhour, J E, Econocrete—what it is and how it is used. Journal American Concrete Institute, Vol 71, December 1974, pp 644-651.

53 Givens, B, Concrete structural member. US Patent No. 3,808,085, assigned to Battelle Development Corporation, Columbus, Ohio, USA, 30 April 1974.

Section 2

Applications of Fibre Cement

2.1 Opening Paper: Applications of fibre cement

J F Ryder
*Building Research Establishment,
Garston, United Kingdom*

Summary

Only two types of fibre cement are at present in commercial production namely asbestos cement and glass fibre reinforced cement. The applications of these two materials are influenced by their properties and by the available production processes. These are briefly described and compared. Descriptions are given of the major applications of the two materials, and of other applications that are of technical interest. It is concluded that at present the two materials are complementary rather than directly competitive.

Résumé

Il n'y a que deux types de ciment de fibres en production en ce moment, notamment le ciment d'amiante et le ciment armé de fibres de verre. Les applications de ces deux matériaux sont influencées par leurs propriétés et par les procédés de production en vigueur; on les décrit brièvement et on établit une comparaison. On décrit les applications principales de ces deux matériaux et d'autres applications d'intérêt technique. On conclut qu'à présent les deux matériaux sont complémentaires plutôt qu'en concurrence directe.

Copyright—Building Research Establishment, Department of the Environment

INTRODUCTION

For the purpose of this review, fibre cement composites are defined as materials consisting essentially of a matrix of cement, or cement plus filler or fine aggregate not exceeding 5 mm, reinforced with fibres. These fibres may be completely random or have various types and degrees of orientation, and more than one type of fibre may be included in one composite material. The fibres may be of any length between 1 mm and 40 mm, or even longer when they have a considerable degree of orientation. The matrix may be modified by various additives.

Only two materials covered by this definition are at present in commercial production, namely asbestos cement, which has been manufactured in the United Kingdom since 1908, and glass fibre reinforced cement (grc), which was first produced for a commercial application in 1970. This review is not, however, entirely restricted to commercial applications of these two materials, but also includes some proposed applications of fibre cement composites that have been demonstrated by means of prototype components. Applications of wood-wool cement composites have not been included, since the coating of cement slurry that binds the wood-wool fibres together does not form a continuous matrix.

The types of application for which a material is used depend not only on its physical properties, its cost and durability, but also on the methods of production that are available for it, since these may impose restrictions on the shapes or sizes that can be made, besides affecting the final cost of a component. It is therefore relevant briefly to consider and compare the properties of asbestos cement and grc, and their methods of production, before the applications of these two materials are described.

ASBESTOS CEMENT

Asbestos cement is composed of finely divided asbestos fibre and Portland cement. Some cellulose fibre or fine silica may also be included. The proportion by weight of the asbestos fibre is normally within the ranges 9 to 12 per cent for flat or corrugated sheet, 11 to 14 per cent for pressure pipes and 20 to 30 per cent for fire resistant boards. Three basic methods of manufacture are in use: the wet transfer roller or Hatschek process, the semi-dry or Magnani process and the dry or Manville process. Various modifications of these basic processes are used to make different products.

The wet or Hatschek process

This is the most widely used method of manufacture, and can be adapted to make a great variety of products. The process developed from one used to make paperboard and, as shown diagrammatically in Fig 1, it begins with the preparation of a dilute suspension of asbestos fibre and cement in water. This contains about 6 per cent by weight of solids. After agitation the suspended solids are picked up as a thin film on the surface of a rotating drum of wire mesh, and are transferred from this to an endless conveyor band of permeable felt. This passes over a vacuum box to remove excess water from the film, which is transferred to a steel assimilation drum, on which it is compacted and further dewatered by a pressure roller and is plied to the required thickness. When the accumulated layers of the 0.6 to 1.4 mm thick film reach the required total thickness, the deposit on the assimilation drum is cut and peels off on to an endless rubber conveyor belt. This 'wet flat' is pliable and has considerable tear strength, and can be moulded, either by mechanical means or by hand, to form flat or profiled sheet or quite complex shapes. The length of this flat, and therefore the maximum length of the final product, is determined by the diameter of the assimilation drum.

The process of forming the original thin layer of asbestos fibre and cement from their dilute suspension tends to give the fibres a two-dimensional orientation in the plane of this layer, with an appreciable degree of preferred orientation in the direction of rotation of the mesh drum. In addition, the method of building up these layers to the final thickness on the assimilation drum produces a laminar structure.

The Mazza process is a modificatoin of the transfer roller method which is used to make asbestos cement pressure pipes. The dewatered film on the fabric belt is

Figure 1 Schematic representation of the manufacture of asbestos cement.

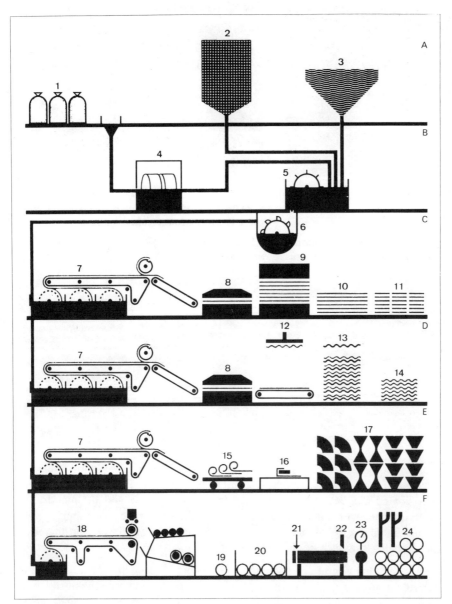

A Raw materials
B Preparation for the manufacture of
C Flat sheets
D Corrugated sheets
E Moulds
F Pipes

1 Crude asbestos from Canada and South Africa, delivered to the factory in bags
2 Portland cement pumped into the factory silos from railway containers
3 Saturated limewater for mixing asbestos fibres and cement
4 In the kollergang the asbestos fibres are finely opened
5 The asbestos fibres and cement are mixed with ample water in the beater to form a homogeneous slurry
6 From the stirring tank, where the asbestos cement slurry is kept in continuous motion, the mixture is fed into the filter boxes of the machines which produce sheets or pipes
7 Rollers form the asbestos cement web into sheets 3 to 20 mm thick
8 The wet flats are cut into the required shapes on the cutting table
9 The wet flats are pressed under high pressure to form roofing and facing components
10 The sheets are stored for curing
11 Tiles for roofing and facing, cut into the various shapes and sizes required, are stored ready for dispatch
12 Wet flats are formed into corrugated sheets
13 The corrugated sheets are deposited in oiled steel molds
14 Corrugated sheets for roofing and facing are stacked in the storeroom for air-curing
15 The still unhardened flats are brought to the molding shop
16 By using different dies, articles of all shapes and sizes are produced
17 Flower boxes, plant pots, ventilation shafts and all the other various articles molded from asbestos cement are air-cured prior to delivery
18 A felt conveyor deposits layer after layer of the thin asbestos cement web on the steel mandrel, until the wall of the pipe reaches the desired thickness
19 The steel core can be removed only a few hours after manufacture
20 Before being subjected to final processing, asbestos cement pipes are treated with water in large tanks
21 The asbestos cement pipe is cut to the required length
22 The ends of the pipes are calibrated to avoid unnecessary work when they are coupled up during laying
23 All pressure pipes are carefully tested at operating pressure before leaving the factory
24 The pipes are stored in the open for air-curing

Figure 2 Hard moulding of asbestos cement 'wet flat'.

Figure 4 Industrial use of corrugated asbestos cement roofing and cladding.

Figure 3 Pilot scale plant for the continuous production of grc sheet by a spray-suction process.

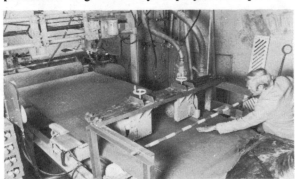

Figure 5 Corrugated asbestos cement sheeting used as a river bank retaining wall.

transferred to a rotating steel mandrel, instead of the assimilation drum. Successive layers are consolidated by pressure until the desired wall thickness has been built up. The coated mandrel is then removed from the machine and the mandrel withdrawn from the pipe, which is matured in a water tank. This process of manufacture imparts a considerable degree of circumferential orientation to the fibres, and thereby increases the bursting pressure of the pipes.

The semi-dry or Magnani process

For the manufacture of corrugated sheet this process has the advantage that it can provide a greater thickness of material at the peaks and troughs of the corrugations, and so increase the bending strength. This profile also enables lapped sheets to fit more closely together.

The mix, which has a solids to water ratio of about 0.5 and is heated to facilitate the dewatering process, is pumped on to a fabric belt and spread and levelled by reciprocating rollers. Both the belt and the rollers may be shaped to form corrugated or profiled sheet. Vacuum boxes under the belt move with it to suck excess water from the hot mix, and the dewatered sheet or profiled material is transferred to pallets to mature.

The Manville extrusion process

Quite complex and sharp edged profiles can be formed by this process. The materials; asbestos fibres, cement, fine silica and a plasticiser such as polyethylene oxide, are fed from a hopper into a mixer, with just sufficient water to produce a stiff mix. This is transferred to the extruder, where a worm drive forces it through a steel die of the desired profile. The resulting extrusion is cut to lengths which are moved by a take-off belt to pallets on rollers. After drying, the extruded sections are autoclaved and finally cut to the required lengths.

Injection moulding

This process is now tending to replace the hand moulding of 'wet flat' from the Hatschek process for the manufacture of special fittings for use with corrugated sheets. It is also used for other products, such as decking tiles. A slurry of asbestos fibre and cement, containing about 45 per cent of solids, is pumped into a permeable mould and then subjected to a hydraulic pressure of about 26 atmospheres via a rubber diaphragm. The mix is dewatered by this process of pressure filtration, and then has sufficient green strength for the product to be demoulded by means of a suction lifting pad, and transferred to a pallet for curing. The complete cycle, from injection to demoulding, takes about one minute.

GLASS FIBRE REINFORCED CEMENT (GRC)

GRC consists of a cement matrix reinforced with glass fibres of a composition that is resistant to attack by the alkalis associated with hydrating cement. The proportion of fibre is usually about 5 per cent of the weight of the matrix. The cement is generally a Portland type but may be a slag or other cement. Up to 50 per cent of the Portland cement can be replaced by pozzolanic material such as pulverised fuel ash or by fine sand. The mix may also include additives to plasticise it or to modify the properties of the hardened matrix.

The production processes for grc have some similarities to those for asbestos cement, but there are also fundamental differences between them. Glass fibre cannot be dispersed in a dilute slurry, as asbestos fibre is for the Hatschek process, but it is available as a continuous roving or bundle of fibres, that can be mechanically handled and orientated and chopped to any required length.

Spray processes for grc

The mechanised spray-suction process can produce either flat sheet or 'wet flat' that can be shaped to form components. The glass fibres in grc made by this process have a two dimensional or planar orientation, although there is normally only a small degree of preferential orientation within the sheet.

In this process a combined spray of cement slurry and strands or bundles of glass fibre, 20-40 mm long, cut from a roving by a rotary chopper in the spray head, traverses a moving mould with a permeable base. This mould acts first as a suction filter to extract excess water from the layer of sprayed mix, and subsequently as a vacuum lifting pad to transport the dewatered sheet to a stack, or to handle the 'wet flat' whilst it is shaped to form a component. Typical water/cement ratios for a neat Portland cement mix are 0.55 for the initial slurry, when sprayed, and 0.25 to 0.30 for the mix after dewatering by suction.

More complex shapes can be made by hand spraying the slurry and chopped roving on to a shaped mould, on which the mixture is left to harden before it is demoulded. The moulds for this process are usually impermeable, but a plasticiser can be added to the slurry mix to enable it to be sprayed at a water/cement ratio as low as 0.30 to 0.35 for a neat Portland cement mix.

Mixing process for grc

Glass fibres can be incorporated in cement mixes by means of pan and paddle mixers or planetary type mixers. The tendency of the fibres to tangle and ball up increases with their length and quantity, and also with the ease that the roving strands separate into filaments. A hard coating on the strands, that does not soften quickly when wetted, therefore aids mixing, and the stiffer bundles of fibre also tend to take up a two dimensional orientation if the mix is cast as a thin layer, whereas the filamentised soft coated fibres retain the random orientation that is produced by mixing. Additives that increase the viscosity of the mixing water and lubricate the fibres facilitate mixing, particularly with soft coated rovings that tend to filamentise.

GRC mixes produced in this way can subsequently be formed on a suction madrel, or extruded. Both processes tend partially to orient the fibres.

COMPARISON BETWEEN ASBESTOS CEMENT AND GRC

The main processes for the production of asbestos cement and of grc are summarised in Table 1, with an indication of typical products from each process, and Table 2 shows some properties of various types of asbestos cement and of grc made by the spray-suction process. Comparison of the properties of the two materials and of the processes available for their production is necessary to understand how, despite the

Table 1 Comparison of production processes for fibre cement products

Material and process	Mix	Method	Primary product and fibre orientation	Further processing	Curing (normal or autoclave)	Typical products
ASBESTOS CEMENT						
Wet process Hatschek	Asbestos fibre and cement suspension (c. 6% solids)	Mesh drum picks up thin (0.6–1.4 mm) ply which is calendered and compacted by rolling until required thickness is built up	'Wet flat' having a laminar structure and 2D fibres partially aligned	None / Pressure / Machine moulding / Hand moulding	N / N or A / N / N	Semi-compressed flat sheet / Fully- " " / Corrugated or profiled " / Specials, of complicated shape or limited market
Mazza	"	Initially as above. The thin ply is wound on a steel mandrel until required wall thickness is built up	The fibres have a mainly circumferential orientation	—	N	Pressure pipes
Semi-dry process Magnani	A hot slurry of asbestos fibre and cement (c. 25% solids)	Compaction on a fabric belt by rolling and suction	The fibres have a largely random orientation	—	N	Corrugated or profiled sheet
Injection moulding	Pumpable slurry of asbestos fibre and cement (c. 45% solids)	Compaction by pressure filtration	"	—	N	Decking tiles / Fittings for corrugated sheets
Dry process Manville	Plastic mix of cement, asbestos fibre, fine silica and plasticiser	Extrusion through a die by means of a worm drive	Some fibre orientation in extrusion direction	—	A	Hollow sections, eg decking, sills and copings. Sharp edged profiles
G R C						
Spray-suction	Fluid slurry of neat cement or cement plus PFA or fine sand (Water/solids c. 0.5)	A combined spray of slurry and chopped roving traverses a moving suction mould or belt	'Wet flat' with 2D fibres	Folding	N	Flat sheet / Shaped single skin components may have textured finish
Hand spray	As above, with water reducing plasticisers (Water/solids c. 0.3)	The combined spray is applied directly to shaped moulds	2D fibres	Sets and hardens on mould	N	Shaped products, eg cladding panels may have textured finish and or sandwich structure
Mixing	Lubricating additives may be necessary with filamentising fibre	Normal pan and paddle or planetary motion mixers	The initial random orientation may be changed by processing	Spread over moulds / Suction mandrel / Extrusion		Coffer units / Fence posts / Experimental sections only

Table 2

Properties of fibre cement

Material	Curing (normal or autoclaved)	Property				Strength			Remarks
		Density g/cc	24 hour water absn %	Thermal expn per °C	Impact Nmm/mm²	Tensile MN/m²	Flexural MN/m²	Compressive MN/m²	
ASBESTOS CEMENT									
Flat sheet fully compressed	A	2.10	14	9×10^{-6}	2	15.7	48.4	25.8	Tensile and flexural strengths
”	A	1.84	14	9×10^{-6}	–	–	37.0	–	Perpendicular to fibre direction
semi compressed	N	1.40	30	–	–	–	22.0	–	
Insulation board	N	0.96	75	5×10^{-5}	–	3.9	11.6	8.0	do.
Extruded material	A	1.76	10-15 max	–	–	–	13.1	56-61	
GRC									
Spray-suction sheet	N	2.10	11	14×10^{-6}	18	14.0	37.0	50-55	28-day ultimate strength values
neat cement, 5% Cemfil						7.0	13.0	50-55	28-day LOP† ” ”

Notes: The data for GRC is based on tests at the Building Research Station. The data for asbestos cement is taken from manufacturers' literature.

† LOP strength values correspond to the strength at the limit of proportionality of the load-deflection curve.

many similarities of the two materials and the present greater cost of grc, it has been possible to exploit this new material to fill gaps in the field of applications of asbestos cement. The similarities between the two materials are at first more apparent than their differences. Table 2 shows that the one outstanding difference in the properties of the two materials is that the impact strength of the grc is, at least initially, an order of magnitude better than that of asbestos cement. Although the fracture toughness of ordinary grc exposed to weathering or to wet conditions may ultimately fall to a value similar to that of asbestos cement, the high resistance to impact of the fresh material can be of considerable value during transport, handling and erection and offers a considerable advantage for uses such as permanent shuttering where the material may be subjected to considerable impacts during construction.

GRC is free from the health hazards associated with asbestos fibres. Since it is a new material, the time dependence of its strength properties are only just becoming understood (1). The loss in strength and pseudo-ductility depend upon the environmental conditions to which it is exposed and designers need to make allowance for these. The properties of asbestos cement are more familiar to designers and they make allowance for its brittleness in deciding its use.

Table 1 enables the Hatschek process for asbestos cement to be compared with the mechanised spray-suction process for grc. Both produce relatively thin 'wet flat' containing fibres with a planar orientation, that can be shaped to form components whose size is limited only by the diameter of the assimilation drum or the dimensions of the suction mould or belt. The main difference between the two processes is in the ease with which they can be modified to make special products. Although 'wet flat' from Hatschek production can be hand moulded to make a wide range of products, this method is now mainly used to make a limited range of standard products by an automated continuous process. By contrast, the grc spray-suction process is, at least in its present stage of development, a batch process that can easily be adapted to make any required number of special shapes. The hand spray process is even more versatile, and can be used to make more complex shapes or sandwich panels with a lightweight core. Both grc processes can be modified to produce material with various textured finishes.

The present differences between the types of products made with asbestos cement or grc reflect not only the differences in properties and production methods of the two materials, but also differences in the historical development of their respective industries.

APPLICATIONS OF ASBESTOS CEMENT

Since its invention by Ludwig Hatschek at the end of the 19th century, asbestos cement has been used for a very wide range of applications that were made possible by its strength in thin sections and the ease with which it could be moulded to quite complex shapes. With the development of other mouldable materials, such as plastics, and the increasing cost of labour for hand moulding, the range of asbestos products has been rationalised, at least in the United Kingdom, and is now largely restricted to products that fully utilise the best properties of the material and can be economically made by the available processes of production.

Table 3 shows that the UK market is dominated by corrugated sheeting of various profiles. This sheeting, together with the various accessories used with it, such as ridges, flashings and closures, accounted for 73 per cent of the total usage of 587 000 tonnes in 1973. The main uses of this sheeting are for roofing and cladding of agricultural and industrial buildings. It is used both in single skin and double skin forms of construction and the latter may incorporate mineral wood or other insulation to provide a U value of less than 1 W/m^2 degrees C. The structural uses of corrugated sheeting include retaining walls for river banks and permanent shuttering for concrete. Examples of the latter use include shuttering for over 20 000 m^2 of cast-in-situ reinforced concrete floors for the new Royal Infirmary Hospital in Hull, and formwork for 50 mm thick concrete acoustic barriers over ceilings in the Sydney Opera House.

Table 3 UK usage of asbestos cement products in 1973

Type of product and application	Quantity (1000 tonnes)	Percentage of total
Corrugated sheeting and accessories	429	73
(1) in agricultural buildings	229	39
(2) in industrial buildings	165	28
(3) other uses	35	6
Flat sheeting	45	8
Pressure pipes	83	14
Rainwater goods	12	2
Other products	18	3
Total	587	100

The trends towards a rationalised range of products has reduced the number of profile classes in the British Standard to five, with corrugation heights ranging from 15 mm to 90 mm and a maximum purlin spacing of 1.67 m. In other countries very deep profiled sheet is used on greater spans; the probable maximum is the 7 m span of the Canalete trough shaped roofing unit, which has a corrugation height of 245 mm, a width of 1008 mm and a wall thickness of 8 mm. These units are made up to 9.2 m long, with a weight of 167 kg, and must therefore be the largest standard asbestos cement component.

Corrugated sheeting is now available in various colours, besides its natural light grey colour. Integral coloured sheet is produced by adding pigments during manufacture, but a wider range of colours is available from factory applied surface finishes, which can also provide protection against acid attack.

A special type of corrugated sheet has been made with reinforcement of steel mesh or high tensile steel wires sandwiched in it to increase its flexural and impact strength. A major use of this material was for the roofs of cement works, where cement dust tends to build up to a coating of sufficient weight to cause failure of unreinforced sheets. It is of interest that some of this material examined after many years showed little sign of corrosion of the wire, except at the cut ends of the sheets, despite the very limited depth of cover.

Although flat sheeting acounts for only 8 per cent of the 1973 UK usage of asbestos cement products, this category includes a considerable variety of products with densities ranging from 1840 kg/m^3 for fully compressed sheet down to 960 kg/m^3 for fire-resistant insulating board. The range also includes material containing cellulose fibre in addition to asbestos fibre and cement. Some properties of various materials in this range are shown in Table 2. In addition to the integral and applied colour finishes that are used for corrugated sheet, flat sheet is also made with a considerable range of coloured mineral enamel finishes and with coatings of sand or aggregate up to 15 mm diameter. It is also produced with a variety of moulded textures, including some with a high relief. These decorative panels are extensively used in the USA and on the continent of Europe for cladding houses and multi-storey buildings, including prestige buildings.

An increasingly important application of flat sheet is as composite panels in prefabricated housing. The 60 mm thick panels, 2.5 m by 1m, used in Israel for the construction of low cost dwellings by the Isabest system, are typical of this application. The sandwich panels consist of timber studding faced with 6 mm asbestos cement sheet, with a core of exfoliated vermiculite in honeycombed paper.

Asbestos cement slates or diagonal tiles are another form of flat sheet. They are coloured by integral pigments and normally are surface treated with silicone resins to prevent lime bloom and algal growth. With a weight of about 20 kg/m² when laid they provide a relatively light roof covering that is suitable for roof pitches down to 20°. The growing scarcity and cost of natural slates has led to their increasing use, not only for roofs of prestige buildings but also for re-roofing historic buildings such as Schloss Tambach in Bavaria and Schloss Rosenberg in Austria.

Table 3 shows that pressure pipes are an important application of asbestos cement, accounting for 14 per cent of the UK usage in 1973. The method of forming these pipes on a steel mandrel ensures that they have a uniform bore with a smooth surface that offers a low hydraulic resistance. Their freedom from the formation of internal deposits means that frictional losses remain constant and a pipe of smaller section than is normal with other materials can therefore often be used. In addition to their use as water mains and sewers they are also used for carrying gas, sea water and many types of slurries and industrial liquors. They have also been used as piles. One example of this use is for the foundations of the El Campin Stadium in Bogota, where 400 mm pressure pipes were driven to a depth of 40 m to provide a bearing capacity of 100 tonnes.

The asbestos cement products manufactured by an extrusion process in the USA and Belgium include not only items such as sills and copings, which exploit the ability of the process to make long lengths of hollow sections having complex profiles and clean edges, but also wall panels and decking that have considerable structural strength. The 80 mm decking units, for example, are available in lengths up to 6 m, weigh 53 kg/m² and have a recommended design load of 1000 N/mm² for a span of 3.7 m.

Despite the competition from plastics gutters and downpipes, rainwater goods still account for a large proportion of the minor applications of asbestos cement. Other applications in the UK include decking tiles for flat roofs, flue pipes, cable ducts and conduits, tanks and cisterns. The larger sized tanks have been used for fish breeding.

In other countries the versatility and mouldability of asbestos cement has led to an even greater variety of applications, that range from 3000 litre barrels for port wine in Portugal, to sculpture in Germany.

APPLICATIONS OF GLASS FIBRE REINFORCED CEMENT

The first prototype grc component developed by the Building Research Station was a cladding panel, of 9 mm thick spray-suction material with a surface texture of crushed stone. However, it was not merely this demonstration of a potential use for the new material that has led to cladding being at present the major commercial application of grc. The combination of strength and light weight and in particular the flexibility of the hand spray process gave the architect great freedom of design, so that he could specify panels of various colours, textures or complex shapes, whilst the relatively low cost of the moulds made small runs practicable. The ability to incorporate an insulating lightweight core in panels of sandwich construction was another advantage.

The early cladding panels tended to be flat, with an applied finish or a small-scale texture. Bolder textures and more complex shapes were soon developed, and door and window openings incorporated in the panels with windows glazed by gasket techniques adopted from the car industry. The size of the panels also tended to increase, the largest so far being 5½ m x 2½ m. The light weight of grc cladding, about 20 kg/m² for panels of sandwich construction with a core of 50 mm expanded polystyrene encapsulated with

Figure 6 GRC cladding panels and window frames. The panels are of sandwich construction and have a reeded texture.

Figure 7 Shaped grc cladding panels with a 17 mm aggregate finish.

Figure 8 GRC coffer units for a mill span cast-in-situ reinforced concrete floor.

Figure 9 10 mm thick grc segments being installed as a lining in a 1.75 m diameter live sewer.

Figure 10 Acoustic hood for an air compressor, made of 8 mm grc sheet lined with 50 mm acoustic foam.

6 mm grc, is of particular advantage for multi-storey buildings.

The first major commercial use of grc cladding was in fact on an 11-storey block of Greater London Council flats, but its use soon extended to other types of construction, including prestige buildings and housing. The use of grc cladding for one type of factory built timber framed house enabled the prototype to be erected on site in one day.

The ease with which blocks of plastic foam can be encapsulated with grc, and the strength and impermeability to water of such coatings, has led to the production by this technique of pontoons, buoys and other items for marine use. Prototype workboats and even a 12 m schooner with hulls of double skinned grc have also been built, but more extensive use of the material for boat building must await the outcome of the detailed assessment that is now in progress of its suitability for such use.

The initial toughness and resistance to impact of grc gives it an appreciable advantage over asbestos cement for use as permanent shuttering, and its potential use for this purpose was first demonstrated by the BRE tests on a reinforced concrete column cast in a 200 mm square box section of 9 mm thick grc. This greatly improved the fire-resistance of the column by preventing spalling of the concrete covering the corner bars, so that in a fire-resistance test to BS 476, carried out a year after the column was cast, it sustained a load of 48 tonnes for 100 minutes, whereas a similar column from a timber mould failed after only 48 minutes.

Examples of the commercial use of grc as permanent shuttering include coffer units for a wide span cast-in-situ reinforced concrete floor for a brewery, and permanent formwork for a bridge over the river Arun. Another example is the use of 10 mm thick grc segments to line a 1.75 m diameter live sewer in London. Three segments, each weighing 50 kg, were bolted together inside the sewer to form a ring 1.2 m long. Up to six of these rings could be laid in a nine hour shift; the completed lining was grouted in 18 m lengths at pressures reaching 0.7 N/mm^2.

The very first commercial application of grc was for ventilation ducts in the three-storey underground car park of a major development scheme in Liverpool. The 12 mm thick ducts, thickened to 25 mm at the base, were designed to withstand the impact of a car at 8 km/h.

An interesting application of the material was for an earth-retaining wall on the M62 motorway. Hollow hexagonal grc units, formed by casting and weighing only 16 kg, were linked by steel bars passing through holes in their flanges and tied to the embankment by the reinforced earth method. Another application is for an experimental noise barrier on the M4 motorway utilising panels of 6 mm grc sheet. The 11 kg/m^2 weight of this material enables the required sound reduction of 20 dB to be achieved. Another example of the use of the material as a noise barrier is an acoustic hood made house an air compressor. The hood was fabricated from a single skin of 8 mm sheet lined with 50 mm flexible acoustic foam. All external fastenings were riveted or screwed with jacknuts directly into the grc sheet.

A variety of minor applications have resulted from the ability to mould grc 'wet flat' and the versatility of the hand spray process. They include garden and street furniture, such as planters, litter bins and street signs, paint marking wheels for white lining machines and junction boxes for electric cables.

In view of the very recent development of grc it is inevitable that many applications are still in the prototype stage and have not yet reached commercial production. For example BRS demonstrated that grc pipes could be made thinner and lighter than reinforced concrete pipes, and that they could be made on existing pipe spinning equipment. Further development by a pipe manufacturer has shown that the present 1 m diameter RC pipe, which has a wall thickness of 100 mm and a collar that increases the diameter by 200 mm, could be replaced by a grc pipe with a wall thickness of only 55 mm that is jointed within this wall thickness. However, before grc pipes can be generally accepted it is necessary to provide sufficient evidence of durability under the conditions to which the pipe would be exposed; research on grc formulation to provide greater durability is proceeding.

A recently developed BRS prototype component that demonstrates a potential major application is a tubular fence post with a wall thickness of 12 mm and an internal diameter of 100 mm tapering to 80 mm over a length of 1 m. This is formed on a rotating suction mandrel. Continuous longitudinal glass fibre was sandwiched in the wall thickness, thus considerably increasing the flexural strength of the post and demonstrating one advantage of grc in comparison with asbestos cement, namely the ability to incorporate continuous fibre and to orient and position it mechanically to provide the most effective reinforcement. This prototype post was only a quarter of the weight of a RC post of similar size, yet was stronger and stiffer and of comparable cost.

Thin renderings of grc can be applied on site, by spray or trowel, without any of the health hazards associated with the spraying of mixes containing asbestos fibre. Trials and tests by BRS have shown that only 1 per cent by weight of Cemfil* fibre premixed in a slurry of a 1:1 mix of cement and fine sand and applied as a rendering by spraying, is sufficient to prevent any shrinkage cracking. A grc rendering only 6 mm thick is therefore completely impermeable to rain. Larger proportions of fibre, added at the spray nozzle to impart a planar orientation, can appreciably increase the flexural strength and stability of the wall.

CONCLUSIONS

In the course of compiling this review of the applications of fibre reinforced cement the author was led to the conclusion that asbestos cement and grc are, at least at present, complementary materials rather than directly competitive ones. Asbestos cement is supreme in the field of standard roofing and cladding at an economical price, whilst grc offers the architect greater freedom of design and initially better resistance to impact. At present the significant loss in strength under some exposure conditions limits the range of applications for grc, but in the event that a second generation of grc becomes available, with modifications to the fibre or matrix that enable the present restrictions on structural uses of the material to be eased, then the scope of its potential applications would be greatly extended.

ACKNOWLEDGEMENT

The work described has been carried out as part of the research programme of the Building Research Establishment of the Department of the Environment and this paper is published by permission of the Director.

REFERENCE

1 Majumdar, A J and Nurse, R W, Materials Science and Engineering, 15 (1974) pp 107-127.

*Registered trade mark of Pilkington Brothers Limited.

Section 3

Theories of Fibre Concrete and Fibre Cement

3.1 Opening Paper: Theories of fibre cement and fibre concrete

A S Argon and W J Shack
*Department of Mechanical Engineering,
Massachusetts Institute of Technology, Cambridge, Mass., USA*

Summary *Theoretical models of cementitious materials reinforced with either continuous or discontinuous, strong and stiff elastic fibres are discussed. Conditions for improved first-crack strength and post-cracking behaviour in such composites are considered. Some uncommon problems encountered in the fracture toughness testing of such composites are elucidated.*

Résumé *Des modèles théoriques de composites de ciment et de béton contenant des fibres à limite élastique et module d'élasticité élevés sont présentés. Les conditions correspondant à l'amélioration de la résistance à l'initiation des fissures ainsi que le comportement en présence de fissures ont été étudiées. Des problèmes particuliers rencontrés dans les essais de tenacité de tels composites ont été clarifiés.*

INTRODUCTION

The reinforcement of relatively brittle construction materials with energy absorbing components is not new. Many examples, such as the use of straw to reinforce mud-bricks, and the present wide-spread practice of the use of steel lath with plaster in interior walls, can be given. Even the use of fibres in concrete is not a novel practice. In fact, the current pre-occupation with reinforcement of cement and concrete with metal and glass fibres is for the purpose of developing an alternative to asbestos fibres which have been used in the past on a large scale (1). As for all applications of composite materials the use of fibres in concrete is largely governed by considerations of economics. Krenchel (1) has recently reviewed the potential of the most promising fibres for application in concrete from both the points of view of technical performance and economics. Taking due note of this constraint, it is nevertheless of interest to assess the theoretical merits of the concepts of fibre reinforcement of brittle materials with high strength elastic fibres. The subject has seen considerable development over the past decade. Romualdi and Batson (2) have shown by an asymptotic fracture mechanics argument that closely spaced fine and very stiff fibres, when incorporated into concrete, can increase the stress at which cracks first appear in a brittle matrix. The same effect has received a different explanation from Aveston, Cooper and Kelly (3) as being due to an insufficiency of releasable elastic energy when the fibre diameter becomes very small while the fibre volume fraction is maintained constant. Most investigators considered effects of fibre aspect ratio, orientation distribution, efficiencies of fibres, effects of interfacial bond strength and many other practical considerations, such as: mixing fibres into wet cement and concrete uniformly, effects of curing time, fibre degradation in wet concrete, etc. These considerations, which are too numerous for us to list here, can be found referred to in several recent reviews and conference proceedings (4-6).

Our considerations here will be limited only to some of the more important theoretical notions in the reinforcement of brittle cementitious matrixes with strong elastic fibres.

Figure 1   **Slender circular fibre in infinite matrix parallel to the principal stress direction in matrix.**

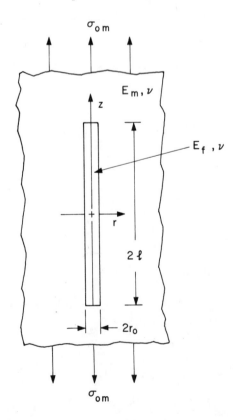

STRESS INHOMOGENEITIES IN FIBRE COMPOSITES AND LAMINATES

The stresses in a composite material are inhomogeneously distributed. Since failure starts from such stress inhomogeneities which are usually accentuated at interfaces, a better understanding of these inhomogeneous stress distributions is essential. In a multi-component material in which each component has different physical and mechanical properties local interfacial stresses arise primarily due to differences in thermal expansion and elastic properties when the temperature of the composite is changed or when it is subjected to external stress. In fibrous composites the stress inhomogeneities of principal importance are those which appear at ends of discontinuous fibres either when the fibre is entirely surrounded by a matrix material or when initially continuous reinforcing fibres fracture under stress. We will be primarily interested here in the interfacial stresses resulting from externally applied stresses for the case of an elastic fibre with a higher modulus embedded in an elastic matrix of somewhat lower modulus—with and without slippage between fibre and matrix.

The geometry of the fundamental problem is shown in Fig 1. A slender fibre of length 2ℓ and radius r_o having a Young's modulus E_f and Poisson's ratio ν is embedded into an infinite elastic matrix of modulus E_m and the same Poisson's ratio ν. The system is subjected to a uniform tensile stress σ_{om} parallel to the fibre axis. This problem has been considered in various forms with different boundary conditions by many investigators in the past (7-9). An exact solution is difficult to obtain, but a very useful approximate solution is readily obtainable by means of Eshelby's (10) method of analysis of transformation induced stresses. Consider the thought experiment outlined in Fig 2a. One starts by removing the fibre from the matrix and replacing it with perfectly bonded matrix material. A uniaxial tension is then applied producing a uniaxial strain of $\epsilon_{om} = \sigma_{om}/E_m$ and a transverse strain $-\nu\epsilon_{om}$. The matrix material occupying the position of the fibre is now removed but a set of tensile tractions σ_{om} are applied at the two ends of the hole to maintain the strain in the matrix uniform. A stress of $\epsilon_{om} E_f$ is applied to the

Figures 2a and 2b   Eshelby's thought experiment for computing the stresses in and around the cylindrical fibre.

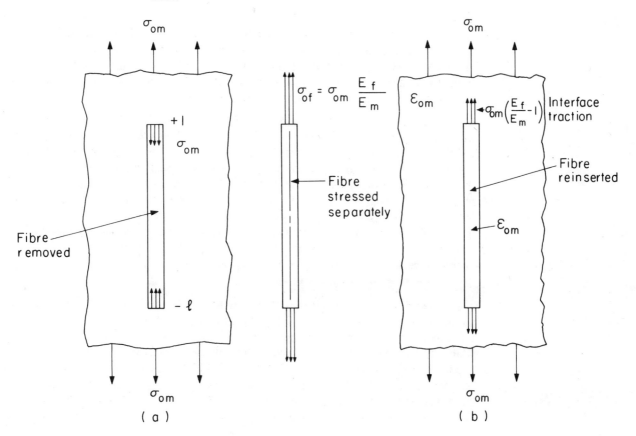

ends of the fibre which stretches the fibre to make it fit perfectly into the existing hole (Fig 2b). This state of uniform strain in matrix and fibre, however, requires application of internal interface tractions of $\epsilon_{om}(E_f-E_m)$ over an area of πr_o^2 at $z = \pm \ell$. These must be released to obtain the superposition stress field to be added to the applied stress σ_{om}. A very adequate solution is obtained from the theory of elasticity (11) by considering these interface tractions as two equal and opposite point forces applied at $z = \pm \ell$ and acting toward the centre against the joint stiffness of the matrix and the fibre. The full details of this solution can be found elsewhere (12). The important result of interest to us here is that the final interface tractions across the end of the fibre at $z = \pm \ell$.

$$\bar{\sigma}_{zz} = \sigma_{om} \frac{E_f}{E_m} \left[1 - \left(1 - \frac{E_m}{E_f}\right) \left(\frac{K_f}{K_m + K_f}\right) \right] \qquad (1)$$

The distributed shear traction along the cylindrical surface $r = r_o$ near the end of the fibre is given by

$$\sigma_{rz} \simeq \sigma_{om} \frac{(r_o/\ell)^3}{8(1-\nu)} \left(\frac{K_m}{K_m + K_f}\right) \left(\frac{E_f}{E_m} - 1\right) \times$$

$$\left[\frac{(1-2\nu)}{[(r_o/\ell)^2 + (1-z/\ell)^2]^{3/2}} + \frac{3(1-z/\ell)^2}{[(r_o/\ell)^2 + (1-z/\ell)^2]^{5/2}} \right] \qquad (2)$$

where

$$K_m = \frac{4\pi r_o E_m (1-\nu)}{4(1-\nu^2) - (1+2\nu)} \qquad (3a)$$

$$K_f = \frac{\pi r_o^2 E_f}{\ell} \qquad (3b)$$

are the appropriate stiffnesses of the matrix and fibre to a pair of concentrated forces at $z = \pm \ell$. The distribution of radial traction on the cylindrical surfaces can also be computed according to the above solution. This traction is smaller than the fibre end tractions by a factor of (r_o/ℓ), it varies by a factor of two from end to end, and in the slender rod case we are considering is negligibly small.

For the following parameters appropriate to the case of steel fibre reinforced concrete or mortar, $E_f/E_m \approx 6$ (1, 13), $\nu \approx 1/3$, and a fibre aspect ratio of 100, *a large fraction* $\alpha = 0.966$ of the initial interfacial traction $\sigma_{om}(E_f/E_m)$ gets transmitted across the end of the fibre into the matrix while the average radial traction appearing across the cylindrical surface is only $5.995 \times 10^{-3} \sigma_{om}$. The shear traction along the side surface at $r = r_o$ is plotted in Fig 2c. We consider this shear traction distribution only as approximate, as it may be subject to considerable distortion due to the replacement of the uniform end tractions with point forces. Figure 2c shows that the maximum shear traction is at $z = \ell$ and is nearly $0.3\sigma_{om}$. The integral of the shear tractions along the fibre increases the average normal stress σ_{zz} in the fibre. This distribution of average fibre stress is also shown in Fig 2c.

It would be natural to expect that failure would occur first at the end of the fibre by tearing away of the matrix. This would, as a first approximation, raise the peak shear traction on the cylindrical surface by a factor of $1/(1-\alpha)$ to a value of $8.823\sigma_{om}$, which is even higher than the level of normal traction across the end of the fibre. Again, it is natural to expect that the fibre end will slip and that a *shear failure zone* will propagate down the fibre end along which the shear traction will be limited to the shear strength τ_b of the interfacial bond. Since the "elastic tail" of the shear traction distribution should be as confined as that of the distribution given in Fig 2c, the well known shear lag estimate for the length λ of the shear failure zone

$$\lambda = \left(\frac{r_o}{2} \frac{\sigma_{om}}{\tau_b} \frac{E_f}{E_o} \right) . \quad (4)$$

gives a rather accurate value. The relative magnitudes given above remain largely unaffected by the fibre aspect ratio as long as the latter is above 50. If σ_{om} is taken as the first crack strength, in the better examples of fibre reinforced concrete the ratio of σ_{om}/τ_b is around unity (1, 13). This gives a shear failure zone $\lambda \approx 1.5d$ only. Quite often, however, the interfacial bond strength is much lower, giving ratios σ_{om}/τ_b as high as 10, which in turn, result in $\lambda \approx 15d$. The length of the shear failure zone at the first-cracking of the matrix is an important parameter. Thus, if the fibre length is so short or the interfacial bond strength so low that the entire length of the fibre becomes unbonded from both ends, the concrete is likely to fail altogether at the first-cracking strength.

The stresses at fibre ends which occur upon cracking of a reinforcing fibre will be, to a first approximation, similar to the stresses discussed above. These have, however, been investigated in more detail in anisotropic, elastic-plastic situations by McClintock (14). Since such fractures in fibres are of importance primarily in composites with ductile matrixes they will not be discussed here.

Figure 2c   **Distribution of shear tractions σ_{rz} on cylindrical end surface of fibre, and distribution of normal stress σ_{zz} along fibre at its end.**

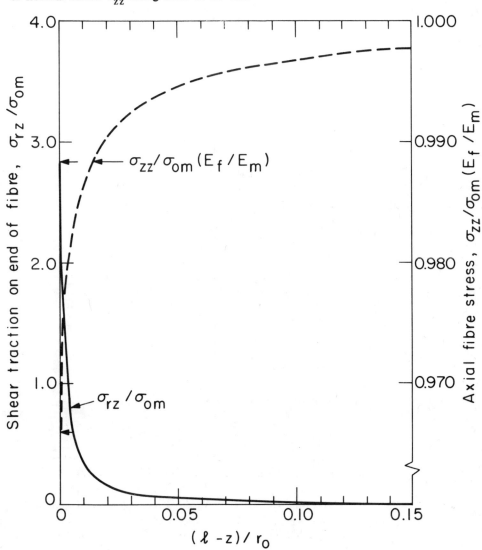

FIBRE COMPOSITES WITH CONTINUOUS ELEMENTS

Non-propagating cracks

Although continuous reinforcing fibres do not lend themselves well for reinforcement of concrete in most conventional applications, some of the more important behaviour can be readily demonstrated in this configuration. One of the most important effects due to incorporation of high strength, high stiffness fibres in concrete (or any other low strength matrix) is the increase they can provide in the first-crack strength of the composite. Romualdi and Batson (2) first pointed out this effect by an argument that strong and stiff fibres tend to decrease the stress intensity factor of matrix cracks and therefore require higher stress to propagate them. Their analysis, which assumed no slippage of fibres, drew attention only to the effect of fibre spacing but failed to indicate the effect of other parameters. A more thorough approximate analysis was given by Aveston, Cooper, and Kelly (3) which we will consider here in a somewhat simplified setting.

Figure 3   **Equivalent three element laminate representing the details for non-propagating cracks in continuous fibre elastic composites.**

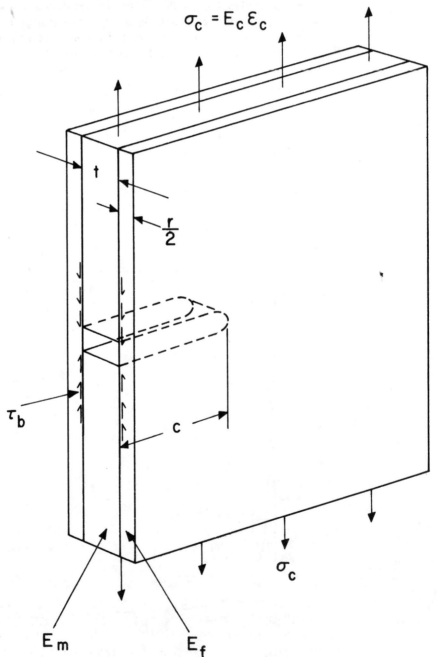

When matrix cracks first form and have to propagate around fibres, which we take for this idealization to be continuous and aligned parallel to the direction of stress, they will locally release some stored elastic energy in the fractured matrix. Such cracks will also: a) propagate shear failure zones up along the fibres and thereby do both dissipative "plastic" sliding work and increase the local elastic energy stored in the surrounding reinforcing elements, b) form new surfaces in the cracked matrix, and finally c) due to the increased compliance of the cracked composite, permit the applied stresses to do additional work on the composite. For the quantitative description of this process consider the equivalent laminate shown in Fig 3, where we have lumped together all of the weak, elastic matrix of modulus E_m and volume fraction V_m into an interior layer and lumped all of the reinforcement as two symmetrically placed surface layers of modulus E_f and combined volume fraction V_f. In this model the cracks in the matrix can only run in the matrix along the width of the laminate. The figure shows that in the region where the crack has run a shear failure zone has propagated vertically up and down between the matrix and the reinforcement. All the components of energy and work consisting of the released energy in the fractured matrix, the additional external work on the laminate, the local dissipative plastic shear work at the sliding interlayer, the additional elastic energy stored in the overstressed portions of the reinforcement, and the energy of the fractured surfaces can be readily evaluated by elementary analysis (for the details of the equivalent treatment for fibres see Aveston et al (3)) and the total free energy change for the laminate with a crack of total length c can be written as:

$$\Delta G = c \left[-\frac{E_m^2 t^2 \epsilon_c^3}{3\tau_b} - \frac{E_m^2 t^2 \epsilon_c^3}{2\tau_b}\left(1 + \frac{t}{r}\frac{E_m}{E_f}\right) + \frac{E_m^2 t^2 \epsilon_c^3}{6\tau_b}\left(1 + \frac{t}{r}\frac{E_m}{E_f}\right) \right.$$

$$\left. + \frac{E_m^2 t^2 \epsilon_c^3}{2\tau_b}\left(1 + \frac{1}{3}\frac{t}{r}\frac{E_m}{E_f}\right) + 2t\alpha \right] \quad (5)$$

where the individual terms are those identified immediately above, and where ϵ_c is the distant uniform strain in the laminate and α the specific fracture work in the matrix alone. The crack runs when:

$$\left(\frac{\partial \Delta G}{\partial c}\right)_{\epsilon_c} = 0 \quad (6)$$

which is only possible when all the terms in the brackets sum to zero. This determines immediately the required elastic strain ϵ_c for crack propagation as

$$\epsilon_c = \left(\frac{12\alpha \tau_b E_f V_f}{E_m^2 E_c t}\right)^{1/3} = \left(\frac{12\alpha \tau_b E_f V_f^2}{E_m^2 E_c r(1-V_f)}\right)^{1/3} \quad (7)$$

where

$$E_c = E_f V_f \left(1 + \frac{t}{r}\frac{E_m}{E_f}\right) \quad (8)$$

is the effective Young's modulus of the laminate. Clearly then cracks in the matrix cannot propagate before the stress on the laminate reaches

$$\sigma_c = E_c \epsilon_c. \quad (9)$$

The strength of the composite given by eqns. (9) and (7) goes to zero when the fibre volume fraction goes to zero. In reality the tensile strength of the composite will level off at the inherent strength f_m' of the concrete itself which is thought to be governed by pre-existing flaws which might be either pre-existing pores, weak interfaces or shrinkage cracks. Since the introduction of very small volume fractions of fibre is not likely to alter the nature of these inherent flaws the first crack strength f_{fc} would not be merely the sum of f_m and σ_c but the bigger of the two. Hence,

$$f_{fc} = f_m \quad (\text{for } f_m > \sigma_c) \tag{10a}$$

$$f_{fc} = E_c \epsilon_c \quad (\text{for } \sigma_c > f_m). \tag{10b}$$

We see from eqns. (10b) and (7) that for given properties of matrix and reinforcement and composition, the first-crack strength of a fibre reinforced concrete increases proportional to the negative 1/3 power of the thickness dimensions of the reinforcement in qualitative support of the discovery of Romualdi and Batson (2), but increases also with increasing interface shear strength and modulus of reinforcement. A hidden assumption in the above analysis is that the reinforcement is strong enough to support the external stress even after the matrix has fully separated. We will discuss this condition more fully in the next section below.

The statement given in eqn. (5) differs in an important respect from the famous crack instability condition of Griffith (15). In fact, since all energy and work terms are linear in the crack length there is no crack length dependent instability at all. At a stress below that given in eqn. (9) cracks of any length in a matrix, uniformly bridged by unfractured reinforcing elements will not propagate, and if the matrix is initially free of cracks there is no reason for cracks to appear of any length greater than the mean spacing between reinforcing elements.

Post-cracking behaviour

In the developments below both for continuous and discontinuous reinforcement we will ignore variability in the strength of the reinforcing elements. This is not an important restriction for work hardened metal fibres which can be expected to neck and rupture in a rather narrow stress range. The restriction is also not important for unprotected brittle glass fibres which can be expected to severely degrade in strength down to a uniformly low level as a result of attack by the alkali constituents of the cement matrix. On the other hand, the cement or concrete matrix in a fibre reinforced concrete composite is a composite itself of cement, sand, and other coarser aggregates and must therefore exhibit some variability. Although we will not consider such variability in a rigorous manner, we will note that different portions of the matrix can crack at somewhat different stresses. Hence we will take the strength of the reinforcement f_f as a constant but admit some range for the strength f_m of the matrix.

As has been discussed by Aveston et al (3), certain requirements have to be met if the composite with continuous, aligned fibres has to have any post-cracking behaviour. When the stressed composite reaches its first crack strength the uniform strain in it just prior to cracking is given by eqn. (7), and the following equation holds.

$$V_f \epsilon_c E_f + (1 - V_f) \epsilon_c E_m = f_{fc} \tag{11}$$

where the first term gives the portion of the stress supported by the fibres just prior to the formation of the first crack. If the fibres are not to fracture before even the first-crack strength is reached they must have a strength

$$f_f > E_f \left(\frac{12 \alpha \tau_b E_f V_f}{E_m^2 E_c t} \right)^{1/3}. \tag{12}$$

Furthermore, if there is to be any post-cracking behaviour at all the fibre strength must be high enough to support the entire post-cracking load without the benefit of any assistance from the matrix, i.e.

$$f_f > \frac{E_c}{V_f} \left(\frac{12 \alpha \tau_b E_f V_f}{E_m^2 E_c t} \right)^{1/3}. \tag{13}$$

Figure 4a   Model of a progressively cracking continuous element, elastic, fibrous composite.

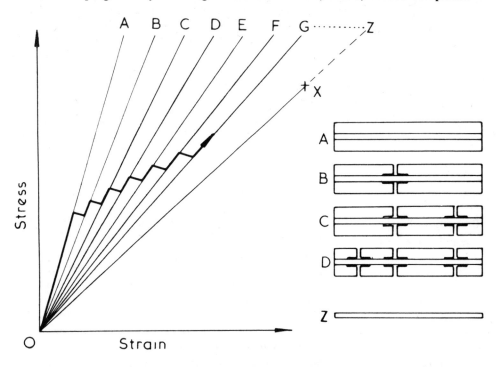

Figure 4b   Behaviour of a PVC-gypsum plaster composite under cyclic tensile extensions of progressively increasing stress (from Allen 1971, courtesy IPC Science and Technology Press).

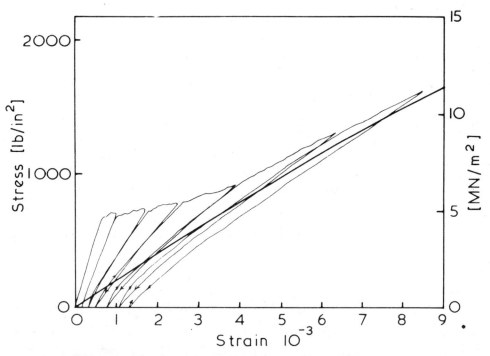

Once the above conditions are satisfied, the composite will continue to support additional load in the presence of increasing fragmentation of the brittle matrix. The process is characterized in Fig 4a, showing on the right inset four stages of the cracking matrix (16). In state A the composite is still intact and its effective modulus is high, giving a loading behaviour shown as line OA. The first crack appears in the matrix as shown in inset B, producing the dark shear failure zone and resulting in a reduction of the effective modulus to a line given by OB. A slight increase in the applied force produces a slight extension in the first shear failure zones along the fibres next to the first crack, this increases the stress in the remaining matrix of the composite in the regions away from the unloaded portions around the first crack until a second crack appears as shown in inset C, resulting in a further reduction of modulus to the line OC. The process of increased fragmentation continues under a slightly increasing load until all shear failure zones touch each other along the fibres. Beyond this point, where the matrix is uniformly cracked with a crack spacing equal to about twice the extent of the shear failure zone, the matrix cannot be stressed any further as long as the bond strength in shear τ_b remains constant. Beyond this point any further increase in load carrying capacity must come entirely from the fibres as the modulus has dropped to its lowest value given by line OZ. Final fracture occurs at point X where the fibre strength is reached. Figure 4b shows an actual set of stress strain curves in a composite made of gypsum plaster and PVC fibres (16).

FIBRE COMPOSITES WITH DISCONTINUOUS ELEMENTS

Properties of discontinuous fibres

The more commonly considered application of fibres in concrete is in discontinuous form where fibres of a given aspect ratio are mixed directly into the concrete prior to pouring. In this practice problems are encountered in the proper wetting, uniform mixing and distribution of fibres which have been widely discussed in the literature. Generally it is found that small volume fractions of fibre in the range of several percent and of aspect ratios in the range of 50-100 are readily incorporated into the concrete. As the volume fraction and aspect ratio are increased fibres tend to bundle-up and become non-uniformly distributed. Volume fractions of the order of 0.05-0.1 can, however, still be accommodated in concrete panels by special spray-laying techniques where, of course, only planar isotropy and reinforcement is achieved (17).

In relatively small volume fractions of fibre where proper mixing can be achieved, the fibres tend to have a random orientation in space. Their site of intersection with any randomly passed plane tend to be distributed according to a Poisson distribution (18) where the probability of finding r or fewer intersections in an area having an overall average number of intersections n_o, is given by

$$P(r) = \sum_{\rho=o}^{\rho=r} \frac{n_o^\rho \exp(-n_o)}{\rho!} . \qquad (14)$$

As was pointed out by Romualdi and Mandel (19) the tensile force carrying efficiency in any one direction of a collection of fibres having random orientations in space is $\eta = 0.41$ as long as the composite remains intact. Furthermore, while the composite is intact the extent of the shear failure zones at the ends of fibres due to displacement incompatibilities between matrix and fibre, which was discussed in detail above, will also decrease with increasing departure of fibre orientation away from the tensile axis. Fibres oriented normal to the tensile axis may have no shear failure zones at their ends as they can transmit the compressive tractions acting across their ends.

For randomly oriented fibres several length dimensions have to be defined. For fibres to have any effect at all on the first-crack strength of a concrete composite they must have a minimum length exceeding four times the length of the shear failure zone λ at the first-crack stress f_{fc} where the matrix fractures. This is the *minimum effective length*.

$$\ell_m = 4\lambda = 2r_o \left(\frac{f_{fc}}{\tau_b}\right)\left(\frac{E_f}{E_c}\right) = 2r_o \frac{E_f}{\tau_b} \epsilon_c \qquad (15)$$

It results from the following consideration. When the matrix cracks and fibres bridge across the crack, the shorter of the two lengths of the fibre threading through the plane of the crack becomes of importance. A little reflection shows immediately that the most probable value of the shorter portion of the fibre defined by its intersection with the plane of the crack is $\ell/4$. If this length is entirely covered with the *shear failure zone* this end of the fibre will be pulled out upon cracking of the matrix. The composite will then fracture at the first-crack strength regardless of the actual tensile strength of the fibres themselves.

If the fibres are longer than this minimum length then upon intersection by a matrix crack they can carry additional load. This happens by the propagation of two new sets of shear failure zones along both sides of the fibre, starting from the plane of the crack as shown in the inset of Fig 4a. If the composite is to have any post-cracking behaviour at all, the portion of the stress $(1 - V_f)\epsilon_c E_m$ previously carried by the matrix will have to be transmitted to the fibres by the tractions along these new shear failure zones, giving a total shear failure zone length λ' which must not exceed one-fourth of the fibre length. This defines a minimum effective fibre length ℓ_{mpost} for post-cracking behaviour

$$\ell_{mpost} = 2r_o \frac{f_{fc}}{\tau_b} \frac{1}{V_f} = 2r_o \frac{\epsilon_c E_c}{\tau_b V_f} . \qquad (16)$$

To achieve this performance the fibre strength f_f must be high enough to support the entire load on the cracked composite, ie

$$f_f > \frac{f_{fc}}{V_f} . \qquad (17)$$

Equations (16) and (17) together prescribe the necessary fibre strength and aspect ratio for post-cracking behaviour.

Finally, when one quarter of the fibre length

$$\frac{\ell}{4} > \frac{r_o}{2} \frac{f_f}{\tau_b} , \qquad (18)$$

fibres will be fractured instead of being pulled out when the composite fails. This length has been called the *critical length* for ultimate performance of a fibre reinforced composite.

Strength of brittle composites with discontinuous fibres

When a composite such as concrete with discontinuous strong elastic or work hardened ductile fibres with random orientation is stressed, the first crack strength which will be obtained will still be given by eqn. (9) with minor modifications.

$$f_{fc} = \epsilon_c E_c \qquad (9)$$

where, however, now

$$\epsilon_c = \left(\frac{12\alpha \tau_b E_f \eta^2 V_f^2}{E_m^2 E_c r_o (1 - \eta V_f)} \right)^{1/3} \qquad (19)$$

$$E_c = E_f \eta V_f + E_m (1 - \eta V_f) \qquad (20)$$

where $\eta = 0.41$ is the fibre efficiency factor discussed above. This strength will be reached, provided the fibre aspect ratio exceeds the value given in eqn. (15), and the fibre strength is high enough so that premature fibre fracture does not occur.

After formation of cracks in the composite the additional extension of the composite due to its increased compliance will straighten out the parts of all fibres bridging the cracks (except perhaps those which were nearly parallel to the crack). This will produce a marked increase in the fibre efficiency factor from 0.41 to a value much nearer to unity. Hence the cracked composite will be able to support additional load without much increased cracking by a more efficient sharing of the load between fibres. Thus, upon cracking, the discontinuous fibre composite tends to improve its performance by acting somewhat like a composite with aligned fibres. If the conditions given by eqns. (16) and (17) are met, the composite will demonstrate post-cracking behaviour with rising stress. For composites having fibres of aspect ratio larger than the critical value given by eqn. (18) the post-cracking behaviour will tend to approach that of composites with continuous and aligned elements, giving an ultimate strength for the composite

$$f_{cu} = \eta' V_f f_f \qquad (21)$$

where, as mentioned above, $0.41 < \eta' < 1.0$.

Traction displacement relations for fibre pull-out

When tension experiments are performed on a typical steel fibre bearing concrete sample in a relatively stiff testing machine load extension curves of the type shown in Fig 5 are obtained. When the first crack strength is reached and the specimen gains in compliance there is a drop of load as part of the testing system unloads. Upon further extension the curve may rise again. If the conditions in eqns. (16) and (17) are met, a new maximum load exceeding the first-crack strength can be obtained. Once the maximum load point is reached in a composite where the fibre length is less than the critical value given in eqn. (18), the shorter ends of the fibres will begin to be pulled out under a decreasing load for a displacement range not exceeding half the fibre length —with the initial slope of the decreasing portion of the force-displacement curve cutting the axis at a point near one-fourth of the fibre length.

Crack propagation and fracture toughness

One of the ultimate reasons for incorporating fibres into concrete is to increase its fracture toughness. In materials which have substantially a linear behaviour it is customary to represent the fracture toughness by either the critical mode I stress intensity factor K_{IC} or the specific work for fracture γ_f. The two are related by the well known expression

$$K_{IC} = \sqrt{\frac{2E \gamma_f}{1 - \nu^2}} \qquad (22)$$

for plane strain. Generally the critical stress intensity factor is obtained from standard notched bar tests, while the specific work for fracture can be taken to be the integral of the traction-displacement law discussed in the previous section. Several investigators (see eg Harris et al (20)) have reported that the K_{IC} and γ_f values obtained for fibre

Figure 5 A typical traction displacement curve for a short steel fibre reinforced concrete specimen (from Naaman et al, 1973, courtesy Pergamon Press).

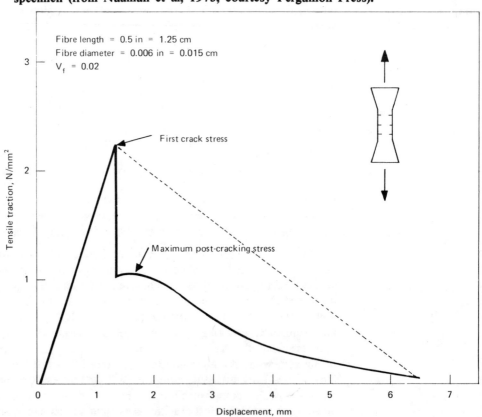

reinforced concrete are not related by eqn. (22). In general, the improvement in K_{IC} due to incorporation of fibres in concrete is much less than the measured improvement in γ_f. The problem appears to be one of specimen size.

Andersson and Bergkvist (21) have recently considered the problem of the instability of a crack in a material having a triangular traction displacement law plotted by the dotted lines in Fig 5 for comparison with the actual law. Their numerical solution shows that as a sample with a crack is loaded, the stress distribution ahead of the crack first starts rising according to linear elastic theory until the crack opening displacement brings the traction to its maximum value. Any further increase in applied stress producing further increases in crack opening displacement will produce a decrease in the traction transmitted across the crack front as it steadily moves the point of maximum traction away from the tip of the crack into the interior. The traction at the tip of the crack finally drops to zero when the crack opening displacement reaches the terminal value shown in Fig 5 where fibres are fully pulled out. At this stage the traction maximum has been pushed ahead of the crack to a distance Δ, and the traction acting across the extension of the plane of the crack between the crack tip and the point a distance Δ away is an inverted and somewhat distorted map of the decreasing portion of the traction displacement law. Andersson and Bergkvist demonstrate that at this point the crack becomes unstable and can propagate under decreasing applied stress. Clearly, the zone of extent Δ ahead of the crack acts in the manner of a "plastic process zone", and the critical crack opening displacement CCOD is somewhere between one quarter and one half of the fibre length. Proceeding further, Andersson and Bergkvist demonstrate that

$$K_{IC} = C_1 \sqrt{(CCOD)} \qquad (23)$$

where C_1 is a constant of proportionality. Since the specific fracture work is a product of one half the maximum of the fibre pull-out traction and the CCOD, it is clear that eqn. (22) also holds. Furthermore, however, an approximate expression between the process zone Δ and the CCOD can be developed from the published results of these authors, which is of the form

$$\frac{\Delta}{CCOD} = \frac{1}{A - (CCOD)/c_o} \qquad (24)$$

where $A = 0.45$ is a constant for one set of numerical studies of the above authors. It is clear, therefore, that for crack lengths of the order of fibre lengths the "plastic process zone" may be many times the CCOD or the fibre length. For the fracture mechanics approach, or the more proper analysis of Andersson and Bergkvist to be applicable, the width dimension of the notched specimens should be many times the combined length of the crack and the "plastic process zone". This condition has not been achieved by any investigator. It is therefore not very surprising that the reported discrepancy occurs.

Naaman et al (18) have considered the size of the largest of the randomly occurring regions of zero fibre density in a fibre reinforced concrete as an initial crack in the concrete, and together with the traction displacement law, prescribe its strength. Although this is an attractive notion which brings in some understanding of a possible size effect, our arguments above show that this concept too would be useful only in very large concrete structures.

DISCUSSION

Above we reviewed some basic concepts which govern the strength of brittle substances reinforced by strong, stiff, and non-brittle fibres, such as work hardened steel. Other, more ductile and extensible fibres, among them polymers, have also been considered as reinforcement for concrete. Our discussion, which does not allow for plastic extension of fibres would, of course, not apply to these systems. The problems of cost and manufacturing which have been dealt with extensively by other investigators are likely to be of as great importance as the purely technical aspects which we have discussed above.

ACKNOWLEDGEMENT

The author's research on fracture is supported in general by the US Army Research Office and the work on fibre reinforced concrete in particular by the Materials Research Laboratory of the Allied Chemical Corporation of Morristown, New Jersey. We are grateful to Mr George Hawkins in helping us review the developments in this field.

REFERENCES

1. Krenchel, H, "Fibre Reinforced Brittle Matrix Materials", *Fibre Reinforced Concrete,* Publ. SP-44 (Detroit: American Concrete Institute) 1974, pp 45-77.
2. Romualdi, J P and Batson, G B, "Mechanics of Crack Arrest in Concrete", *Journal of the Engineering Mechanics Division Proc. ASCE,* vol 89, No EM3, 1963, pp 147-168.
3. Aveston, J, Cooper, G A and Kelly, A, "Single and Multiple Fracture", *The Properties of Fibre Composites,* Proc. of a Conf. at the Nat. Phys. Lab. (Guildford, Surrey: IPC Science and Technology Press) 1971, pp 15-24.
4. ACI Committee 544, "State-of-the-Art Report on Fibre Reinforced Concrete", *ACI Journal,* November 1973, pp 729-743.
5. National Physical Laboratory, *The Properties of Fibre Composites,* Proc. of a Conf. at the Nat. Phys. Lab. (Guildford, Surrey: IPC Science and Technology Press) 1971, pp 1-90.
6. American Concrete Institute, *Fibre Reinforced Concrete,* Publ. SP-44 (Detroit: American Concrete Institute) 1974, pp 1-554.
7. Dow, N F, "Study of Stress Near a Discontinuity in a Filament Reinforced Composite", *Report R635D61* (Philadelphia: Space Sciences Laboratory, Missile and Space Division, General Electric Co) 1963.
8. Spencer, A J M and Smith, G E, "Interfacial Tractions in Fibre-Reinforced Composites", *The Properties of Fibre Composites,* Proc. of a Conf. at the Nat. Phys. Lab. (Guildford, Surrey: IPC Science and Technology Press) 1971, pp 87-89.
9. Agarwal, B D, Lifshitz, J M and Broutman, L J, "Elastic-Plastic Finite Element Analysis of Short Fibre Composites", *Fibre Science and Technology,* vol 7, 1974, pp 45-62.
10. Eshelby, J D, "The Determination of the Elastic Field of an Ellipsoidal Inclusion and Related Problems", *Proc. Roy. Soc.* (London) vol A241, 1957, pp 376-396.
11. Timoshenko, S and Goodier, J N, *Theory of Elasticity* (2nd Ed) (New York: McGraw-Hill) 1951, p 354.
12. Argon, A S, "Stresses in and Around Slender Elastic Rods and Platelets of Different Modulus in an Infinite Elastic Medium under Uniform Strain at Infinity", submitted to *J. Comp. Materials.*
13. Takagi, J, "Some Properties of Glass Fibre Reinforced Concrete", *Fibre Reinforced Concrete,* Publ. SP-44 (Detroit: American Concrete Institute) 1974, pp 93-111.
14. McClintock, F A, "Problems in the Fracture of Composites with Plastic Matrixes", unpublished, 1969.
15. Griffith, A A, "The Phenomena of Rupture and Flow in Solids", *Phil. Trans. Roy. Soc.* (London), vol A221, 1921, pp 163-198.
16. Allen, H G, Comment to paper of Aveston, Cooper, and Kelly, *The Properties of Fibre Composites,* Proc. of a Conf. at the Nat. Phys. Lab. (Guildford, Surrey: IPC Science and Technology Press) 1971, pp 25-26.
17. Majumdar, A J, "Glass Fibre Reinforced Cement and Gypsum Products", *Proc. Roy. Soc.* (London), vol A319, 1970, pp 69-78.
18. Naaman, A E, Argon, A S and Moavenzadeh, F, "A Fracture Model for Fibre Reinforced Cementitious Materials", *Cement and Concrete Research,* vol 3, 1973, pp 397-411.

19. Romualdi, J P and Mandel, J A, "Tensile Strength of Concrete Affected by Uniformly Distributed and Closely Spaced Short Lengths of Wire Reinforcement", *ACI Journal,* June 1964, pp 657-671.
20. Harris, B, Varlow, J and Ellis, C D, "The Fracture Behavior of Fibre Reinforced Concrete", *Cement and Concrete Research,* vol 2, 1972, pp 447-461.
21. Andersson, H and Bergkvist, H, "Analysis of a Non-Linear Crack Model", *J. Mech. Phys. Solids,* vol 18, 1970, pp 1-28.

3.2 Fibre pull-out in multiply-cracked discontinuous fibre composites

D K Hale
National Physical Laboratory, United Kingdom

Summary

The process of fibre pull-out in multiply-cracked aligned discontinuous fibre composites is examined and it is shown that, with fibres of uniform cross-section, pull-out will normally take place preferentially at a single crack. The conditions required for multiple pull-out and uniform extension of the composite are examined and defined. It is found, for example, that with fibres of circular cross-section an extension of 10% at a nearly constant pull-out stress should be achieved when the crack spacing is small compared with the fibre length, if the local value of the frictional stress τ_q obeys the relationship

$$\tau_q = \tau_o (1 + 8.25\, x/\ell_f)$$

where τ_o is the initial frictional stress, x is the relative displacement of fibre and matrix, and ℓ_f is the fibre length. It is shown that the required increase in the local value of τ can be obtained if the radius of the fibre increases uniformly from the centre towards the ends so that the local interfacial pressure increases during fibre pull-out. Expressions are derived relating the required increase in radius to the elastic constants of fibre and matrix, τ_o, μ the coefficient of sliding friction and V_m the volume fraction of the matrix. The conditions required to avoid premature failure of the composite through matrix-splitting are examined and fibre geometries suitable for practical application are discussed.

Résumé

Nous analysons l'arrachement (pull-out), c'est à dire le processus par lequel les fibres sont arrachées de la matrice dans les matériaux composites qui contiennent des fibres alignées et discontinues, ayant subi des fissures multiples. On démontre que s'il s'agit des fibres d'une section uniforme, l'arrachement aura lieu d'abord à une seule fissure. Les conditions sous lesquelles le composite fibreux subira des arrachements multiples et d'une extension uniforme sont examinées et définies. On trouve, par exemple, qu'avec des fibres d'une section circulaire, on devrait atteindre 10% d'extension sous contrainte à peu près constante, quand l'espacement entre les fissures est faible auprès de la longeur d'une fibre, à condition que

$$\tau_q = \tau_o (1 + 8.25\, x/\ell_f)$$

τ_q *signifie le chiffre local de la contrainte de frottement,*

τ_o *la contrainte de frottement initiale,*

x *le déplacement de la fibre par rapport à la matrice,*

ℓ_f *la longeur de la fibre.*

On démontre qu'on pourrait réaliser l'augmentation requise du chiffre local τ à condition que le rayon du fibre s'accroisse constamment depuis le centre aux extrémités, de façon à augmenter la pression locale interfaciale pendant l'arrachement. On établi des rapports liant l'augmentation requise du rayon aux modules d'élasticité de la fibre et de la matrice, à τ_o, à μ, (le coefficient de frottement glissant) et à V_m, la proportion volumétrique de la matrice. On examine les conditions requises pour éviter la panne prématurée par rupture de la matrice, et on discute des configurations de fibres convenables aux applications en pratique.

Two of the more important energy absorbing mechanisms in the failure of fibre composites are (a) fibre "pull-out" and (b) multiple cracking of the matrix (1). The first can lead to a substantial increase in the work of fracture and the second, with brittle matrix composites, to a marked degree of pseudoductility and an increased capacity for energy absorption during failure. In this paper, the process of fibre pull-out from multiply-cracked composites is examined and ways in which the maximum energy absorption might be achieved are discussed.

We consider the behaviour of an aligned discontinuous fibre composite subject to a tensile stress acting in the fibre direction. It is assumed that (a) the fibres are of uniform strength σ_{fu}, length ℓ_f, and radius r, and randomly distributed in the matrix with a fibre volume fraction V_f and a matrix volume fraction V_m, (b) both fibres and matrix are elastic with the fibres having the higher failure strain and (c) stress is transferred uniformly between fibre and matrix at a rate depending on τ the maximum shear stress which can be sustained by the interface so that, during fibre pull-out, τ is determined by the frictional forces at the fibre-matrix interface. The rate of stress transfer from fibre to matrix will then be given by

$$\frac{d\sigma}{dz} = \frac{2\tau}{r} \qquad (1)$$

where $d\sigma$ is the stress transferred in a distance dz.

Now consider pull-out of fibres at a single transverse crack in the matrix; the discussion of this simple case follows that given by Kelly (2). When a fibre is pulled out from one half of the matrix or the other, the tensile stress σ_f in the fibre at the crack surface is, from equation (1),

$$\sigma_f = \frac{2\tau y}{r} \qquad (2)$$

where y is the length of the fibre being pulled out of the matrix. If y is equal to a critical length $\frac{1}{2}\ell_c$, then the stress in the fibre will become equal to σ_{fu}, the fibre breaking stress and we have

$$\sigma_{fu} = \frac{\tau \ell_c}{r}. \qquad (3)$$

It follows that no fibres will be broken if the fibre length is less than ℓ_c and we restrict the following discussion to the case where $\ell_f < \ell_c$.

With a random distribution of fibres each of length ℓ_f, the maximum embedded length which will be pulled out will be $\ell_f/2$ and the mean pull-out length will be $\ell_f/4$. The mean tensile stress in the fibres at the crack surface $\bar{\sigma}_f$ will, from equation (2), be given by

$$\bar{\sigma}_f = \frac{\tau \ell_f}{2r}. \qquad (4)$$

Multiplying this by V_f and substituting for τ/r from equation (3), we find that pull-out will initially occur at a composite stress σ_p^o given by

$$\sigma_p^o = \tfrac{1}{2} \sigma_{fu} V_f \frac{\ell_f}{\ell_c}. \qquad (5)$$

As pull-out continues, both the mean length of fibre being pulled out and the number of fibres bridging the crack will decrease. When the crack width is p, the mean length of the fibres being pulled out will be $\frac{1}{2}\left(\frac{\ell_f}{2} - p\right)$ and the fraction of fibres still bridging the crack will be $\left(1 - \frac{2p}{\ell_f}\right)$.

The stress in the composite will be

$$\sigma_p = \tfrac{1}{2} \sigma_{fu} V_f \frac{\ell_f}{\ell_c} \left(1 - \frac{2p}{\ell_f}\right)^2$$

or

$$\sigma_p = \sigma_p^o \left(1 - \frac{2p}{\ell_f}\right)^2 . \tag{6}$$

The essential condition for multiple fracture of the matrix (1) is that the fibres bridging a crack should be able to sustain the additional load resulting from the failure of the matrix. Immediately before the matrix cracks, the stress in a composite reinforced with discontinuous fibres will be

$$\sigma_c = \bar{\sigma}_f V_f + \sigma_{mu} V_m$$

$$= \epsilon_{mu} E_f V_f (1 + \alpha) \tag{7}$$

where ϵ_{mu} is the failure strain of the matrix, E_f and E_m the Young's moduli of fibre and matrix, and $\alpha = E_m V_m / E_f V_f$.

The stress which can be sustained by the fibres alone is the pull-out stress given by equation (5) so that, for multiple fracture, we have the condition

$$\tfrac{1}{2} \sigma_{fu} V_f \frac{\ell}{\ell_c} > \epsilon_{mu} E_f V_f (1 + \alpha) . \tag{8}$$

This condition for multiple fracture in a discontinuous fibre composite has been derived previously by McLean (3).

Multiple fracture of the matrix can therefore be expected when the fibre length exceeds a value, ℓ_m, given by

$$\frac{\ell_m}{\ell_c} = 2 \frac{\epsilon_{mu}}{\epsilon_{fu}} (1 + \alpha) . \tag{9}$$

If ℓ_m is less than ℓ_c, multiple cracking followed by fibre pull-out can therefore occur with discontinuous fibre composites. With, for example, a wire-reinforced cement, we may take $E_m = 10$ GN/m², $E_f = 200$ GN/m², $V_m = 0.97$, and $V_f = 0.03$ to give $\alpha = 1.6$. If $\epsilon_{mu} = 2 \times 10^{-4}$ and $\epsilon_{fu} = 1 \times 10^{-2}$ then $\ell_m/\ell_c = 0.1$ so that if ℓ_c is 100 mm then ℓ_m will be 10 mm.

We now consider an aligned discontinuous fibre composite in which $\ell_m < \ell_f < \ell_c$ and in which multiple cracking has taken place so that the matrix is broken into blocks of length a. We assume that the composite is built up from elements such as that shown in Fig 1(a). To simplify the discussion we have assumed that the matrix cracks coincide with the fibre ends and that each fibre is embedded in a discrete number (n) of matrix blocks so that $\ell = na$. Fig 2 represents an assembly of the multiply-cracked elements.

We first assume that, when the composite is extended, the blocks of matrix move apart to leave cracks of the same width and that the fibre ends are pulled back through the blocks against the shear stresses as shown in Fig 1(b). In the assembly of nine fibres shown in Fig 2, frictional forces will act in the directions shown by the arrows to oppose fibre pull-out. It will be noted that, in each block of matrix, the frictional forces acting in one direction are balanced by those acting in the other.

We now examine what happens initially in the plane XY when the composite is extended. Fibre 5 will be pulled out of the blocks on either side of XY against shear stressses acting on a fibre length of $\ell_f/2$ in each case. Fibre 2 will be pulled out of block B against a shear stress acting on a length of fibre equal to the block length. At

Figure 1   Pull-out with multiply-cracked composite element.

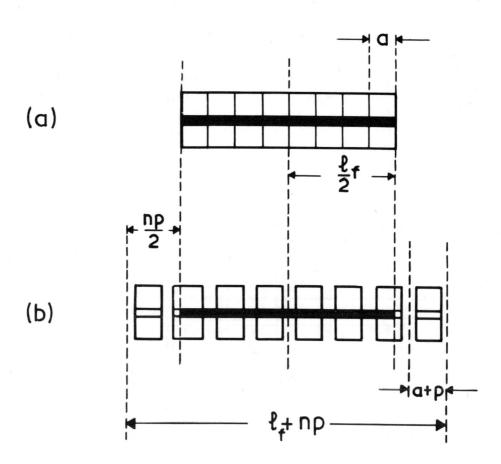

the same time it will be pulled *into* block C on the other side of the plane XY. The net shear stress acting on the fibre on one side of the plane will however be equal and opposite to that on the other. The shear stresses acting initially on the fibres crossing plane XY and determining the pull-out stress will, as with the singly-cracked system, clearly correspond to the shear stresses acting on fibre lengths ranging from zero to $\ell_f/2$.

When the crack width is p the length of the composite element shown in Fig 1 will be (ℓ_f + np). The length of fibre in the half-element which is still in contact with the matrix will be

$$\frac{\ell_f}{2}\left(\frac{\ell_f}{\ell_f + np}\right) \quad \text{or} \quad \frac{\ell_f}{2}\left(1 - \frac{np}{\ell_f + np}\right).$$

This will be the maximum length of fibre on which shear stresses can act in the same direction to resist widening of the crack. The mean effective length will be

$$\frac{\ell_f}{4}\left(1 - \frac{np}{\ell_f + np}\right).$$

Figure 2   Pull-out with multiply-cracked composite.

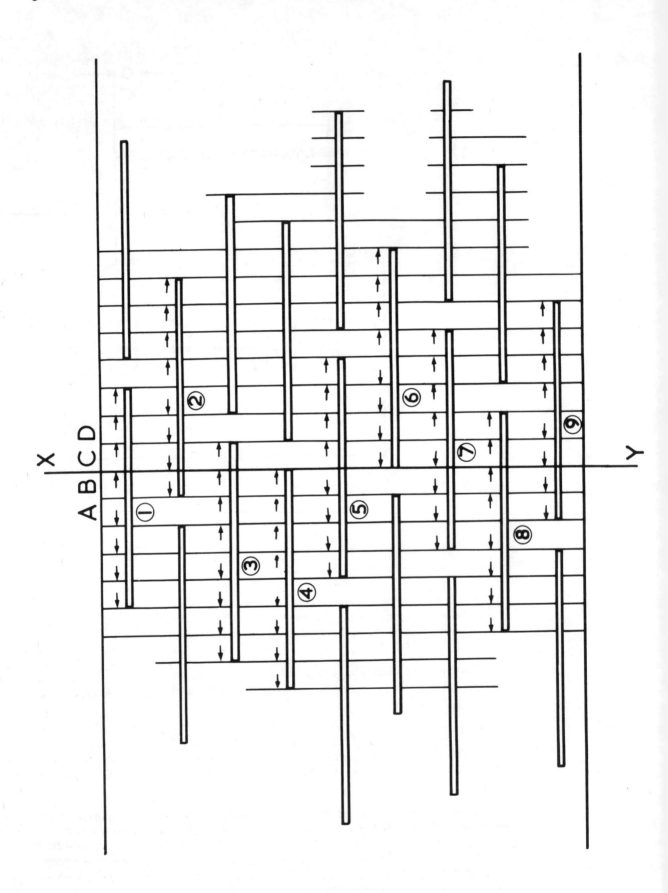

The fraction of the extended composite element length still bridged by fibre will be

$$\frac{\ell_f}{\ell_f + np} = \left(1 - \frac{np}{\ell_f + np}\right).$$

Since we have a random distribution of fibres, each part of the composite element will be represented in a thin section of the composite element assembly such as that including the crack at XY and half the blocks B and C on either side (Fig 2). This means that the fraction of fibres bridging a crack in the composite will also be

$$\left(1 - \frac{np}{\ell_f + np}\right).$$

From equations (2) and (3) we can now obtain the tensile stress σ_{mp} in the multiply-cracked composite:–

$$\sigma_{mp} = \tfrac{1}{2} \sigma_{fu} V_f \frac{\ell_f}{\ell_c} \left(1 - \frac{np}{\ell_f + np}\right)^2, \quad \ell_m < \ell_f < \ell_c . \tag{10}$$

The same expression is obtained if we evaluate the mean fibre stress in the extended composite element shown in Fig 1 and then multiply the mean fibre stress by V_f. When $p = 0$

$$\sigma_{mp}^o = \tfrac{1}{2} \sigma_{fu} V_f \frac{\ell_f}{\ell_c} \tag{11}$$

and σ_{mp}^o is therefore equal to σ_p^o for a single crack so that we can rewrite equation (10) as

$$\sigma_{mp} = \sigma_p^o \left(1 - \frac{np}{\ell_f + np}\right)^2, \quad \ell_m < \ell_f < \ell_c . \tag{12}$$

Equation (12) can be expressed in the form

$$\frac{\sigma_{mp}}{\sigma_p^o} = \left(\frac{1}{1 + \gamma}\right)^2, \quad \ell_m < \ell_f < \ell_c \tag{13}$$

where $\gamma = np/\ell_f$, the relationship between σ_{mp}/σ_p^o and the fractional extension γ of the multiply-cracked composite is shown in Fig 3.

The relationship between σ_p/σ_p^o and p/ℓ for the singly-cracked system is also shown in Fig 3. The composite stress corresponding to a given extension is always less with the singly-cracked system than with the multiply-cracked system (provided the length of the composite is greater than $\ell_f/2$) and less work is needed when extension takes place at a single crack. Consequently, in the absence of any constraint, we would expect pull-out to occur preferentially at a single crack under normal tensile loading. To achieve a uniform extension of the multiply-cracked composite along its length under normal tensile loading conditions, fibre pull-out must become progressively more difficult as a crack widens so that pull-out does not take place preferentially at a single crack.

An increase in the forces resisting pull-out can be obtained by an increase in the value of τ. If τ is determined by frictional forces at the interface then, with a fibre of circular cross-section, an increase in the local value of τ as pull-out proceeds should be achieved if the diameter of the fibre increases from the middle towards the ends (Fig 4a). With increasing pull-out, the increase in the radial pressure at the fibre-matrix interface will lead to a corresponding increase in τ.

Figure 3 **Extension during pull-out.**
(a) *Singly-cracked system (equation (6))*
(b) *Multiply-cracked system (equation (13))*
(c) *Multiply-cracked system (equation (21))*
with $A = 8.25\ \tau_o/\ell_f$.

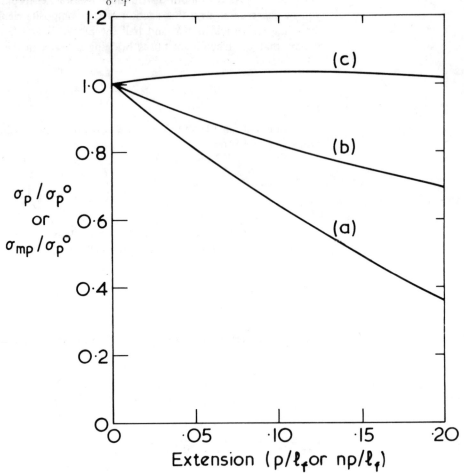

Figure 4 **Fibre geometries.**

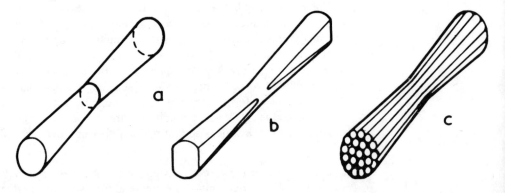

We now consider the rate at which the local value of τ has to increase with p, the crack width, in order to achieve a pull-out stress which increases significantly with p, at least in the initial stages of pull-out.

We consider a composite element such as that shown in Fig 1, but in this case we assume that the fibre has the form shown in Fig 4a and that the local value of τ within a block of matrix increases as the block moves away from its initial position towards the end of the fibre. We also assume that the radius of the fibre at the ends does not differ greatly from that in the middle so that r in equation (1) can be considered constant. We have to remember that the rate of movement of the block relative to the fibre will depend on its distance from the centre. Consider the qth block from the centre. When the crack width is p, this block will have moved through a distance

$$x = p(q - \tfrac{1}{2}). \tag{14}$$

Now suppose

$$\tau_q = \tau_o + Ax \tag{15}$$

where τ_q is the local value of τ in the qth block, τ_o is the initial value of τ and A is a constant. The stress $\Delta\sigma_q$ transferred between fibre and matrix in the qth block will be

$$\Delta\sigma_q = \frac{2a}{r}\left(\tau_o + Ap(q - \tfrac{1}{2})\right). \tag{16}$$

The total stress transferred in the half-element will correspond to the maximum tensile stress in the fibre so that, if there are n' blocks of matrix in contact with the fibre

$$\sigma_{max} = \sum_{q=1}^{q=\frac{n'}{2}} \Delta\sigma_q = \frac{n'a}{2r}\left(2\tau_o + \frac{n'Ap}{2}\right). \tag{17}$$

From equations (16) and (17) we find that the average stress in the fibre in the first block from the centre will be

$$\frac{n'a}{2r}\left(2\tau_o + \frac{n'Ap}{2}\right) - \frac{a}{r}\left(\tau_o + \frac{Ap}{2}\right) \tag{17a}$$

and the average stress in the qth block will be given by

$$\bar{\sigma}_q = \frac{n'a}{2r}\left(2\tau_o + \frac{n'Ap}{2}\right) - \frac{2a}{r}\left(q\tau_o + \frac{Apq(q+1)}{2} - \frac{qAp}{2}\right) + \frac{a}{r}\left(\tau_o + Ap(q - \tfrac{1}{2})\right)$$

$$= \frac{a}{r}\left\{\tau_o(n' + 1) + \frac{Ap}{2}\left(\frac{(n')^2}{2} - 1\right) + q(Ap - 2\tau_o) - Apq^2\right\} \tag{17b}$$

so that the average stress in the fibre in the $\frac{n'}{2}$ blocks will be

$$\frac{2}{n}\sum_{q=1}^{q=\frac{n'}{2}} \bar{\sigma}_q = \frac{an'}{2r}\left(\tau_o + \frac{Ap(n' - 1)(n' + 1)}{3n'}\right). \tag{17c}$$

This will also be the average stress in the fibre (including that between the blocks) in a length $\frac{\ell_f}{2}$ from the centre. If we put

$$n' = \frac{\ell_f^2}{(\ell_f + np)a}, \quad n = \frac{\ell_f}{a}, \quad \text{and} \quad \gamma = \frac{np}{\ell_f} \qquad (17d)$$

we find that the average stress in the fibre over the length ℓ_f is

$$\frac{\ell_f}{2r(1+\gamma)} \left\{ \tau_o + \frac{A\ell_f\gamma}{3n} \left(\frac{n}{1+\gamma} - \frac{1+\gamma}{n} \right) \right\} \qquad (17e)$$

and over the length $\ell_f + np$ is

$$\frac{\ell_f}{2r} \left(\frac{1}{1+\gamma} \right)^2 \left\{ \tau_o + \frac{A\ell_f\gamma}{3n} \left(\frac{n}{1+\gamma} - \frac{1+\gamma}{n} \right) \right\}. \qquad (17f)$$

Multiplying by V_f we obtain the stress for multiple pull-out,

$$\sigma_{mp} = \tfrac{1}{2} V_f \frac{\ell_f}{r} \left(\frac{1}{1+\gamma} \right)^2 \left\{ \tau_o + \frac{A\ell_f\gamma}{3n} \left(\frac{n}{1+\gamma} - \frac{1+\gamma}{n} \right) \right\}. \qquad (18)$$

When $A = 0$, this equation reduces to equation (10). When $\gamma = 0$

$$\sigma_o = \tfrac{1}{2} V_f \frac{\ell_f}{r} \tau_o \qquad (18a)$$

so that

$$\frac{\sigma_{mp}}{\sigma_o} = \left(\frac{1}{1+\gamma} \right)^2 \left\{ 1 + \frac{A\ell_f\gamma}{3n\tau_o} \left(\frac{n}{1+\gamma} - \frac{1+\gamma}{n} \right) \right\} \qquad (19)$$

Differentiating equation (19) we find that $d\left(\frac{\sigma_{mp}}{\sigma_o}\right) / d\gamma$ will be positive if

$$A > \frac{6\tau_o}{\ell_f} \left(\frac{n^2(1+\gamma)}{n^2(1-2\gamma)-(1+\gamma)^2} \right). \qquad (20)$$

If n is large, equation (19) can be simplified to give

$$\frac{\sigma_{mp}}{\sigma_o} = \left(\frac{1}{1+\gamma} \right)^2 + \frac{A\ell_f\gamma}{3\tau_o} \left(\frac{1}{1+\gamma} \right)^3 \qquad (21)$$

and differentiating we find that $d\left(\frac{\sigma_{mp}}{\sigma_o}\right) / d\gamma$ will be positive if

$$A > \frac{6\tau_o}{\ell_f} \left(\frac{1+\gamma}{1-2\gamma} \right) \qquad (22)$$

so that to obtain controlled multiple pull-out up to an extension of 10%, A must be equal to or greater than $8.25\, \tau_o/\ell_f$. Values of σ_{mp}/σ_o calculated using equation (21) and this value of A are plotted in Fig 3.

Figure 5   Composite element.

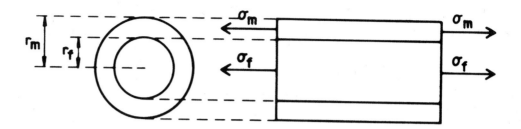

Since we have identified τ with the friction stress, we may write

$$\tau = P\mu \qquad (23)$$

where μ is the coefficient of sliding friction and P is the radial pressure at the interface. The radial pressure will be determined by the difference δ between the radius of the fibre and that of the cavity in the block of matrix if the fibre were removed. The relationship between δ and P for a simple composite element such as that shown in Fig 5 is evaluated in the Appendix where it is shown (equation (4a)) that

$$\frac{\delta}{r} = P\left(\frac{(1-\nu_f)}{E_f} - \frac{(1-\nu_m)}{E_m} + \frac{2}{V_m E_m}\right) + \left(\frac{\nu_f \sigma_f}{E_f} - \frac{\nu_m \sigma_m}{E_m}\right). \qquad (24)$$

P is the interfacial pressure, ν_f and ν_m are the Poisson's ratios for fibre and matrix, E_f and E_m the corresponding Young's moduli, and σ_f and σ_m the longitudinal tensile stresses in fibre and matrix.

The final term in brackets in equation (24) represents the "Poisson contraction" effect of the longitudinal tensile stresses on fibre pull-out. Its effect will be greatest when $\sigma_m = 0$, and it will then have a significant effect on the interfacial pressure P when $\nu_f \sigma_f / E_f$ is significant compared with δ/r. In the case of a steel wire reinforced Portland cement we may take $\nu_f = 0.29$, $E_f = 200$ GN/m² and if we then take $\delta/r = 20-30 \times 10^{-4}$, the ultimate drying shrinkage value of the cement paste (4), we find that the Poisson contraction effect is not important if σ_f is less than 140–200 N/mm² or, with $V_f = 0.03$, the composite stress is less than 4–6 N/mm². The interfacial pressure will only be reduced to zero when σ_f reaches 1.4–2 GN/m².

If, for the present, we disregard the Poisson contraction effect we may write

$$\frac{d\delta}{dP} = r\left\{\frac{(1-\nu_f)}{E_f} - \frac{(1-\nu_m)}{E_m} + \frac{2}{V_m E_m}\right\}. \qquad (25)$$

From equations (24) and (16) we have

$$\frac{dP}{d\tau} = \frac{1}{\mu} \quad \text{and} \quad \frac{d\tau}{dx} = A \qquad (25a)$$

so that

$$\frac{d\delta}{dx} = \frac{Ar}{\mu}\left\{\frac{(1-\nu_f)}{E_f} - \frac{(1-\nu_m)}{E_m} + \frac{2}{V_m E_m}\right\}. \qquad (26)$$

For multiple pull-out, A will be given by equation (20) or, if n is large, by equation (22).

For controlled multiple pull-out up to an extension of 10%, when n is large, we have previously shown that A must be equal to or greater than 8.25 τ_o/ℓ_f, so that in this case the required value of $d\delta/dx$ is given by

$$\frac{d\delta}{dx} = \frac{8.25 \tau_o r}{\ell_f \mu} \left\{ \frac{(1-\nu_f)}{E_f} - \frac{(1-\nu_m)}{E_m} + \frac{2}{V_m E_m} \right\} \qquad (27)$$

and the required percentage increase in the radius at the fibre end will be

$$\frac{d\delta}{dx} \frac{100 \ell_f}{2r} = \frac{412.5 \tau_o}{\mu} \left\{ \frac{(1-\nu_f)}{E_f} - \frac{(1-\nu_m)}{E_m} + \frac{2}{V_m E_m} \right\} \qquad (28)$$

For steel wire reinforced Portland cement, we may take $\nu_f = 0.29$, $\nu_m = 0.25$ (4), $E_f = 200$ GN/m², $E_m = 10$ GN/m² and $V_m = 0.97$. If we then assume $\mu = 0.5$ we find that, for $\tau_o = 1$ N/mm² a percentage increase in radius of 0.11% is required. For $\tau_o = 10$ N/mm² the required increase is 1.1%.

With the higher values of τ_o there is the possibility of failure through matrix-splitting. From equation (14), the n'/2 th block from the centre will have moved through a distance

$$p \left(\frac{n'}{2} - \frac{1}{2} \right) = \frac{\ell_f \gamma}{2} \left(\frac{1}{1+\gamma} - \frac{1}{n} \right). \qquad (28a)$$

If n is large and $\gamma = 0.1$, this distance will correspond to 0.045 ℓ_f so that the local value of τ calculated from equations (15) and (22) will be 1.37 τ_o. The corresponding interfacial pressure will be 1.37 τ_o/μ or 2.74 τ_o if we assume $\mu = 0.5$. The tangential stress in the matrix at the fibre surface is shown in the Appendix (equation (5a)) to be given by

$$\sigma_t = \frac{(1 + V_f)}{(1 - V_f)} P \qquad (28b)$$

so that if $V_f = 0.03$, we have $\sigma_t = 1.06P$. It follows that the maximum tangential stress, up to an extension of 10%, with A = 8.25 τ_o/ℓ_f will be 2.90 τ_o. To avoid premature failure through matrix-splitting the tangential stress should not exceed the matrix failure stress so that if, for example, the matrix failure stress is 5 N/mm² the value of τ_o should not exceed 1.72 N/mm².

This analysis indicates that it should be possible to obtain large extensions (eg up to 10%) through multiple pull-out provided failure by matrix-splitting can be avoided. Although a fibre of the form shown in Fig 4a may not be a practical possibility, similar behaviour should be observed with fibres or fibre bundles such as those shown in Figs 4b and 4c. In the case of a metal fibre, the increase in the effective radius can be achieved, for example, by slightly flattening the fibre towards the ends (Fig 4b). This will mean that the effective area of contact during pull-out will not only be decreased but will also decrease more rapidly with pull-out than with a fibre of circular cross-section. This effect will, however, be small compared with the increase in the value of τ. If we assume that a fibre of radius r_0 is deformed as shown in Fig 4b and the cross-sectional area remains constant, then it can be shown that

$$\frac{\pi}{2} \left(\frac{r_o}{r_1} \right)^2 = \frac{b}{2r_1} + \sin \frac{b}{2r_1} \cos \frac{b}{2r_1} \qquad (28c)$$

where r_1 is the radius and b the length of each arc bounding the new cross-section. If $r_o/r_1 = 0.99$ then b = 2.415 r_1, so that the effective perimeter will have been reduced from 6.22 r_1 to 4.83 r_1. This corresponds to a reduction of 23% compared with the more than eight-fold increase in the local value of τ which may be achieved with, for example, wire reinforced Portland cement, by a 1% increase in the effective radius.

ACKNOWLEDGEMENTS

The author is indebted both to Dr D McLean whose interest in this problem prompted this analysis and to Mr J Aveston for helpful discussions. The work was carried out as part of the research programme of the National Physical Laboratory.

APPENDIX

Interfacial pressure in a simple composite element

We consider a simple element such as that shown in Fig 5 in which a fibre of circular cross-section and radius r is surrounded by a sleeve of matrix with an external radius r_m such that $r^2/r_m^2 = V_f$, the volume fraction of fibre in the composite.

We wish to know the interfacial pressure which will be developed when fibre and matrix are subject to longitudinal tensile stresses σ_f and σ_m, and the difference between the external radius of the fibre and the internal radius of the sleeve of matrix (before engagement) has a known value δ.

The stresses in a circular cylinder are related by

$$\frac{u}{h} = \frac{1}{E}\left(\sigma_t - \nu(\sigma_\ell + \sigma_r)\right) \tag{1a}$$

where u is the radial displacement, h is the distance from the axis, E is Young's modulus, ν is Poisson's ratio, and σ_t, σ_ℓ, and σ_r are the tangential, longitudinal and radial stresses.

If we now consider the sleeve of matrix as a thick tube in which the surface (h = r) is subject to a pressure P while the surface (h = r_m) is unstressed we have

$$\sigma_r = \frac{r^2 P}{r_m^2 - r^2}\left(1 - \frac{r_m^2}{h^2}\right) \tag{2a}$$

$$\sigma_t = \frac{r^2 P}{r_m^2 - r^2}\left(1 + \frac{r_m^2}{h^2}\right). \tag{3a}$$

Substituting equations (2a) and (3a) in (1a) we have

$$u = \frac{r^2 P}{(r_m^2 - r^2)E_m}\left\{(1-\nu_m)h + (1+\nu_m)\frac{r_m^2}{h}\right\} - \frac{\nu_m \sigma_m h}{E_m} \tag{3a1}$$

where E_m is Young's modulus, ν_m Poisson's ratio, and σ_m the longitudinal tensile stress in the matrix.

If the matrix is disengaged from the fibre, the (negative) increase in its internal radius on removing the pressure P and the tensile stress σ_m will be equal to that produced by applying a pressure $-P$ and a stress $-\sigma_m$ so that

$$u_m = \frac{r P}{(r^2 - r_m^2) E_m}\left[(1-\nu_m)r^2 + (1+\nu_m)r_m^2\right] + \frac{r \nu_m \sigma_m}{E_m} \tag{3a2}$$

where u_m is the radial displacement of the matrix at the surface ($h = r$).

Similarly, it can be shown for the fibre that

$$u_f = \frac{rP}{E_f}(1-\nu_f) + \frac{r\nu_f \sigma_f}{E_f} \qquad (3a3)$$

where u_f is the radial displacement of the fibre at the surface ($h = r$) when fibre and matrix are disengaged, E_f is Young's modulus and ν_f Poisson's ratio for the fibre.

It follows that

$$\delta = u_f - u_m = rP\left\{\frac{(1-\nu_f)}{E_f} - \frac{(1-\nu_m)r^2}{(r^2-r_m^2)E_m} - \frac{(1+\nu_m)r_m^2}{(r^2-r_m^2)E_m}\right\}$$

$$+ r\left(\frac{\nu_f \sigma_f}{E_f} - \frac{\nu_m \sigma_m}{E_m}\right). \qquad (3a4)$$

If we now put $r_m^2 = r^2/V_f$ we have

$$\frac{\delta}{r} = P\left\{\frac{(1-\nu_f)}{E_f} - \frac{(1-\nu_m)}{E_m} + \frac{2}{V_m E_m}\right\} + \left(\frac{\nu_f \sigma_f}{E_f} - \frac{\nu_m \sigma_m}{E_m}\right). \qquad (4a)$$

From equation (3a) the tangential stress in the matrix at $h = r$ will be given by

$$t = \left(\frac{r_m^2 + r^2}{r_m^2 - r^2}\right)P = \left(\frac{(1+V_f)}{(1-V_f)}\right)P. \qquad (5a)$$

REFERENCES

1. Aveston, J, Cooper, GA and Kelly, A, "Single and multiple fracture", "The Properties of Fibre Composites", Conference Proceedings, National Physical Laboratory, IPC Science and Technology Press Ltd, 1971, Paper 2, pp 15–24.
2. Kelly, A, "Strong Solids", 2nd Ed., Clarendon Press, Oxford, 1973, p 181.
3. McLean, D, private communication.
4. Newman, K, "Composite Materials", Ed. Holliday, L, Elsevier, 1966, p 336.

3.3 Fibre spacing and specific fibre surface

Herbert Krenchel
*Structural Research Laboratory,
Technical University of Denmark*

Summary   *The average fibre spacing and the total fibre surface per unit volume of composite material are two vitally important factors governing the behaviour of a fibre-reinforced material both in the green stage, during the mixing and building-up of the composite, and in the final product.*

The average fibre spacing governs the rheological properties of the composite during mixing and also, to some extent, affects the mechanical properties of the final product, whereas the specific fibre surface is a parameter primarily governing the way in which cracking-up of the matrix takes place when the composite is strained.

The average fibre spacing is calculated from the number of fibres crossing a unit area in an arbitrary cross-section of the composite. This number is a function of the cross-sectional area of the fibre and the fibre concentration and of the type of orientation of the reinforcement. The specific fibre surface, on the other hand, is a function of the perimeter and the cross-sectional area of the fibre and of the fibre concentration, but it is independent of the orientation of the reinforcement.

Résumé   *L'écartement moyen des fibres et leur surface totale par unité de volume d'un matériau composé sont deux facteurs d'une importance vitale qui régissent le comportement d'un matériau renforcé de fibres, à la fois au cours de la phase préliminaire pendant le mélange et l'élaboration du composé, aussi bien que dans le produit final.*

L'écartement moyen des fibres gouverne les propriétés rhéologiques du composé pendant le mélange et affecte aussi, dans une certaine mesure, les propriétés mécaniques du produit final, tandis que la surface spécifique des fibres est un paramètre qui régit principalement le procédé de fissurage de la matrice quand le composé est tendu.

L'écartement moyen des fibres est calculé à partir du nombre de fibres qui se croisent par espace-unité à une intersection quel conque du composé. Ce nombre est une fonction de la zone d'intersection de la fibre et de la concentration des fibres et du type d'orientation du renforcement. La surface spécifique des fibres, d'autre part, est une fonction du périmètre et de la zone d'intersection de la fibre et de la concentration des fibres, mais elle est indépendante de l'orientation du renforcement.

FIBRE SPACING

In a composite material containing more or less uniformly distributed and closely spaced reinforcing fibres, the fibre spacing affects the behaviour of the material in a number of different ways.

During the mixing and casting, or building-up, of the composite, the rheological properties of the mix will naturally depend on how close the fibres get to each other, greater energy being required to place and distribute the matrix material in a large number of narrow fibre spaces than in a few, open ones. And when the final product comes under strain, the ability of the fibres to act as crack arrestors will also depend on how much an individual crack can open and how far it can run before it encounters a fibre.

Looking at the matter in this simple way, it will be realized that the average fibre spacing is a purely geometrical factor than can be determined statistically by considering the number of fibres crossing an arbitrary plane cross-section of the composite. The more fibres that pass a unit area of this plane section, the more densely they will naturally be packed, and it is difficult to see how the distance between the centroids of the individual stable fibres in an FRC-material can be of any significance to the material properties discussed here, (1, 2). An arbitrary crack in a brittle matrix will primarily develop along a plane through the material and will at any rate never be found running from fibre-midpoint to fibre-midpoint of randomly dispersed short pieces of fibre unless these midpoints happen to lie in a single plane, which is of course not normally the case.

It is also difficult to see why the fibre spacing should have anything to do with the efficiency of the reinforcement, which is a function of the orientation of the fibres and of their anchorage (3). The fibre spacing is a purely geometrical quantity—a length. If we produce two different composite materials with the same volume concentration of fibres V_f, the same fibre diameter and the same fibre orientation, and with exactly the same structure, although with different matrix materials, one, say, Portland cement, and the other, an epoxy, the fibre spacing in these two materials will naturally be the same even though the efficiency of the reinforcement differs considerably—at any rate in the case of short-fibred reinforcement.

Calculations

The following calculations are all based on the useful statistical rule about summed cross-sectional areas versus volume concentrations, which holds true for all well homogeneous and heterogeneous materials provided sufficiently large cross-sectional areas of the material are considered.

When an arbitrary, plane section is made through a heterogeneous material the cut face will normally look like a map of a sea with a great many islands in it. The sea is the matrix, and the plane cut through the other components of the composite—grains, particles, air-bubbles, fibres, etc—gives the islands.

We will now consider a unit area of this plane section in the composite. The area must be large enough to contain a statistically acceptable quantity of each of the different components of the material. *The summed area of all "islands" representing a given type of inclusion within the unit area will then be equal to the volume concentration of this component in the composite.*

In the case of a fibre-reinforced material this means that the summed area S_f of all cut fibre surfaces appearing within a unit area of an arbitrary, plane section of the material will always be equal to the volume concentration of fibres in the composite, regardless of the orientation of the reinforcement, see Fig. 1. Further, if one group of the fibres is selected, say, all fibres forming the angle ϕ with the normal to the section, then the summed areas $S_{f\phi}$ of their cut faces within the unit area will be equal to the volume concentration of these specific fibres within the composite, $V_{f\phi}$ ((3), pages 13–14).

It will thus be seen that, for an arbitrary, fibre-reinforced material, the number of fibres n passing a unit area of a given plane section through the material can be determined as follows:

Figure 1   Sections of a fibre-reinforced material.

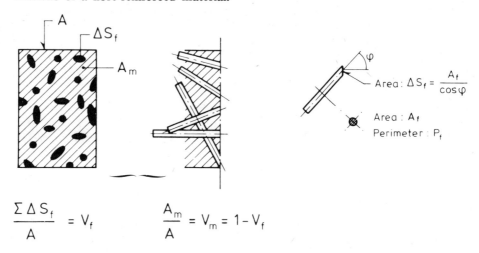

$$\frac{\Sigma \Delta S_f}{A} = V_f \qquad \frac{A_m}{A} = V_m = 1 - V_f$$

1. The entire reinforcement is divided into groups of fibres, which form the angle ϕ with the normal to the plane section investigated.

2. A unit area of the plane section is considered. The total area of the cut faces of the selected fibres within the unit area is known, see above:

$$S_{f\phi} = V_{f\phi}. \qquad (1)$$

3. The cross-sectional area A_f of each fibre (at right-angles to its longitudinal axis) is known. The area of the fibre appearing in the arbitrary plane section in the composite depends on the angle ϕ:

$$\Delta S_f = \frac{A_f}{\cos \phi}. \qquad (2)$$

4. The number of fibres passing the unit area at the angle ϕ with the normal is then:

$$n_\phi = \frac{S_{f\phi}}{\Delta S_f} = \frac{V_{f\phi}}{A_f} \cos \phi. \qquad (3)$$

5. Hence, the total number of fibres passing the unit area is obtained by summation over all fibre-angles represented:

$$n = \Sigma n_\phi. \qquad (4)$$

We can then calculate an average fibre spacing $\bar{s}$, for example as proposed by Romualdi and Mandel in 1964 (1):

$$\bar{s} = \frac{1}{\sqrt{n}}. \qquad (5)$$

However, it must be realized that the spacing thus determined will, in most cases, only be an approximation of the correct value, as the distance is not, in fact, determined from the number n alone. It also depends on the mutual placing of the fibre-faces in the plane section, as demonstrated in Fig 2a-c.

The figure shows three different, regular patterns of fibre sections in a plane: a square pattern, a triangular pattern and a hexagonal pattern. The scale used for the figures is such that they all have the same number of fibre sections within the same area (same d and same V_f, i.e. same n); however, as will be seen, the fibre spacing differs in the three cases. In Fig 2a the spacing accords with formula (5), in Fig 2b it is about 7% bigger, and in Fig 2c about 12% smaller.

Figure 2 Fibre sections in regular square pattern, triangular pattern and hexagonal pattern, respectively. All three materials shown have the same fibre concentration ($V_f \sim 0.08$) but different fibre spacing.

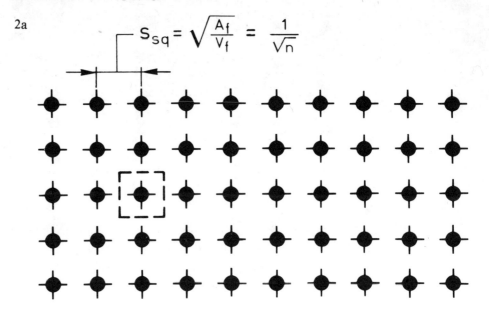

2a $\quad S_{sq} = \sqrt{\dfrac{A_f}{V_f}} = \dfrac{1}{\sqrt{n}}$

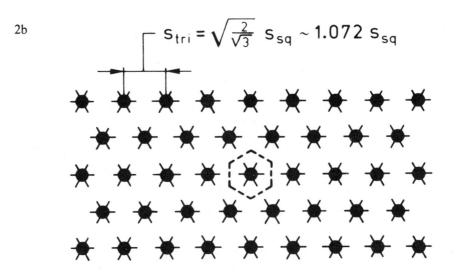

2b $\quad S_{tri} = \sqrt{\dfrac{2}{\sqrt{3}}}\; S_{sq} \sim 1.072\, S_{sq}$

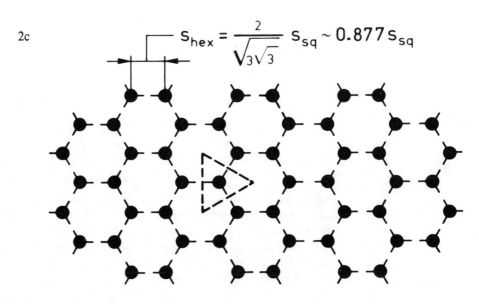

2c $\quad S_{hex} = \dfrac{2}{\sqrt{3\sqrt{3}}}\; S_{sq} \sim 0.877\, S_{sq}$

However, in most cases in practice, it will be very difficult to predict the pattern formed by the fibres in the section and consequently impossible to carry out an exact determination of the average fibre spacing. In the following we shall therefore limit ourselves to determining the approximate quantity $\bar{s}$ by means of the simple formula (5) for the square pattern.

Example 1 *All fibres parallel, plane section perpendicular to the direction of orientation*

For all the reinforcement: $\phi = 0$

of which $\Delta S_f = A_f$

$$(F3): \quad n_\phi = n = \frac{V_{f\phi}}{A_f} \cos \phi = \frac{V_f}{A_f} \tag{6}$$

$$(F5): \quad \bar{s}_{1.0} = \sqrt{\frac{A_f}{V_f}}. \tag{7}$$

In the special case of cylindrical fibres, we get:

$$\sqrt{A_f} = \frac{\sqrt{\pi}}{2} d$$

$$\bar{s}_{1.0.cyl} = \frac{\sqrt{\pi}}{2} \frac{d}{\sqrt{V_f}} \sim 0.885 \frac{d}{\sqrt{V_f}}. \tag{8}$$

Example 2 *2-d randomized fibre orientation, plane section perpendicular to the plane of orientation*

The conditions relating to the entire reinforcement are best determined by considering a representative selection of the fibres, viz, a collection of fibres all intersecting at the same point and uniformly distributed over all directions in the plane of orientation. The plane section is inserted perpendicular to this and outside the point of fibre intersection.

Figure 3 Selection of fibres in 2-d random orientation, all intersecting at the same point.

Figure 4 Selection of fibres in 3-d random orientation, all intersecting at the same point.

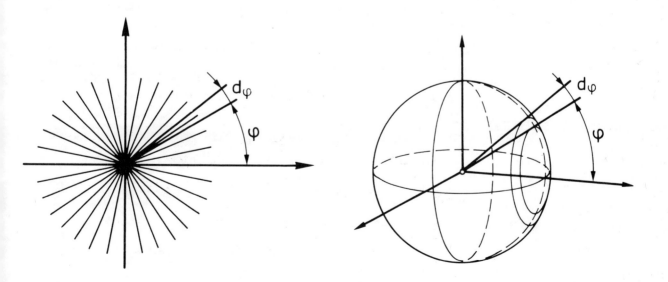

The part of the reinforcement forming the angle ϕ with the normal to the plane section lies within the angle ϕ to $\phi + d\phi$. The ratio between these fibres and the total reinforcement is then given by the ratio between the angle $d\phi$ and π, see Fig 3:

$$\frac{V_{f\phi}}{V_f} = \frac{d\phi}{\pi}$$

from which

(F3) $$n_\phi = \frac{V_{f\phi}}{A_f} \cos \phi = \frac{V_f}{A_f} \frac{1}{\pi} \cos \phi \, d\phi$$

hence

(F4) $$n = \int_{-\frac{\pi}{2}}^{\frac{\pi}{2}} n_\phi = \frac{V_f}{\pi A_f} \int_{-\frac{\pi}{2}}^{\frac{\pi}{2}} \cos \phi \, d\phi = \frac{2V_f}{\pi A_f} \qquad (9)$$

and

(F5) $$\bar{s}_{2.0} = \frac{1}{\sqrt{n}} = \sqrt{\frac{\pi}{2}} \sqrt{\frac{A_f}{V_f}} = \sqrt{\frac{\pi}{2}} \, \bar{s}_{1.0}. \qquad (10)$$

In the special case of cylindrical fibres, we get:

$$\bar{s}_{2.0.\text{cyl}} = \sqrt{\frac{\pi}{2}} \frac{\sqrt{\pi}}{2} \frac{d}{\sqrt{V_f}} = \frac{\pi}{2\sqrt{2}} \frac{d}{\sqrt{V_f}} \sim 1.11 \frac{d}{\sqrt{V_f}}. \qquad (11)$$

Example 3

3-d randomized reinforcement

The conditions relating to the entire reinforcement are also determined here by considering a representative selection of fibres, all of which intersect at the same point and are uniformly distributed over all directions in space. The plane section is inserted outside the point of fibre intersection.

The part of the reinforcement forming the angle ϕ with the normal to the plane section lies between the cones from ϕ to $\phi + d\phi$, see Fig 4. The ratio between these fibres and the total reinforcement is then given by the ratio between the area of the zone of the sphere (ϕ to $\phi + d\phi$) and the area of the hemisphere:

$$\frac{V_{f\phi}}{V_f} = \frac{2\pi r^2 \sin\phi \, d\phi}{2\pi r^2} = \sin\phi \, d\phi$$

from which

(F3) $$n_\phi = \frac{V_{f\phi}}{A_f} \cos\phi = \frac{V_f}{A_f} \sin\phi \cos\phi \, d\phi$$

hence

$$(F4) \qquad n = \int_0^{\frac{\pi}{2}} n_\phi = \frac{V_f}{A_f} \int_0^{\frac{\pi}{2}} \sin\phi \cos\phi \, d\phi = \frac{1}{2} \frac{V_f}{A_f} \qquad (12)$$

and

$$(F5) \qquad \bar{s}_3 = \frac{1}{\sqrt{n}} = \sqrt{2}\sqrt{\frac{A_f}{V_f}} = \sqrt{2}\, \bar{s}_{1.0} \, . \qquad (13)$$

In the special case of cylindrical fibres, we get:

$$\bar{s}_{3.\text{cyl}} = \sqrt{2}\, \frac{\sqrt{\pi}}{2}\, \frac{d}{\sqrt{V_f}} = \sqrt{\frac{\pi}{2}}\, \frac{d}{\sqrt{V_f}} \sim 1.25\, \frac{d}{\sqrt{V_f}}. \qquad (14)$$

A comparison of formulae (8), (11) and (14) above with formula (14) in reference (1) shows that the expression derived by Romualdi and Mandel in 1964 for the average fibre spacing:

$$S = 13.8\, d \sqrt{\frac{1}{p}} = 1.38\, \frac{d}{\sqrt{V_f}}$$

gives values that are 56% too high in the case of 1-d fibre orientation, 24% too high in the case of 2-d randomized orientation, and about 10% too high in the case of 3-d randomized orientation according to our premises.

SPECIFIC FIBRE SURFACE

The specific fibre surface (SFS) is the total surface area of all fibres within a unit volume of the composite (excluding the area of the cut fibre ends). This specific surface also provides some information on the rheological properties of the material during mixing and casting but it is primarily a parameter governing the way in which the matrix cracks when the composite is strained beyond the elongation at rupture of the matrix material. The greater the specific fibre surface the closer the cracks will be and the narrower their widths.

For so-called Ferro-cement materials this was demonstrated by Shah and Key in 1972 (4), and even for ordinary reinforced concrete and prestressed concrete, the basic crack parameter, as suggested for instance by Efsen and Krenchel in 1959 (5), is, in fact, the same as the specific surface of the reinforcement in the tensile zone of a reinforced concrete member (more correctly: our parameter β is the reciprocal value of SFS).

For all three types of materials—fibre-reinforced materials, Ferro-cement and reinforced concrete—we still have the problem of determining the exact size of a safe crack width, i.e. one that will not harmfully affect the material or its reinforcement during their useful lifetime. But, anyway, the smaller the cracks, the better, and that means as big a specific surface of the reinforcement as possible.

Calculation

The specific fibre surface can be calculated along the same lines as the average fibre spacing, except that we now imagine not one, but two parallel plane sections inserted in the composite, and calculate the surface of the fibres within a unit volume between these two sections. However, the calculations can be simplified considerably when it is realized that the specific fibre surface must be independent of the orientation of the fibres.

Let us imagine a fibre-reinforced material with a given fibre concentration V_f, a given type of fibre with a specific ℓ/d ratio and a fibre orientation chosen at random. We consider a unit volume of this material and, by way of experiment, suppose that the matrix material is a liquid, so that the fibre orientation can be altered without significant resistance (eg by means of magnetism in the case of steel fibres). It is obvious that such an alteration of the orientation of the fibres will not change the total volume of the sample as the volume of the matrix and the total number of fibres remain constant. The specific surface of the reinforcement is then directly determined from the number of fibres within the unit volume, multiplied by the curved surface of each individual fibre—in the case of infinitely long fibres: the total fibre length within the unit volume, multiplied by the periphery of the fibre.

In the following calculations we will neglect the end surface of the fibres in the determination of the specific fibre surface, partly because the end surfaces do not act in the same way as the remaining fibre surface in the determination of the anchorage of the fibres and thus have no significant influence on the formation of cracks in the matrix material, and partly because the area of these end surfaces is infinitessimal compared with the remaining fibre surface in the case of fibres with a reasonably high aspect ratio (less than 1% at $\ell_f/d > 50$ for ordinary cylindrical fibres).

Let us consider a material with a fibre concentration V_f. Each fibre in the material has the cross-section A_f and the length ℓ_f. The volume of the fibre is thus $v_f = A_f \ell_f$, whereby the number of fibres per unit volume of the composite is expressed by:

$$N = \frac{V_f}{v_f} = \frac{V_f}{A_f \ell_f}.$$

The periphery of the fibre—the outer circumference of a plane section perpendicular to the axis of the fibre—is denoted by P_f, see Fig. 1. The curved surface of the fibre is then $0_f = P_f \ell_f$.

The total fibre surface in a unit volume of the composite material is thus

$$SFS = 0_f \, N = P_f \ell_f \frac{V_f}{A_f \ell_f} = \frac{P_f}{A_f} V_f. \tag{15}$$

In the special case of cylindrical fibres, we get:

$$SFS = \frac{\pi d}{\pi/4 \, d^2} V_f = \frac{4}{d} V_f. \tag{16}$$

Example 1

Calculations on some typical FRC-materials:
All fibres parallel, plane section perpendicular to the direction of orientation.

Material 1.0.1 $\qquad\qquad$ d = 0.15 mm, V_f = 0.20

(F6): $\quad n = \dfrac{0.20}{\pi/4 \; 0.15^2} = 11.3$ fib/mm²

(F8): $\quad \bar{s} = \dfrac{\sqrt{\pi}}{2} \dfrac{0.15}{\sqrt{0.20}} = 0.297$ mm

(F16): $\quad SFS = \dfrac{4}{0.15} \, 0.20 = 5.33$ mm²/mm³

Material 1.0.2 $\qquad$ $d = 0.010$ mm, $V_f = 0.10$

(F6): $\quad n = \dfrac{0.10}{\pi/4 \; 0.01^2} = 1.273$ fib/mm^2

(F8): $\quad \bar{s} = \dfrac{\sqrt{\pi}}{2} \dfrac{0.01}{\sqrt{0.10}} = 0.028$ mm

(F16): $\quad \text{SFS} = \dfrac{4}{0.01} \; 0.10 = 40.0$ mm^2/mm^3.

Example 2 — 2-d randomized fibre orientation, plane section perpendicular to the plane of orientation.

Material 2.0.1. Fibre: glass strands with a cross-section as shown in Fig 5.

Figure 5 — **Cross-section of a glass strand consisting of 204 monofilaments each of a diameter $d = 0.010$ mm (see also (6), Figure 3).**

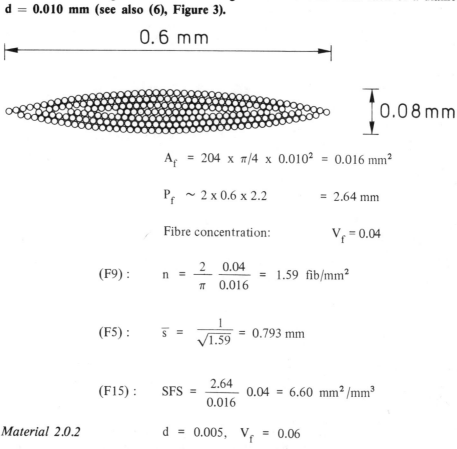

$A_f = 204 \times \pi/4 \times 0.010^2 = 0.016$ mm^2

$P_f \sim 2 \times 0.6 \times 2.2 \qquad = 2.64$ mm

Fibre concentration: $\qquad V_f = 0.04$

(F9): $\quad n = \dfrac{2}{\pi} \dfrac{0.04}{0.016} = 1.59$ fib/mm^2

(F5): $\quad \bar{s} = \dfrac{1}{\sqrt{1.59}} = 0.793$ mm

(F15): $\quad \text{SFS} = \dfrac{2.64}{0.016} \; 0.04 = 6.60$ mm^2/mm^3

Material 2.0.2 $\qquad$ $d = 0.005$, $V_f = 0.06$

(F9): $\quad n = \dfrac{2}{\pi} \dfrac{0.06}{\pi/4 \; 0.005^2} = 1.945$ fib/mm^2

(F5): $\quad \bar{s} = \dfrac{1}{\sqrt{1.945}} = 0.023$ mm

(F16): $\quad \text{SFS} = \dfrac{4}{0.005} \; 0.06 = 48.0$ mm^2/mm^3

Example 3 3-d randomized reinforcement.

Material 3.1: Same type of fibre as in material 2.0.1, $V_f = 0.02$

$$(F12): \quad n = \frac{1}{2} \frac{0.02}{0.016} = 0.63 \text{ fib/mm}^2$$

$$(F13): \quad \bar{s} = \sqrt{2}\sqrt{\frac{0.016}{0.020}} = 1.26 \text{ mm}$$

$$(F15): \quad SFS = \frac{2.64}{0.016} \, 0.02 = 3.30 \text{ mm}^2/\text{mm}^3$$

Material 3.2 $d = 0.15 \text{ mm}, \quad V_f = 0.02$

$$(F12): \quad n = \frac{1}{2} \frac{0.02}{\pi/4 \; 0.15^2} = 0.566 \text{ fib/mm}^2$$

$$(F5): \quad \bar{s} = \frac{1}{\sqrt{0.566}} = 1.33 \text{ mm}$$

$$(F16): \quad SFS = \frac{4}{0.15} \, 0.02 = 0.53 \text{ mm}^2/\text{mm}^3$$

Material 3.3 $d = 0.005 \text{ mm}, \quad V_f = 0.012$

$$(F12): \quad n = \frac{1}{2} \frac{0.012}{\pi/4 \; 0.005^2} = 306 \text{ fib/mm}^2$$

$$(F5): \quad \bar{s} = \frac{1}{\sqrt{306}} = 0.057 \text{ mm}$$

$$(F16): \quad SFS = \frac{4}{0.005} \, 0.012 = 9.6 \text{ mm}^2/\text{mm}^3$$

It is interesting to see how small is $\bar{s}$ and how big SFS is, even in the case of a comparatively low fibre concentration, when d is as small as it is in materials 1.0.2, 2.0.2 and 3.3.

For the purposes of comparison it can be mentioned that the specific surface of the reinforcement in Ferro-cement normally lies between 0.120 and 0.500 mm²/mm³ (ie 120 to 500 m²/m³) (4, 7), and in ordinary reinforced concrete members, between 0.005 and 0.020 mm²/mm³ (ie 5 to 20 m²/m³) (5). The crack spacing at failure lies between 5 mm and 20 mm in Ferro-cement (4) and between 50 mm and 200 mm in ordinary reinforced concrete members (5).

If the linear relationship between the number of cracks at failure and the specific surface of the reinforcement found by Shah and Key for Ferro-cement ((4), Fig. 4) holds good even at much bigger values of SFS, we will find crack spacings at failure in the FRC-materials of between 0.05 mm (for SFS = 50 mm^2/mm^3) and 5.0 mm (for SFS = 0.5 mm^2/mm^3).

REFERENCES

1. Romualdi, J P and Mandel, J A, "Tensile Strength of Concrete Affected by Uniformly Distributed and Closely Spaced Short Lengths of Wire Reinforcement", ACI Journal, Proceedings, Vol 61, No 6, June 1964, pp 657-670.
2. Abolitz, A L, Agbim, C C, Untrauer, R E and Works, R E, Discussion of the paper by Romualdi and Mandel (1), ACI Journal, Proceedings, Vol 61, No 12, Disc 61-38, Dec 1964, pp 1651-1656.
3. Krenchel, H, "Fibre Reinforcement", Dissertation, Technical University, Copenhagen, Akademisk Forlag, 1964.
4. Shah, S P and Key, W H, "Impact Resistance of Ferro-Cement", Journal of the Structural Division, ASCE, Vol 98, No ST1, Proc. Paper 8640, Jan 1972, pp 111-123.
5. Efsen, A and Krenchel, H, "Tensile Cracks in Reinforced Concrete", Ingenioren (Copenhagen), No 3, Feb 1959, pp 101-110.
6. Majumdar, A J, "A Method for Assessing the Quantity and Distribution of Glass Fibre in an Opaque Matrix", Journal of Material Science, 9, 1974, Letters, pp 512-514.
7. Nervi, P L, "Aesthetics and Technology in Building", Harvard University Press, Cambridge, Mass., 1965.

3.4 Mechanics of glass fibre reinforced cement

N G Nair
*Pilkington Brothers Limited, Research and Development Laboratories,
Lathom, Ormskirk, Lancashire*

Summary Glass fibre reinforced cement made by the spray moulding technique has a complex microstructure which is very different from the classically assumed 'single filament in a homogeneous matrix' model. This paper presents the results of a micromechanics analysis which takes into account strand geometry, cement penetration into strands and matrix structure.

A new unit cell has been defined and the stress distributions and elastic properties prior to matrix cracking have been theoretically determined by a shear lag type analysis. A fracture mechanics analysis has been used to study the propagation of microcracks and to interpret the transitional region of the stress-strain curve prior to the onset of multiple cracking.

Résumé Le ciment renforcé de fibres de verre, manufacturé par moulage par projection a une microstructure complexe très différente du modèle de l'hypothèse classique du "filament simple dans une matrice homogène". Cette communication présente les résultats d'une analyse des micromécanismes qui tient compte de la géométrie des fils, de la pénétration du ciment dans les fils et de la structure de la matrice.

On a défini une nouvelle cellule-unité et on a établi théoriquement les répartitions des contraintes et les caractéristiques élastiques avant fissuration de la matrice par une analyse du genre temps de cisaillement. On a fait une analyse des mécanismes de fissuration pour étudier la propagation des microfissures et interpréter la zone de transition de la courbe contrainte/déformation avant le commencement de la fissuration multiple.

INTRODUCTION

Glass fibre reinforced cement (GRC) made by incorporating an alkali-resistant glass fibre sold under the trade-mark "Cem-FIL" into a cementitious matrix is a relatively young member of the composite family of materials, but it is already seen as a promising material for civil engineering applications. The wide scale use of any material, however, requires among other things a considerable understanding of its mechanical and physical characteristics, and with this objective in mind an extensive investigation into the basic mechanics of GRC is being carried out at Pilkington Research Laboratories. This paper summarises some of the results of the theoretical studies carried out as a part of this overall research programme.

The purpose of these theoretical studies is to highlight certain microscopic characteristics and structural features of the GRC made by the spray moulding method (1) and to establish their influence on the mechanical behaviour, in particular on the tensile stress-strain curve of the material. The work dealing with the stress-strain curve up to the bend point (see Figure 1) has been completed and forms the basis of this paper. Oakley and Proctor (2), in a separate paper, give certain behavioural aspects of the material beyond the onset of multiple fracture.

MICROSCOPIC CHARACTERISTICS

GRC is a composite in which a strong brittle fibre is used to strengthen and toughen a weak brittle matrix. The basic mechanics of this material is no different from that of other ceramic and brittle matrix composites, or reinforced plastics and metals. But, both in its microscopic structure and in its characteristics, it is one of the most complex composites ever made. Some of the special characteristics are given below.

The purpose of reinforcing the cement matrix is not quite the same as reinforcing a plastic or metal. For example, in the case of plastics, fibres are added to strengthen and stiffen the matrix, while cement is relatively stiff but brittle and fibres are added to inhibit crack propagation, increase the tensile strength and fracture energy and to provide a degree of ductility in the behaviour of the material and a local stress relieving mechanism. The localised cracking of the cement gives the required energy absorbing

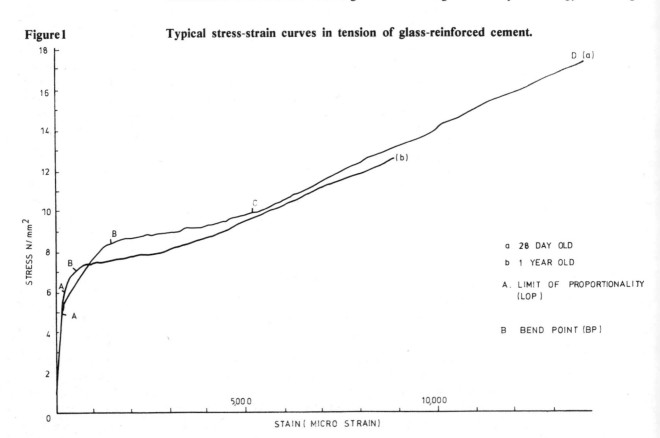

Figure 1   **Typical stress-strain curves in tension of glass-reinforced cement.**

mechanism, but the catastrophic propagation of these cracks is prevented by the presence of unbroken bridging fibres which hold the matrix together and confer a load-bearing function on the locally cracked material. For this type of function neither very stiff nor very high fibre content is required. Glass fibre which is three times stiffer and 300 times stronger than the cement matrix is an ideal reinforcing material and 5 to 10% of glass by volume is adequate for this function.

The microscopic characteristics of cement are also different from those of plastics and metals. Hardened cement has a very low tensile strength (1.4–7 N/mm²) and failure strain ($200 - 600 \times 10^{-6}$). Microscopically, it has a porous structure which itself changes with age due to continuing hydration of cement. The size and extent of the porosity depends on the w/c ratio and the fabrication and curing methods. The Young's modulus, shear modulus and compressive strength of cement paste vary with the porosity.

Like concrete and cement mortar, plain cement paste contains, albeit to a lesser extent, randomly oriented and distributed microcracks. Differential shrinkage of cement paste and the presence of unhydrated cement particles or additives like sand determine the size and density of the microcracks. Wetting and drying shrinkage, alternate freezing and thawing, together with the constraining effects of fibres, could increase the microcracking of cement paste within the composite. The failure of cement paste is considered to be by the slow growth and merging of such microcracks to form macroscale cracks which then catastrophically propagate through the section. The total effect of all the voids, inclusions and microcracks is to make the stress-strain relation somewhat non-linear and the strength and failure strain erratic. In other words, unreinforced cement paste possesses neither a stable stress-strain curve nor a well-defined tensile strength, and, therefore, cannot be used to bear tensile, bending or cyclic loads.

Glass fibre, unlike cement paste, has a well-defined Young's modulus and the stress-strain behaviour is linear up to the point of failure. The spray method of fabrication inflicts relatively little damage on the fibre.

The bonding characteristics of the glass-cement interface play a significant role in determining the composite behaviour (3). The interface bond in a 28 day old material is predominantly frictional, but with age, it may remain frictional or change to an adhesive bond followed by frictional slip after debonding. There are indications that the bond strength increases with the age of the composite (3), however, the relative increase in the magnitudes of the two forms of bond has not yet been conclusively determined.

MICROSTRUCTURE OF GRC

The major difference of GRC from most other composites is in its special microstructure. Typical microphotographs of spray moulded GRC given in Figure 3 of Reference 2 and Plate 1 of Reference 4 reveal these special features. Because the fibres are sprayed as strands of 204 filaments and because the cement particles with their 20 μm average size are unable to penetrate into the strand, the strands remain integral as short bundles or fibre rich layers separated by cement as schematically illustrated in Figure 2. The low glass content and the lack of filament dispersion make the proportion of such fibre rich layers small compared to the cement paste. The thickness of cement paste between two strands stacked one above the other is about 560 μm compared to the 40 μm between fibres that are uniformly dispersed.

The fibre rich regions are oval or nearly rectangular in cross-section and are randomly packed in two dimensions. However, in many spray moulded GRC products, a preferential orientation of strands in certain directions has been observed.

The continuous change of microstructure with age is another feature that is of some significance. The fibre bundle of 28 day old material has practically no cement hydrates within it and, therefore, tends to act as a single reinforcing element. As chemical reaction continues in the cement and to some limited extent at the interface, the reaction products penetrate into the strand and those filaments in the strand that are covered by cement

Figure 2   Idealised unit cell of glass fibre reinforced cement.

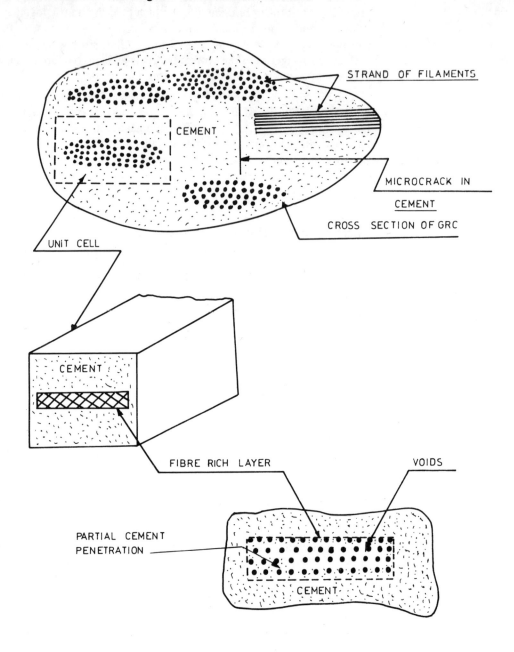

products this way tend to act themselves as independent reinforcements. In 6–12 months' time, the entire strands of a water-stored GRC are filled by the reaction products (4), and the fibre bundle may be effectively modified from a single reinforcement to a 'miniature composite'. The bonded surface area increases from the initial surface area of the bundle to the total surface area of all the filaments in the bundle.

MICROSTRUCTURAL PARAMETERS

The commonly used microstructural parameters such as volume fraction, fibre length, fibre spacing etc, are insufficient to represent the special microstructure of GRC correctly in a theoretical analysis. Four new parameters have, therefore, been defined and by varying these, the material behaviour for a wide range of microstructural conditions and ageing characteristics can be studied. The parameters are listed as follows:-

Filament dispersion factor, f_d

Filament dispersion factor, f_d, indicates the degree of dispersion of filaments for a given fibre volume fraction and is defined as:

$$f_d = \frac{\text{Volume of fibre rich regions}}{\text{Total volume of composite}} \quad (1)$$

Multiplying the numerator and the denominator by the total volume of glass, f_d can be shown as

$$f_d = \frac{V_f}{V'_f} \quad (2)$$

where V_f and V'_f are the fibre volume fractions of the whole composite and the fibre rich regions respectively. f_d varies from $1.1 V_f$ for the closely packed bundle to 1.0 for a uniform dispersion of filaments.

State of filamentisation

The fibres within the composite are said to be in a *non-filamentised* state if all the filaments in a strand act collectively as one reinforcement. On the contrary, fibres are in a *fully-filamentised* state if each of the filaments can act as independent reinforcement. Depending upon the degree of penetration of cement and the dispersion of filaments, a strand could be in partially filamentised state between the two extremes defined above.

A strand could be in non-filamentised state in the longitudinal direction either when there is no cement within the strand which is the case for closely packed bundles, or when the penetrated cement products have plenty of voids or cracks in them which make them ineffective as a matrix continuum. When there is no cement within the strand, it acts as a void in the transverse direction. But, a non-filamentised fibre bundle with non-uniformly penetrated cement products, depending upon the cement content may offer some transverse rigidity and in the extreme case could act as a composite in the transverse direction. This special case in which the non-filamentised strand acts as a composite transversely is called a *transversely filamentised* state.

It may be noted that although filament dispersion encourages filamentised behaviour, it is the state of the cement within the strand which really determines the state of filamentisation.

Bundle efficiency factor, f_e

The effectiveness of a non-filamentised strand as reinforcement depends on how well it is bonded to the cement around the strand and how well the load is transferred to all the filaments. To represent the effectiveness, a bundle efficiency factor, f_e, is defined as the ratio of the total bonded surface area of all the filaments in a bundle of length equal to the fibre radius r to the cross-section area of all the load bearing filaments in that bundle. Taking N as the total number of filaments and as the number of load bearing filaments, f_e can be mathematically expressed as:

$$f_e = \frac{(\Sigma_{i=1}^{N} \alpha_i \pi r) r}{N' \pi r^2} \quad (3)$$

α_i is a parameter that represents the fraction of the surface area of the ith fibre bonded to the cement. It is 2.0 when the filament is completely bonded, 1.0 when only half is bonded and 0.0 for no bonding. f_e varies from 0.0 to 2.0 and increases with age of the composite.

Fibre orientation function, $P(\theta)$

The strands, although sprayed generally in a random fashion in two dimensions, could be aligned or preferentially oriented in order to improve the property in any required direction. A faulty spray technique may give a preferential orientation instead of random orientation. The parameter $P(\theta)$ accounts for the orientation of strands in a composite and is the probability that a strand can be found at any point in the composite oriented in the direction θ with respect to a chosen axis. $P(\theta)$ is such that

$$\int_{-\frac{\pi}{2}}^{\frac{\pi}{2}} P(\theta)\, d\theta = 1.0$$

and

$$P(\theta) \geq 0.0. \tag{4}$$

For a perfect random orientation $P(\theta)$ is constant for $-\frac{\pi}{2} \leq \theta \leq \frac{\pi}{2}$ and is $\frac{1}{\pi}$

MODEL UNIT CELL

A unit cell is the smallest region of the material that behaves as a composite. All the theoretical studies of GRC published so far (5–8) take a single fibre embedded in

Table 1 **Dimensions and properties of the model unit cell of GRC**

(a) Microscopic dimensions and properties relevant to the analysis

Fibre content by volume, V_f	5%	Bundle efficiency factor, f_e	0.4
Thickness of the unit cell	0.67 mm	Fibre length (also length of unit cell)	30.0 mm
Width of the unit cell	0.75 mm	Fibre diameter	13.0 μm
Width of the fibre rich layer	0.65 mm	Number of filaments in strand	204
Thickness of the fibre rich layer	0.113 mm	Young's modulus of glass fibre, E_f	70000 N/mm²
Effective fibre content V'^1_f of fibre rich layer	34%	Young's modulus of cement, E_m	25000 N/mm²
		Poisson's ratio of glass, ν_f	0.2
Filament dispersion factor, f_d	0.15	Poisson's ratio of cement, ν_m	0.25

(b) Formulae for determining the elastic properties of the unit cell

	Non-filamentised state	Transversely filamentised state	Fully filamentised state
E_ℓ	$E_f V_f \dfrac{N'}{N} + E_m (1-f_d)$	$E_f V_f \dfrac{N'}{N} + E_m (1-f_d)$	$E_f V_f + E_m (1-V_f)$
E_t	$E_m (1-f_d)$	$E_m \left[\dfrac{f_d\, E_f (1+2V'_f) + 2E_m(1-V'_f)}{E_f(1-2V'_f) + 2E_m(1+V'_f)} + (1-f_d) \right]$	
$G_{\ell t}$	$G_m (1-f_d)$	G_m	$G_m \left[\dfrac{G_f(1+V'_f) + G_m(1+V'_f)f_d}{G_f(1-V'_f) + G_m(1-V'_f)} + (1-f_d) \right]$
$\nu_{\ell t}$	$\nu_f V_f + \nu_m (1-f_d)$	$\nu_f V_f + \nu_m (1-f_d)$	$\nu_f V_f + \nu_m (1-V_f)$

cement as the unit cell, which does not truly model the microstructure described previously. In the analysis outlined here, one strand with the cement surrounding it is extracted as a unit cell of the composite as illustrated in Figure 2. The fibre rich layer and the unit cell itself are approximated to be rectangular in shape. The typical dimensions of a unit cell calculated from 34 measurements made by using the Quantimet Image Analising Microscope are tabulated in Table 1 along with other parameters and properties used in this analysis.

The composite with a known fibre orientation function, $P(\theta)$, can be considered in this modelling as a collection of unit cells oriented in the same way as the strands with the same function $P(\theta)$. The packing of enough such unit cells makes a subregion of the composite the behaviour of which is representative of the overall behaviour of GRC.

f_e and $P(\theta)$ have not been experimentally determined. $P(\theta)$ is, therefore, taken as $\frac{1}{\pi}$, and f_e is taken as 0.4 assuming that only the outer surface of the outermost filaments are bonded to the cement.

MICROSCOPIC STRESS DISTRIBUTION

A knowledge of the stress and strain distributions within the micro-region of the composite is essential not only for predicting the elastic properties, but also for assessing the microscopic failure and crack arrest mechanisms.

Shear-lag type analyses have been carried out on the unit cell to find the stress distribution within the material under an external tensile stress. The unit cell has been considered for analysis as a microlaminate with the fibre rich layer sandwiched between two layers of cement. The following assumptions have been made in the analysis:

(i) The fibre rich layer containing a non-filamentised strand behaves as an equivalent homogeneous material.

(ii) The fibre rich layer with a filamentised strand behaves as a composite with fibre volume fraction V'_f.

(iii) Cement does not creep.

(iv) The stresses are so small that the shear stress at the interface is less than static frictional shear. There is no slip and elastic load transfer is valid.

The analysis has shown that the stress distributions on either end of the unit cell are greatly influenced by the microstructural parameters of the material. The load transfer length is not governed by the filaments, but by the strand as well as by the size and stiffness of the fibre rich layer. The peak stresses in the fibre and fibre rich layers and the maximum interface shear stress at the fibre ends are plotted in Figure 3a against a filament dispersion factor. The results for the filamentised and non-filamentised states show that non-filamentised fibres particularly when they are closely packed carry a greater proportion of stress than the filamentised fibres. Shear stress, on the other hand, is higher around the filamentised fibres which means that they will debond at an earlier stage in the stress-strain curve than the non-filamentised fibres would do.

INITIAL ELASTIC PROPERTIES

The stiffness constants of a composite in which the strands are oriented with a function $P(\theta)$, can be derived from the elastic properties of the unit cell. To derive the constants in any direction, the elastic properties of all the unit cellls at different orientations are resolved in the required direction and a weighted integration is carried out between $\theta = -\pi/2$ and $\theta = \pi/2$. A non-random orientation of strands can make the material generally orthotropic. The six stiffness constants Q_{11} to Q_{66} for this general case can be given as follows:-

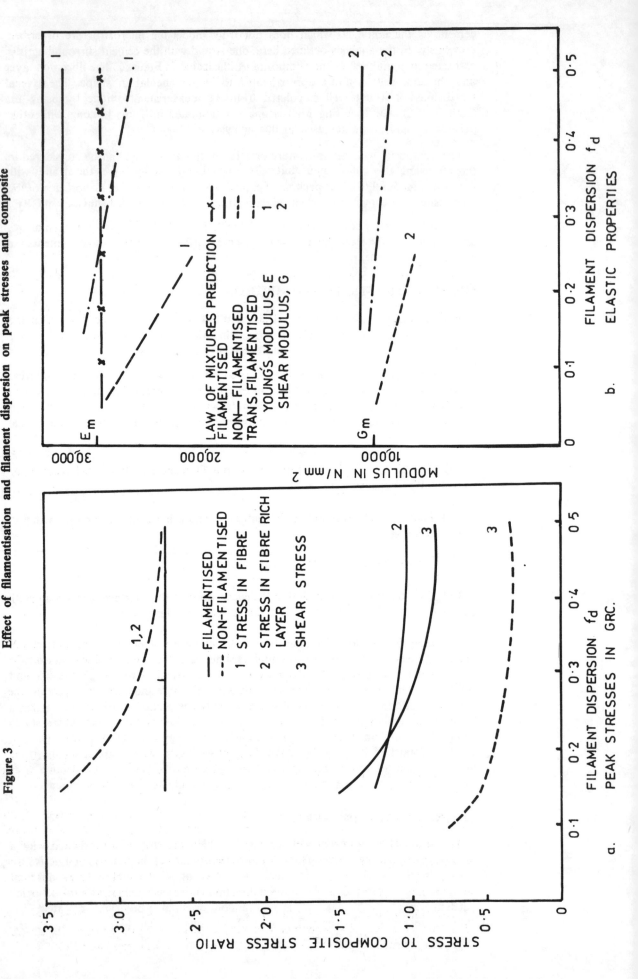

Figure 3   Effect of filamentisation and filament dispersion on peak stresses and composite elastic properties

$$\begin{bmatrix} Q_{11} \\ Q_{22} \\ Q_{12} \\ Q_{66} \\ Q_{16} \\ Q_{26} \end{bmatrix} = \int_{-\frac{\pi}{2}}^{\frac{\pi}{2}} P(\theta) \begin{bmatrix} m^4 & n^4 & 2m^2n^2 & 4m^2n^2 & -- \\ m^4 & m^4 & 2m^2n^2 & 4m^2n^2 & -- \\ m^2n^2 & m^2n^2 & m^4+n^4 & -4m^2n^2 & -- \\ m^2n^2 & m^2n^2 & -2m^2n^2 & (m^2-n^2)^2 & -- \\ m^3n & -mn^3 & mn^3-m^3n & 2(mn^3-m^3n) & -- \\ mn^3 & -m^3n & m^3n-mn^3 & 2(m^3n-mn^3) & -- \end{bmatrix} \begin{bmatrix} \beta E_\ell \\ \beta E_t \\ \beta \nu_{\ell t} E_\ell \\ G_{\ell t} \\ 0 \\ 0 \end{bmatrix} d\theta \qquad (5)$$

where $m = \cos\theta$, $n = \sin\theta$ and $\beta = \frac{1}{1-\nu_{\ell t}\nu_{t\ell}}$. For a perfect random orientation, however, the material is planar isotropic. By taking $P(\theta)$ as $1/\pi$ and integrating Equation 5 and by using the formula (9)

$$E = Q_{11}\left(1 - \frac{Q_{12}}{Q_{11}}\right)^2 \qquad (6)$$

$$\nu = \frac{Q_{12}}{Q_{11}} \qquad (7)$$

$$G = Q_{66} \qquad (8)$$

the three elastic properties E, G, and ν of the isotropic medium can be derived as follows:-

$$E = [\tfrac{3}{8} E_\ell + \tfrac{3}{8} E_t + \tfrac{1}{4}\nu_{\ell t} E_t + \tfrac{1}{2} G_{\ell t}(1-\nu_{\ell t}\nu_{t\ell})] \frac{1-\nu}{1-\nu_{\ell t}\nu_{t\ell}} \qquad (9)$$

$$G = [\tfrac{1}{8} E_\ell + \tfrac{1}{8} E_t - \tfrac{1}{4}\nu_{\ell t} E_t + \tfrac{1}{2} G_{\ell t}(1-\nu_{t\ell}\nu_{\ell t})] \frac{1}{1-\nu_{\ell t}\nu_{t\ell}} \qquad (10)$$

$$\nu = \frac{E_\ell + E_t + 6\nu_{\ell t} E_t - 4G_{\ell t}(1-\nu_{\ell t}\nu_{t\ell})}{3E_\ell + 3E_t + 2\nu_{\ell t} E_t + 4G_{\ell t}(1-\nu_{\ell t}\nu_{t\ell})}. \qquad (11)$$

The terms E_ℓ, E_t, $G_{\ell t}$ and $\nu_{\ell t}$ are the longitudinal modulus, transverse modulus, shear modulus and Poisson's ratio respectively of the unit cell. They differ for different states of filamentisation and the formulae for the non-filamentised, transversely filamentised and the fully-filamentised states are tabulated in Table 1.

Figure 3b shows how the Young's modulus and shear modulus obtained from Equations 9–10 vary with filamentisation and filament dispersion. At the filamentised state E and G do not vary with f_d, but for the other two states shown, E and G decrease slightly as f_d increases. Filamentised fibres improve the Young's modulus to a level slightly above the Young's modulus of the cement matrix. Non-filamentised state, on the contrary, because of the voidage tends to reduce the Young's modulus.

The modulus predicted by the law of mixture's rule which is also shown in Figure 3b does not consider the dispersion of filaments or the states of filamentisation of strands. However, because of the low fibre content, it appears to be a good approximation for the fibre composite. All the predictions give values very close to the Young's modulus of plain cement paste, and because of the variability in the Young's modulus of cement paste itself it is hardly possible to distinguish in an experiment the influence of the microstructure on the measured composite modulus.

MICROSCOPIC FAILURE MECHANISMS AND THEIR INFLUENCE ON THE STRESS-STRAIN CURVE.

The stress-strain relation of the composite remains linear with the slope equal to the Young's modulus predicted by Equation 9 until some form of microscopic failure capable of modifying the average composite stress or strain takes place. The subsequent change in the slope is determined by the nature of failure and the extent of damage which has taken place due to the failure.

Debonding of fibres at their ends followed by frictional slip and the microcracking of cement paste are the two likely failure mechanisms at this stage of material behaviour. Frictional slip causes a redistribution of stress and strain near the fibre ends over a length equal to the frictional load transfer length. Because the fibre content is small and the fibres are several times longer than the length over which changes take place, fibre slip makes very little difference to the Young's modulus and is unnoticeable on a macroscopic scale.

Microcracking, on the contrary, has a much more pronounced effect and has, therefore, been investigated further. Microcracking, as envisaged here, is the slow and stable crack growth within cement paste confined between strands, as distinct from the unstable crack growth considered by Aveston et al (8) which is a macroscopic cracking that propagates across several fibres and eventually through the complete section. Only those existing cracks which are smaller in size than the strand spacing qualify for this treatment. Cracks that are larger than strand spacing and are bridged by fibres have greater resistance to growth because part of the work done by the external loads is dissipated in destroying the bond and in causing frictional slippage and the energy available for creating crack surface is correspondingly less. An energy balancing criterion similar to the one used by Aveston et al (8) is aplicable to this latter case.

In order to establish how the strands and their geometry influence the microcracking of cement, a fracture mechanics analysis of a tunnel-shaped crack situated between two fibre rich layers as illustrated in Figure 4 has been undertaken. The influence is measured in terms of the changes in the stress intensity factor, K_i. One effect of the strand is that the voids within a non-filamentised or partially filamentised strand make the cement matrix discontinuous in the thickness direction and the crack in such a case could behave as if it is in a finite width strip. On the other hand, the stiffness of the fibre rich layer in the direction normal to the plane of the crack can exert pinching forces, on the crack tip, as postulated by Romualdi and Batson (11) which may tend to close the crack. To study the finite width effect, it was first considered that the crack is in a block of plain cement with width equal to the spacing of strands. Fibre rich layers were later superposed on either side of the crack for evaluating the stiffening effect of the strands.

Stress intensity factors, K_i, have been found for both the cases described above by a finite element analysis making use of the distorted element technique (11), and the values of K_i obtained for various crack sizes are plotted in Figure 4. The results for plain cement paste obtained by the finite element technique are in close agreement with the analytical results (also shown in Figure 4) from Paris and Sih (12) for a centrally cracked finite width strip, which confirms the reliability of the former technique. The bottom half of Figure 4 shows the percentage changes in the cracking stress level resulting from the finite width and the stiffening effects.

Because of the stiffening effect of the strands aligned at right angles to the plane of the cracks, K_i decreases and the cracking stress increases. The maximum increase in the cracking stress is about 46% and for most crack sizes it is less than 10%. The low glass-to-cement modular ratio and the wider spacing of strands are the primary reasons

Figure 4 Fracture arrest characteristics of GRC.

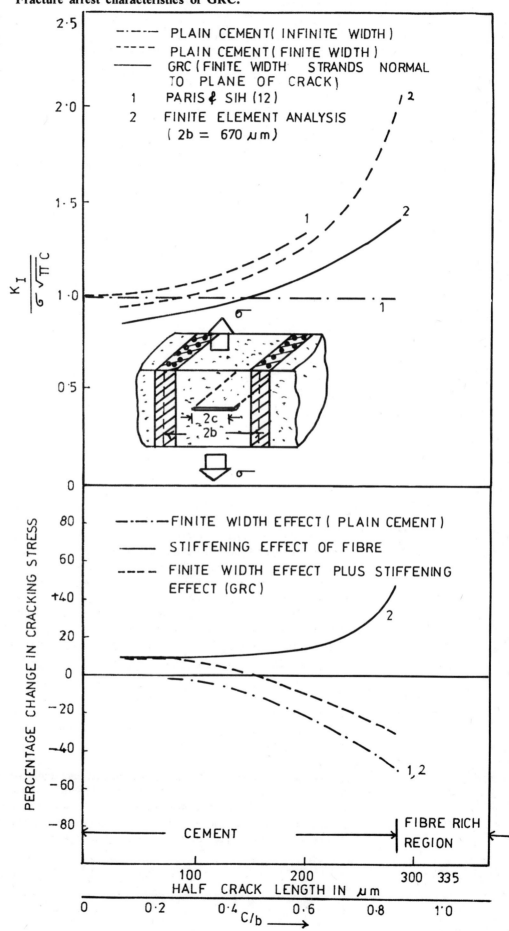

for such small improvements in the cracking stress. Furthermore, a crack lying between the fibre rich layers is influenced not by the stiffness of the fibres, but by the stiffness of the fibre rich layer which is only a fraction of that of glass.

The finite width effect, on the contrary, reduces the cracking stress by as much as 51%, and the effect is so predominant that even with the stiffening effects superposed on the finite width effect there is still a reduction in the cracking stress level for large size cracks. However, the discontinuities or voids in the strand act as barriers to the advancing cracks by preventing them from propagating across the section.

It is appropriate now to consider how the above-described effects can modify the stress-strain curves of a 28 day old and one year old GRC. The effect of microcracking is to decrease the slope of the stress-stain curve, and the change in slope depends on the size and density of cracking. Although isolated cracking could occur at very low stress levels making small changes in the slope, it is at the limit of proportionality (LOP) where the intensity of cracking is sufficiently high to make a marked change in the slope. The level of LOP can be determined by finding the stress at which the intense cracking begins in the composite. The largest cracks in cement paste which are vulnerable to microcracking and which are capable of making a noticeable change in the shape of stress-strain curves are the cracks near the fibre ends, as illustrated in Figure 2. For these cracks, fibre rich layers are spaced at twice the unit cell size distance and the crack can be as large as the thickness of cement between the strands. In a twenty-eight day old material, the strands have no cement penetration and this introduces a finite width effect. At the same time, if the strands are so oriented that their fibre axis is parallel to the plane of the cracks, they do not offer the stiffness effect. Therefore, the cracks of the type considered above in the cement paste close to the fibre ends lying between two transversely oriented strands have the least resistance to propagation. Applying Griffiths' theory for such a crack, the cracking stress can be predicted as:

$$\sigma = \eta(c) \sqrt{\frac{2E\gamma}{\pi(1-\nu^2)C}} \qquad (12)$$

where $\eta(c)$ is the reduction in stress due to the finite width effect obtained from Figure 4 for a crack of length 2c. Taking the surface energy γ as $4 J/m^2$, the crack length as 1.05 mm, crack length to plate width ratio as 0.85 and $\eta(c)$ as 0.49, it can be shown that the crack which would normally propagate in plain cement paste at $11.38 N/mm^2$ will propagate because of the finite width effect at $5.58 N/mm^2$. As the stress level increases, further similar cracking will take place in the cement paste layer between strands. Since the advancing cracks are stopped by the voids within the strands they cannot develop into macroscopic cracks. The process of microcracking will continue up to the turn-over point (TOP) where macroscopic cracking begins. Because of the variability in γ, crack size, crack density and strand spacing the stresses predicted may differ slightly from measured LOP, but the study amply demonstrates the mechanism by which the twenty-eight day old material derives it LOP to TOP transition zone.

In a one year old water-stored GRC, the finite width effect no longer exists and the cracking stress is increased by the stiffening effect of the strands. None of the microcracks considered above can propagate as the critical stresses for their propagation are above the TOP level. The stress-strain curve will remain fairly linear up to the TOP and the LOP to TOP transition zone is very small.

CONCLUSIONS

GRC has a very complex microstructure not generally observed in other cement based composites. The conventional single-fibre-in-cement model is inadequate to represent all the microstructural features. The strand-in-cement model is a much closer representation of the microstructure than the single filament-in-cement model. The investigation shows that the characteristics and geometry of the strands and their packing within the composite, represented by the filamentisation, filament dispersion f_d, bundle efficiency f_c, and orientation functions $P(\theta)$ do not greatly affect the initial elastic properties. But, they have significant effects on the slope of the stress-strain curve, particularly around LOP and TOP, and on the multiple fracture process.

Although crack arresting as postulated by Romualdi and Batson (10) is not very significant, strands because of their bundle effect act as barriers preventing the propagation of cracks into macroscopic cracks. The cracking is thus very much localised by the fibres and the stress-strain curve, therefore, becomes much more stable. Contrary to the general belief that glass fibre is inactive until the multiple fracture appears, the study shows that the presence of fibre is very essential and the fibres play an important role in making the stress-strain curve of cement paste stable in the so called initial elastic range so that products can be designed to operate safely within this range.

ACKNOWLEDGEMENT

The author wishes to thank Mr B A Proctor and Mr H W McKenzie for the many helpful suggestions and discussions, also Pilkington Brothers Limited and Dr D S Oliver, Director, Group Research and Development for permission to publish this article.

REFERENCES

1. Steele, B R, "Glass Fibre Reinforced Cement", Proc. Int. Building Conf., Prospects for fibre reinforced construction materials, 1971, Dept. of the Environment, (1972).
2. Oakley, D R and Proctor, B A, "Tensile Stress-strain Behaviour of Glass Fibre Reinforced Cement Composites", this conference.
3. Majumdar, A J, "The Role of Interface in Glass Fibre Reinforced Cement", Cement and Concrete Research, Vol 4, 1974, pp 247–266.
4. Litherland, K L and Jaras, A C, "Microstructural Features of Glass Fibre Reinforced Cement and Composites", This Conference.
5. Lawrence, P, "Some Theoretical Considerations of Fibre Pull-out from an Elastic Matrix", J. Mat. Science, Vol 7, 1972, pp 1–6.
6. Laws, V, Lawrence, P and Nurse, R W, "Reinforcement of Brittle Matrices by Glass fibres", J. Phys. D.: Applied Physics, Vol 6, 1973, pp 523–537.
7. Allen, H G, "The Strength of Thin Composites of Finite Width with Brittle Matrices and Random Discontinuous Reinforcing Fibres", J. Phys D.: Applied Physics, Vol 5, 1972, pp 331–343.
8. Aveston, J, Cooper, G A and Kelly, A, "Single and Multiple Fracture", The Properties of Fibre Composites, IPC Science and Technology Press, 1971, pp 15–26.
9. Ashton, J E et al, "Primer on Composite Materials: Analysis", Technomic Publishing Co, 1969.
10. Romualdi, J P and Batson, G B, "Mechanics of Crack Arrest in Concrete", Journal of the Engineering Mechanics Div., Proceedings of American Society of Civil Engineering, Vol 89, 1963, No EM3, pp 147–168.
11. Henshell, R D and Shaw, K G, "Crack Tip Finite Elements are Unnecessary", Int. J. Num. Methods Eng. (to appear).
12. Paris, P C and Sih, G C, "Stress Analysis of Cracks", American Society for Testing and Materials, Special Technical Publication 381.

3.5 Strength of concrete beams with aligned or random steel fibre micro-reinforcement

V S Parameswaran and K Rajagopalan
*Structural Engineering Research (Regional) Centre,
Madras, India*

Summary

This paper presents a method of predicting the ultimate flexural strength of steel fibre reinforced concrete beams in which failure is initiated by the pull-out of the micro-reinforcement. The relative efficiency of aligned continuous fibres, aligned discontinuous fibres, and random discontinuous fibres is briefly discussed. The possibility of predicting the ultimate strength of fibre reinforced beams based on an assumed mechanism of failure and with a tessellated configuration of the fibre-matrix packing has also been considered. A prediction equation, without and with inter-fibre interaction, has been postulated and the parameters in the equation obtained by fitting it to the experimental results.

Résumé

Cet exposé présente une méthode pour prédire la résistance à la rupture de poutres de béton renforcé de fibres de verre dans lesquelles la rupture est amorcée par l'arrachement du micro-renforcement. On discute brièvement l'efficacité relative de fibres continues ou discontinues en alignement et de fibres discontinues réparties au hazard. On considére également la possibilité de prédire la résistance à la rupture de poutres de béton de fibres sur la base d'un mécanisme de rupture hypothétique et avec une configuration en damier du tassement fibre-matrice. On a postulé une équation pour la prédiction, en considérant ou en négligeant l'interaction des fibres, et on a obtenu les paramètres de l'équation en l'adaptant aux résultats expérimentaux.

INTRODUCTION

Fibre reinforced cement composite is a relatively new material in which steel or other fibres are introduced as micro-reinforcements. It brings to the concrete industry several advantages such as superior crack control, ductility and energy absorption capacity. Rheological considerations of the fresh mix often demand small aspect ratios of the fibres but with such aspect ratios the improvement in tensile strength is only modest due to the pulling out of the fibres (1,2). This problem is presently preventing the attainment of high strengths that are associated with the fracture of the fibres. The problem has been partially solved by several techniques that result in bond improvement (3). However, until methods are found for capitalizing on the high fracture strength of the fibres, dependency has to be placed only on the pull-out strength of the micro-reinforcement. This paper is, therefore, concerned with a discussion of the factors influencing the pull-out of fibres in fibre reinforced beams, describes an analytical approach based on an assumed failure mechanism and presents a prediction equation based on experimental results.

LONG ALIGNED FIBRES

Distinct post-cracking strengths can be achieved when long, aligned fibres are used as micro-reinforcements as in the case of ferro-cement (4). The cracking strength of such a composite in direct tension can be written by the rule of mixtures (equilibrium) relation

$$\sigma_{cr} = \sigma_{mu} v_m + \sigma'_f v_f \qquad (1)$$

where σ_{cr} is the cracking strength of the composite, σ_{mu} is the ultimate strength of the matrix, σ'_f is the stress in the fibres when the matrix cracks and v_m and v_f are respectively the matrix and the fibre volume ratios.

As soon as the matrix cracks, the stress carried by it until then is thrown onto the fibres. By a detailed analysis of the diffusion of the interfacial stress into the contiguous concrete, it can be shown that debonding occurs with fibres, such as steel, upon cracking of the concrete matrix (5). Though debonding occurs with long, aligned fibres, pull-out would not occur due to the large fibre lengths and successive debonding would culminate only in the fracture of the fibres. Hence the ultimate strength of such a composite is given only by the fracture strength of the fibres.

$$\sigma_{cu} = \sigma_{fu} v_f \qquad (2)$$

where σ_{cu} is the ultimate strength of the composite and σ_{fu} is the fracture strength of the fibres. Understandably, the above equation of equilibrium involves only the fibres and there is no 'mixture' as in the cracking stage. A discussion on crack spacing etc, of such members containing long, aligned fibres is available in reference (4).

Ferro-cement beams containing long, aligned fibres again show a distinct post-cracking strength. The authors have discussed the cracking and ultimate strength characteristics of such beams in reference (6).

DISCONTINUOUS ALIGNED FIBRES

Strengths of composites with discontinuous aligned fibres are lower than those with long aligned fibres due to the probabilistic dispersions of the fibre centres resulting in a weakest link and also due to the steep stress gradient in the fibres. It is generally recognised (7) that the relation between the average and maximum fibre stress can be written as

$$\sigma_{f,av} = \sigma_{f,max} \left(1 - (1-k)\frac{\ell_c}{\ell}\right) \qquad .. (3)$$

where ℓ_c is the critical fibre length (5), ℓ is the fibre length used and k is a factor that depends on the fibre stress gradient. For a linear fibre stress distribution (resulting from a uniform bond stress distribution), it can be verified that k = 1/2. This is normally the situation that exists after debonding and hence for such situations the length efficiency factor, Ψ, can be written as

$$\Psi = (1 - \frac{\ell_c}{2\ell}). \qquad (4)$$

It can be seen that the efficiency increases rapidly with lengths greater than the critical. When the length of the fibres in a composite is critical, the above relation shows that such a composite is only 50% efficient compared to one with long aligned fibres.

Assuming a uniform stress gradient in the fibres, Laws (8) has derived the following expression for the length efficiency factor at the point of incipient cracking of the matrix

$$\Psi = \left(1 - \frac{\ell_c}{2\ell} \frac{\epsilon_{mu}}{\epsilon_{fu}}\right) \qquad (5)$$

where ϵ_{mu} and ϵ_{fu} are respectively the failure strains of the matrix and the fibre. From the above relation it can be seen that, in steel fibre reinforced concrete, due to the large difference between the matrix and fibre failure strains, the length efficiency factor is close to unity. Length efficiency factors after the cracking of the matrix have also been discussed by Laws (8).

DISCONTINUOUS RANDOM FIBRES

Composites with discontinuous fibres not aligned in the stress direction result in lower strengths compared to those with discontinuous aligned fibres. Generally, in fibre reinforced cement products, it is difficult to align short fibres in the direction of stress but an ingenious way of doing this has been recently described by Hannant (9). Cement composites, vibrated on a table while casting, usually contain randomly oriented fibres in planes normal to the direction of vibration (10). Assuming that each fibre has equal probability of orientating itself in any particular direction, simple orientation factors, λ, can be derived for obtaining the effective fibre volume ratio in any particular direction (1). Other deterministic methods of obtaining the orientation factors in the elastic and post-cracking stages are also available (8,11).

Figure 1 **Typical fibre reinforced beam with pull-out mode of failure.**

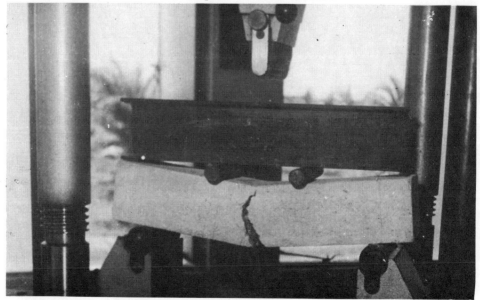

FIBRE PULL-OUT IN BEAMS

Beams containing steel fibre micro-reinforcements with sub-critical aspect ratio fail with the fibres pulling out from the concrete matrix. It has been observed that beams with a fibre pull-out mode of failure fail with a single crack traversing straight up to the neutral axis at the cracking load of the matrix. There is also a coincident increase in strength with a small increase in deflection giving the impression of a rigid body rotation of the cracking halves about the neutral axis. The load-carrying capacity then starts to decline with a gradual pull-out of the fibres in several layers accompanied by a gradual diminution of load, giving the structure a superior post-cracking ductility, integrity and energy absorption. The axis of rotation gradually shifts upward terminating finally at the topmost edge. A typical fibre reinforced beam with a pull-out mode of failure is shown in Fig 1.

Expression for ultimate strength

An expression for the ultimate strength of beams in which the fibres fail by pull-out may be obtained by writing a "rule of mixtures" relation for moment about an axis of rotation. In the following, it has been assumed that fibre centres have been distributed uniformly so as to form tessellated packing arrays throughout the beam as shown in Fig 2. It has also been assumed that at incipient failure, the axis of rotation coincides with the neutral axis of the plain concrete beam. With these assumptions, the ultimate moment of a fibre reinforced beam, M_{uc}, can be written as

$$M_{uc} = M_{uo} + \sum_{k=1}^{k=m} n(\tfrac{2}{3}\tau)\,\pi\,d\,\tfrac{\ell}{2}\,(2k-1)\,\tfrac{R}{2}\,\tfrac{2k-1}{2m-1} \qquad (6)$$

where M_{uo} is the ultimate moment of resistance of the plain concrete beam, d is the fibre diameter and m is the number of fibre layers below the axis of rotation. In the above, the interfacial bond stress is assumed to have a semi-parabolic profile, the bond stress dropping to $\tau/2$ at the end of half the fibre length. A discussion of interfacial bond stress profiles is presented in reference (5).

Figure 2

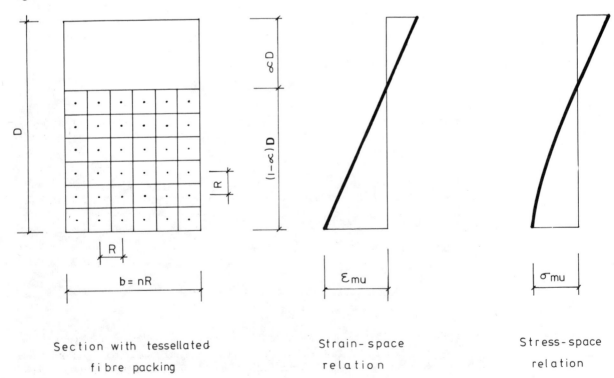

Section with tessellated fibre packing   Strain-space relation   Stress-space relation

For random, discontinuous fibres, the length efficiency factor, Ψ, and the orientation factor, λ, have to be introduced in the second term on the right hand side of equation 6. With this and after further simplification, the above relation can be written as

$$M_{uc} = M_{uo} + \frac{2}{9} \Psi \lambda \tau (1-\alpha)^2 (2+\frac{1}{m}) \, a \, v_f \, bD^2 \qquad (7)$$

where a is the fibre aspect ratio and v_f is the fibre volume ratio. The ultimate moment of a plain concrete beam can be estimated by assuming a linear stress-space relation in the compression zone and a parabolic stress-space relation in the tension zone as shown in Fig 2, for which $\alpha = 0.45$

$$M_{uo} = \frac{bD^2}{4.23} \sigma_{mu} \qquad (8)$$

Using equation 8 in equation 7, the following relation is obtained

$$R = 1 + \left(0.28 \, \Psi \, \lambda \, \eta \, (2+\frac{1}{m}) \right) \, a \, v_f \qquad (9)$$

where R is the ratio of the ultimate moment of the fibre reinforced beam to that of the plain concrete beam and η is the ratio τ divided by σ_{mu}.

The ultimate strength of a fibre reinforced concrete beam in which fibres fail by pull-out can be estimated from equation 9. It has been pointed out that at incipient cracking, the length efficiency factor is around unity. For the assumed semi-parabolic distribution of the interfacial bond stress (which results in a cubic relation for the fibre stress), it can be shown that Ψ is nearly unity. In fact, an exact bond stress distribution is relatively of little value in assessing the length efficiency factor at incipient cracking, although it is highly relevant in deciding the question of fibre debonding (5). Thus equation 9 can be further simplified, for all practical purposes, as

$$R = 1 + \lambda \, a v_f \, . \qquad (10)$$

The above expression, though simple, is only approximate since a cascade of assumptions, through reasonable, have been made in its derivation. It is therefore deemed necessary to postulate a prediction equation, the form of which can be conjectured from equation 9 and to obtain the parameters in the prediction equation by fitting it to the available experimental data.

PREDICTION EQUATION

The analysis presented above has established the form of a prediction equation which must consist of a term signifying the matrix strength and one denoting the strength of the fibres in pulling out. However, there are several unknown effects which pose formidable problems in a theoretical formulation such as the bond behaviour of fibre bundles, the probabilistic nature of fibre dispersion and fibre-aggregate interaction. In such a situation, only a prediction equation whose parameters are obtained for satisfactory agreement with experimental results would be of practical value. The various factors affecting the problem must indeed be kept in mind to know the scope of the equation parameters. The following equation in non-dimensional form with one parameter β can be postulated

$$R = 1 + (\beta a - 1) v_f \, . \qquad (11)$$

In the above equation, the only parameter to be determined by a statistical process is β and this must then encompass the effects of stress gradient in the cross-section of the beam, fibre efficiency and the nature of τ, together with the factors that influence τ itself.

An interesting result can be derived by assuming β as λ times a constant, thus isolating the orientation factor. It can be seen that, since the orientation is the only *extrinsic* factor, the values of β for random two- and three- dimensional fibre orientations must be in the ratio of the respective orientation factors, provided other factors remain constant. Thus

$$\frac{\beta^{2D}}{\beta^{3D}} = \frac{(2/\pi)}{(4/\pi^2)} = \frac{\pi}{2}. \qquad (12)$$

An important factor influencing the pull-out mechanism which cannot be neglected is the fibre–fibre interaction. The magnitude of this effect depends on the spacing of the fibres (which depends on the orientation factor, fibre diameter and fibre volume ratio) as well as on the tensile strength of the concrete between the fibres (which itself depends on a number of parameters such as the matrix-aggregate bond). The penalty on bond stress caused by inter-fibre interaction must therefore depend on the number of fibres per unit area which is $4v_f/\pi d^2$ and hence the effect of inter-fibre interaction on the strength ratio, R, may be assumed to vary as $a^3 v_f^2$ provided other factors remain constant. Thus the prediction equation 11 can be refined to include this effect by adding a non-dimensional penalty function as follows

$$R = 1 + (\beta a - 1) v_f - \alpha\, a^3 v_f^2. \qquad (13)$$

There are thus two parameters α and β that are to be determined statistically. A careful consideration of the effect of inter-fibre interaction would lead to the conclusion that it may be logical to consider this effect only in the case of mortars involving large fibre volume and aspect ratios and two-dimensional random fibre orientations. The effect may perhaps be negligible in cases involving three-dimensional random orientations.

CORRELATION WITH EXPERIMENTS

With the help of a digital computer programme, the parameters in the proposed prediction equations are determined for comparison with the experimental results reported in the literature. In analysing the results, the orientation effect is given the importance it deserves. If a single equation is fitted to all the experimental results, the resulting equation would over-estimate the strength of beams in which the fibre orientations are random in three dimensions, and underestimate the strength of beams with a random two-dimensional fibre orientation. The major factors affecting the fibre orientation are gravity, aspect ratio of the fibres, consistency of the mix and the vibration characteristics used in the process of compaction. Though the effect of table vibration on the fibre orientation has been established (10), the effects of form and other types of vibration still remain unknown.

Equation 11 applied to the experimental results of the authors (those presented in reference (1) with a pull-out fibre failure mode and those obtained subsequently) and those of Luke and Waterhouse (12) gives the following relation

$$R = 1 + (0.69\, a - 1)\, v_f \qquad (14)$$

and when applied to the results of Kar and Pal (13) results in the following relation

$$R = 1 + (0.47\, a - 1)\, v_f. \qquad (15)$$

While it is difficult to state explicitly the reasons for the differences in the values of β in the above relations, the authors believe that it is due to a random, two-dimensional fibre orientation in the former and a random, three-dimensional orientation in the latter. The present authors have used table vibration in their experiments. Luke and Waterhouse state that they have used external vibration and in addition, their mix had a water-cement ratio of 0.52. Kar and Pal do not state the mode of compaction used in their experiments. It is also interesting to note that the β values in the above two equations agree closely with equation 12. Further, it can be seen that the above equations based on experimental data also agree with equation 10.

The results of the authors, Luke and Waterhouse and of Kar and Pal are typical for fibre reinforced concretes with a maximum aggregate size of 10 mm. To ascertain how the proposed equation agrees with the test results on mortars, the experimental results of Romualdi and Mandel (14) and of Lankard (15) are analysed. In some of his experiments, Lankard has used mixed fibres. In such cases, a "weighted aspect ratio", $\bar{a}$, is computed and used in place of a. If a^i is the aspect ratio and v_f^i is the fibre volume ratio of the ith type of fibres, then

$$\bar{a} = \frac{\Sigma a^i v_f^i}{\Sigma v_f^i} \tag{16}$$

and v_f in equations 11 and 13 is taken to mean Σv_f^i. The following relation is then obtained for the results on mortars

$$R = 1 + (0.67\ a - 1)v_f. \tag{17}$$

Lankard has used external vibration with a frequency of 60 Hz. Equation 17 again shows the effect of random two-dimensional orientation. Also by comparing equations 14 and 17, it can be deduced that the effect of fibre-aggregate interaction is negligible in concretes with a maximum coarse aggregate size of 10 mm or less.

To study the effect of inter-fibre interaction, equation 13 is applied to the results of Romualdi and Mandel and of Lankard on mortars. The resulting expression is

$$R = 1 + (0.73\ a - 1)\ v_f - 0.00041\ a^3 v_f^2. \tag{18}$$

The degree of agreement is improved by the inclusion of the effect of inter-fibre interaction as seen from Table 1. However, the effect does not seem to be very significant for the cases studied.

As the fibre-aggregate interaction seems to have an insignificant influence, it seems logical to obtain an equation based on a larger number of results and covering both mortars and concretes. Hence the results of the authors, of Luke and Waterhouse, of Romualdi and Mandel and of Lankard are combined and analysed. The following relations have been obtained without and with inter-fibre interaction.

$$R = 1 + (0.68\ a - 1)\ v_f \tag{19}$$

and

$$R = 1 + (0.69\ a - 1)\ v_f - 0.00009\ a^3 v_f^2. \tag{20}$$

These equations can be considered typical of fibre reinforced mortars and concretes with a random two-dimensional orientation of fibres.

A single and general equation fitted to the whole set of experimental results is also given below.

$$R = 1 + (0.57\ a - 1)\ v_f. \tag{21}$$

This equation represents an average situation wherein the orientation of fibres is random but neither wholly in two dimensions nor completely in three dimensions. It is also clear that the above equation would be an upper bound for the case of random fibre orientation in space and a lower bound for two-dimensional fibre orientations. A single equation reflecting an average situation has also been recently obtained by Swamy (16). His equation reads

$$\sigma_{cu} = 0.97\, \sigma_{mu}\, v_m + 3.41\, av_f. \qquad (22)$$

This equation has also been compared with the experimental results of various investigators in Table 1. In general, the authors' equations, presented in the paper, give better correlation than equation 22.

Table 1   Correlation of prediction equations with experiments

Results of	Prediction equation (reference no in the text)	R(Experiment) / R(Calculated)		
		Mean	Standard deviation (per cent)	Coefficient of variation (per cent)
Authors, Luke and Waterhouse	(14)	0.99	6.0	6.1
	(22)	1.04	7.3	7.0
Romualdi and Mandel, and Lankard	(17)	0.99	11.2	11.3
	(18)	0.98	11.0	11.1
	(22)	1.01	16.5	16.3
Authors, Luke and Waterhouse, Romualdi and Mandel, and Lankard	(19)	0.99	8.9	9.0
	(20)	0.99	8.9	9.0
	(22)	1.03	12.7	12.4
Kar and Pal	(15)	1.00	8.6	8.6
	(22)	0.81	10.4	12.7
All the Investigators	(21)	0.99	10.3	10.5
	(22)	0.89	15.2	17.0

All the aforementioned equations have been based on test results obtained on small beams. In large beams, the probability of the occurrence of the "weakest-link" is greater and hence, such beams, depending possibly on the ratio of fibre length to span length, may suffer a reduction in strength by failing in the weakest-link mode. The magnitude of this size effect is not known at present and research in this area may be needed. It must also be pointed out that the study described in this paper is concerned only with the "pull-out bond" and hence may be applicable only to cases of direct tension or pure bending. The presence of a shear force and its effect on the pull-out mechanism has not been considered and research in this area may be needed. In addition, research is also needed on the mechanics of failure in beams wherein the fibres are used as an adjunct to conventional reinforcements.

CONCLUSIONS

The mechanics of fibre pull-out in beams and the factors influencing it have been discussed and form the theoretical basis for a prediction equation proposed for computing the ultimate flexural strength of fibre reinforced concrete beams in which failure is triggered by fibre pull-out. The proposed equation is then compared with the experimental results of several investigators which show that, in general, the comparison tends to support the validity of the proposed equation. The equation is also refined to include the effect of inter-fibre interaction. It has also been found that the effect of fibre-aggregate interaction is negligible in fibre reinforced concrete with the normally used maximum coarse aggregate sizes of 10 mm or less.

ACKNOWLEDGEMENT

The authors are grateful to Prof G S Ramaswamy, Director, Structural Engineering Researching Centre, Madras, for the encouragement given in the preparation of the paper and also for his comments. The paper is published with his kind permission.

REFERENCES

1. Rajagopalan, K, Parameswaran, V S and Ramaswamy, G S, Strength of Steel Fibre Reinforced Concrete Beams, The Indian Concrete Journal, Vol 48, No 1, January 1974, pp 17-25.
2. ACI Committee 544, State-of-the-Art Report on Fiber Reinforced Concrete, Journal of the ACI, Vol 70, No 11, November 1973, pp 729-744.
3. Mayfield, B and Zelly, B, Steel Fibre Treatment to Improve Bonds, Concrete, Vol 7, No 3, March 1973, pp 35-37.
4. Rajagopalan, K and Parameswaran, V S, Cracking and Ultimate Strength Characteristics of Ferro-cement in Direct Tension and in Pure Bending, The Indian Concrete Journal, Vol 48, No 12, December 1974, pp 387-393 and 395.
5. Rajagopalan, K and Parameswaran, V S, A study on the Mechanics of Fibre Debonding in Concrete with Micro-Reinforcement, to appear in RILEM, Materials and Structures: Research and Testing.
6. Rajagopalan, K and Parameswaran, V S, Analysis of Ferro-cement Beams, Journal of Structural Engineering, Vol 2, No 4, January 1975, pp 155-164.
7. Broutman, L J and Krock, R H, (Ed), Modern Composite Materials, Addison Wesley Publishing Company Inc, USA, 1967.
8. Laws, V, The Efficiency of Fibrous Reinforcement of Brittle Matrices, Journal of Physics, D. Applied Physics, Vol 4, 1971, pp 1737-1746.
9. Hannant, D J and Spring, N, Steel-Fibre-Reinforced Mortar: A Technique for Producing Composites with Uniaxial Fibre Alignment, Magazine of Concrete Research, Vol 26, No 86, March 1974, pp 47-48.
10. Edgington, J and Hannant, D J, Steel-Fibre-Reinforced Concrete: The Effect on Fibre Orientation of Compaction by Vibration, RILEM, Materials and Structures: Research and Testing, Vol 5, No 25, January-February 1972, pp 41-44.
11. Pakotiprapha, B, Pama, R P and Lee, S L, Mechanical Properties of Cement Mortar with Randomly Oriented Short Steel Wires, Magazine of Concrete Research, Vol 26, No 86, March 1974, pp 3-15.
12. Luke, C E and Waterhouse, B L, Steel Fibre Optimisation, Publication M-28, Construction Engineering Research Laboratory, Champaign, Illinois, USA, December 1972, pp 63-81.
13. Kar, J N and Pal, A K, Strength of Fibre Reinforced Concrete, Journal of the Structural Division, Proceedings of the ASCE, Vol 98, No ST5, May 1972, pp 1053-1068.
14. Romualdi, J P and Mandel, J A, Tensile Strength of Concrete Affected by Uniformly Distributed and Closely Spaced Short Lengths of Wire Reinforcement, Journal of the ACI, Vol 61, No 6, June 1964, pp 657-671.
15. Lankard, D R, Summary Report on Cost/Effectiveness Study to WIRAND Licenses, National Standard Company, UK, 1972.
16. Swamy, R N, Fibre Reinforced Concrete: Mechanics, Properties and Applications, The Indian Concrete Journal, Vol 48, No 1, January 1974, pp 7-16 and 29.

3.6 Some results of investigations on steel fibre reinforced concrete

B Schnütgen
*Institute of Structural Engineering,
Ruhr-University, Bochum*

Summary

In this paper a theory for the evaluation of the influence of steel fibres on the tensile strength of steel fibre reinforced concrete has been presented. The theory was based on the assumption that the theory of crack propagation was applicable ie the material is a quasi-brittle material. The different dimensions and orientations of the fibres are considered by means of numerical influence factors.

The expressions obtained numerically were checked experimentally. There was a good agreement between the theoretical and experimental results.

Résumé

Cette communication présente une théorie pour l'évaluation de l'influence des fibres d'acier sur la résistance à la traction du béton armé de fibres d'acier. Cette théorie est basée sur l'hypothèse que la théorie de la propagation des fissures était applicable à ce cas, c'est à dire que le matériau est presque un matériau fragile. Les différentes dimensions et l'orientation des fibres sont prises en considération au moyen de facteurs d'influence numériques.

Les expressions obtenues par voie numérique ont été vérifiées expérimentalement. Il y a un bon accord entre les résultats théoriques et expérimentaux.

NOTATION

Geometrical values:

x,y,z	Cartesian coordinates
ξ	local coordinate
α	angle between fibre and load direction
c	half crack length
c_o	half crack length in unloaded state
ℓ_f	length of fibre
d	diameter of fibre
s	fibre spacing

Strength and stresses:

f_{fc}	strength of composite
f_c	strength of concrete
$f_{c,t}$	strength of concrete in tension
$f_{f,t}$	strength of fibres against pull-out
σ_z	stresses in z-direction
$\sigma_{z,z=0}$	stresses in z-direction in the plane z = 0
σ_{z,ξ_o}	stresses in z-direction in the point

Material constants:

E	elastic modulus
v	Poisson's ratio
K	fracture toughness
n	ratio of the moduli of elasticity

Indices:

c	with regard to concrete
f	with regard to fibre
fc	with regard to composite
t,b,sp	in tension, bending, splitting tensile
z	in z-direction
α,α'	in direction of the angle α and rectangular to that
$\parallel, \perp$	parallel or rectangular to the fibre direction

Other notations:

k	constant factors
η	factors of efficiency

INTRODUCTION

At the Institute of Structural Engineering for the Ruhr-University, Bochum research work on steel fibre reinforced concrete has been carried out since 11970. This research project involves the theory and experimental investigations of the strength of fibre reinforced concrete and its structural behaviour. In particular, emphasis was laid on finding out the effect of the content of the fibre, the dimensions of the fibre and the mix proportions.

The theoretical investigations were based on the assumption that the effect of the fibres on a brittle material is approximately the same as that on concrete. The postulate of brittle behaviour is not exactly complied with in a material like concrete. As a result the theoretical values were checked with experimental tests.

Above all the influence of the maximum size of aggregate on the tensile strength of concrete was checked, because of the great influence of the size of aggregate on the internal stresses of plain concrete.

THEORETICAL INVESTIGATION

Fundamentals

The fibres provide only an improvement on strength of the cement paste. They have no influence on the adhesion between cement paste and aggregate grains. Investigations on plain concrete have shown that for the concrete strength the following order of influence is valid:

1st) adhesion between cement paste and aggregate grain,

2nd) strength of cement paste,

3rd) strength of aggregate.

This consideration shows that a model based on the theory of crack propagation in elastic solids must have a correction to consider the influence of the maximum size of aggregate. The basic material is an elastic material with internal cracks. These cracks are small in relation to the dimensions of the solid. First the influence of fibre on the elastic solid will be investigated. After that the influence of the maximum size of aggregate is taken into consideration.

Internal stress of a cracked elastic solid

On condition that the proportions of the crack are small compared to the proportions of the solid it is permissible to investigate a crack in an infinite elastic solid. The tensile stresses near a crack in an infinite solid resulting from the theory of elastic solid are shown by equation (1)

$$\sigma_{z,z=0} = \frac{2 f_{c,t}}{\pi} \left[\frac{\pi}{2} + \frac{1}{\sqrt{\left(2\frac{\xi}{c} + \frac{\xi^2}{c^2}\right)}} - \frac{\sqrt{\left(2\frac{\xi}{c} + \frac{\xi^2}{c^2}\right)}}{1+\frac{\xi}{c}} \arcsin \frac{1}{1+\frac{\xi}{c}} \right] . \quad (1)$$

This equation tends to infinity for $\xi \to 0$. But in conformity with the theory of crack propagation a transitional area exists (1). This area is an element of the crack but it is not stress-free. The distances between the surfaces of crack are so small in this area, that the adhesion has a measurable intensity.

Irwin (2) has shown that the sphere of plastic deformation should be added to the above-mentioned transitional area. The theory of crack propagation is applicable to plastic materials if the sphere of plastic deformation is small compared to the crack dimensions.

According to Muskhelishvili (3) there are two hypotheses for the area of a crack at the edge. The edge area included the area of plastic deformations and the area of measurable adhesion.

Figure 1 **Stress distribution near a crack in tension (qualitative).**

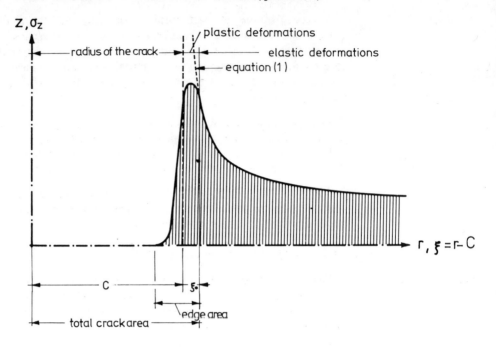

1st) The edge area is small compared to the measurement of the crack.

2nd) The form of the cross-section of the edge area is a material constant.

From equation (1) one can obtain for the crack in unstable equilibrium the corresponding external equation when $\xi_o \ll c$

$$f_{c,t} = \pi \sigma_{z,\xi_o} \frac{\sqrt{\xi_o}}{\sqrt{2c}} . \qquad (2)$$

As a result of the afore-mentioned hypotheses both ξ and σ_{z,ξ_o} are material constants. As a result we obtain the well-known expression (3)

$$f_{c,t} = \frac{K_c}{\sqrt{2c}} . \qquad (3)$$

Effects of fibres

The fibres have a threefold influence on equation (3).

1st) Influence as a conventional reinforcement:

$$\text{factor } A_1 = [1 + V_f(n - 1)] \qquad (4)$$

2nd) Influence on the dimensions of the cracks:

$$c = c\,(s) \qquad (5)$$

3rd) Influence as a result of fibres in the crack area. This is an additional term. We can suppose that the fibres have only a negligible effect on the crack size before they intersect the crack area. The cracks extend as far as the positions of the fibres in the crack area. This allows additional loading. Therefore most of the fibres in the crack area are placed in the edge area of cracks. The application of uniformly-distributed loading of the fibres in the crack area is an approximation with additional safety. The fibres in the crack area have as a result of the high deformation of the concrete the ultimate strength $f_{f,t}$.

After consideration of all the three influences, equation (6) is obtained

$$f_{c,t} - f_{f,t} = \frac{K_c}{\sqrt{2c(s)}} \cdot \qquad (6)$$

The dependence of the values of $f_{f,t}$, $c(s)$ and V_f on the length, diameter and orientation of the fibres is to be determined more exactly.

Influence of fibres in crack area: The ultimate strength of fibres against pull-out is given by equation (7) with τ_{bu} as the ultimate bondstress.

$$F_{f,t} = \eta \, \tau_{bu} \, \pi \, d \, \frac{\ell_f}{2} \cdot \qquad (7)$$

The fibres are randomly distributed. As a result, the average value of embedded length of fibres is $\ell_f/4$. The coefficient of length is shown as follows

$$\eta_{\ell u} = 0.5 \, . \qquad (8)$$

In addition, the coefficient of fibre-orientation is taken into consideration because the fibre-orientation is inclined with angle α to the orientation of loading. The ultimate load of a fibre increases as a result of the different orientation of fibres and loading. In the crack surface there is a change of direction of the fibre and this causes a deflection force. The deflection force produces additional frictional force which causes a greater resistance of the fibre to pull-out.

The factor η_d which considers this effect is to be determined by tests.

The complete factor in equation (7) is given by

$$\eta = \eta_{\ell u} \, \eta_\alpha \, \eta_d \, . \qquad (9)$$

The fibre strength corresponding to the total cross-sectional area of the composite becomes

$$f_{f,t} = 2\tau_{bu} \, \frac{\ell_f}{d} \, V_f \, \eta \, . \qquad (10)$$

Influence of fibres as a conventional reinforcement: In contrast to conventional reinforcement the fibres are randomly orientated in steel fibre reinforced concrete and their length is small. The factors showing these influences are considered in (4).

a) **Factor for fibre orientation.** When the fibres are orientated at an angle α to the direction of uniaxial load the fibre strength in the fibre direction is

$$\sigma_f^\parallel = \sigma_{fc}^\alpha \left(\frac{n \cos \alpha}{1+V(n-1)} - \frac{n(1-V_f) v_c \sin^2 \alpha}{1+V_f(n-1)} \right) \qquad (11)$$

and in the transverse direction to the fibres

$$\sigma_f^\perp = \sigma_{fc}^\alpha \sin^2 \alpha \, . \qquad (12)$$

The shear stress in the fibres is

$$\tau_f^{\parallel \perp} = \sigma_{fc}^\alpha \sin\alpha \cos\alpha \, . \qquad (13)$$

In direction of loading the fibre stresses are

$$\sigma_f^\alpha = \frac{\sigma_{fc}^\alpha}{1+V_f(n-1)} \cdot$$

$$\cdot \left\{(\sin^4\alpha + 2\sin^2\alpha\cos^2\alpha)\,[1+V_f(n-1)] + n\cos^4\alpha + n(V_f^{-1})\,v_f\sin^2\alpha\cos^2\alpha\right\}. \quad (14)$$

Perpendicular to the load direction the fibre stress is

$$\sigma_f^{\alpha'} = \frac{\sigma_{fc}^\alpha \sin^2\alpha}{1+V_f(n-1)}\left\{(n-1)\cos^2\alpha - n\,v_c\sin^2\alpha\right\}. \quad (15)$$

The corresponding shear stress is

$$\tau_f^{\alpha,\alpha'} = \frac{\sigma_{fc}^\alpha (1-V_f)\sin\alpha\cos\alpha}{1+V_f(n-1)}\left\{n\,v_c\sin^2\alpha - (n-1)\cos^2\alpha\right\}. \quad (16)$$

The factor of fibre orientation $\eta_{f,\alpha}$ resulted from integration of (14) over all directions of α. For a three-dimensional distribution of fibre orientation

$$\eta_{f,\alpha} = 1 - \frac{2}{15}(1-V_f)\left[\frac{6(n-1)}{n} + v_c\right] \quad (17)$$

The corresponding expression for a two-dimensional distribution of the fibre orientation for example in the surface of a concrete specimen is

$$\eta_{f,\alpha} = 1 - \frac{1}{8}(1-V_f)\left[\frac{5(n-1)}{n} + V_c\right] \quad (18)$$

The graphical solution of (17) and (18) is shown in Figure 2 for different values of n and a Poisson's ratio of 0.25.

Figure 2   Factor for consideration of the fibre orientation near the concrete surface.

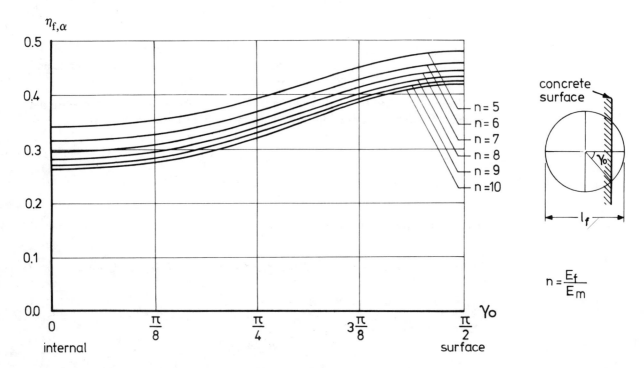

b) **Factor for the fibre length.** On the assumption that total load on the fibre is applied near the end of the fibre and with ultimate bond stress τ_{bu}, the following approximate formula is valid:

$$\eta_{f,\ell} = 1 - \frac{n}{4} \frac{f_{fc,t}}{\tau_{bu}} \frac{d}{\ell_f} \left\{ 1 + \ln\left(\frac{4}{n} \frac{\tau_{bu}}{f_{fc,t}} \frac{\ell_f}{d}\right) \right\}. \qquad (19)$$

This approximation is shown in Figure 3 with the ratio of the Young's moduli of steel and concrete equal to 6. A comparison with the exact solution shows that the formula (19) is a satisfactory approximation.

Figure 3   **Influence coefficient of the aspect ratio of fibres.**

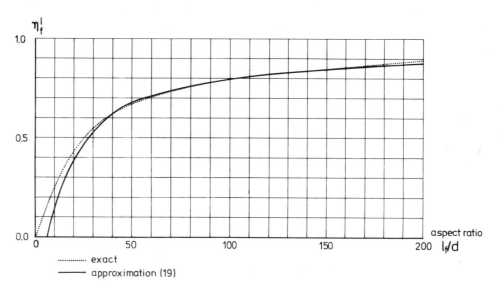

Influence of the crack size on the tensile strength of fibre reinforced concrete: The influence of the crack size on the tensile strength of fibre reinforced concrete is shown in equation (6). The interdependence of the crack diameter and spacing of the fibres is given in the form:

$$c = c_o + k_1 s(\eta). \qquad (20)$$

The term c_o is the crack radius as a result of the concrete production and placement. The factor k_1 is a constant obtained from experiment. The fibre spacing $s(\eta)$ is to be taken as the effective spacing in consideration of the factors $\eta_{f,\ell}$ and $\eta_{f\alpha}$.

The interdependence of the fibre spacing on the diameter and the content of fibres is given by

$$s(\eta) = \sqrt{\frac{\pi d^2}{4 \, \eta_{f,\ell} \, \eta_{f,\alpha} \, V_f}}. \qquad (21)$$

The tensile strength of steel fibre reinforced concrete is shown as follows

$$f_{fc} = \frac{K_c}{\sqrt{2 c_o}} \frac{1 + \eta_{f,\ell} \, \eta_{f,\alpha} \, V_f \, (n-1)}{\sqrt{1 + \frac{k_1}{c_o} \sqrt{\frac{d}{\eta_{f,\ell} \, \eta_{f,\alpha} \, V_f}}}} + 2 \, \tau_{bu} \frac{\ell_f}{d} \, \eta \, V_f. \qquad (22)$$

Equation (22) is applicable only for concretes with a maximum size of aggregate smaller than the effective fibre spacing. In addition the state of cracks in the unloaded concrete must be such that in plain concrete the dimensions of cracks would be greater than the effective fibre spacing.

The usual content of fibres in steel fibre reinforced concrete is 1 to 2% by volume. With that volume of fibres and d = 0.4 mm (diameter of fibres) the effective fibre spacing is 4 to 7 mm. Therefore equation (22) is only applicable to a mortar with an unfavourable state of cracks as a result of the conditions during curing.

The influence of fibres on the crack diameter for concrete with a maximum size of aggregate equal to or greater than 8 mm is only assumed for cracks of diameter greater than the diameter of the aggregate.

In the unloaded state of concrete, cracks of this dimension occur only as an exception.

For this reason equation (22) can be reduced to

$$f_{fc} = f_c [1 + \eta_{f,\ell} \, \eta_{f,\alpha} V_f (n-1)] + 2 \tau_{bu} \frac{\ell_f}{d} \eta V_f . \qquad (23)$$

The validity of (22) and (23) is shown in Figure 4.

Figure 4 **Qualitative relation of strength to fibre spacing.**

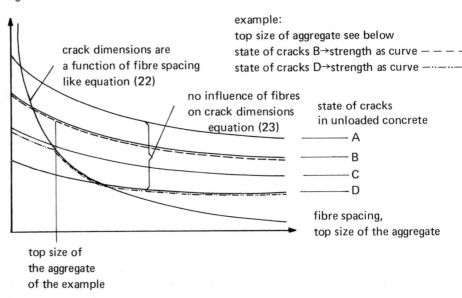

EXPERIMENTAL INVESTIGATION

Programme

The most interesting parameters affecting the tensile strength of steel fibre reinforced concrete are:

the fibre content

the fibre dimensions (aspect-ratio)

the state of cracks in the unloaded concrete.

Influence of fibre content:

Specimens

Bending tests: beams 150 x 150 x 700 mm.

Splitting tests: cubes 150 mm.

Mix-proportions

Cement: Portland cement Z 350 F (German standard) 500 kg/m³

Water-cement ratio: 0.42

Aggregate: fine aggregate with round grain in correspondence with grading curve B8 in the German standard DIN 1045.

Fibres: $d_f = 0.4$ mm

$\ell_f = 25$ mm

Curing

Series A: water cured

Series: B: 7 days water cured ard subsequently air cured.

Age of specimens when tested: 3, 7, 14, 28, 56, and 90 days.

Influence of fibre dimensions: Specimen and mix proportions as above. Curing as above (series B).

Fibre dimensions

Diameter of fibres	Length of fibres			
	15 mm	25 mm	38 mm	50 mm
0.25 mm	X	X	X	–
0.40 mm	X	X	X	X
0.60 mm	–	X	X	X

Figure 5   **Effect of fibre content on flexural strength.**

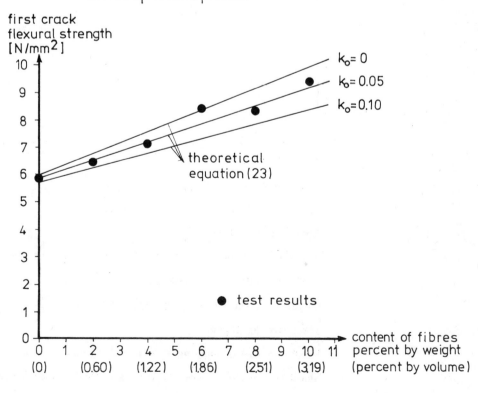

Figure 6   Flexural strength in relation to the age of specimen.

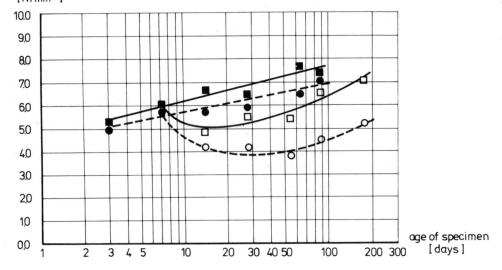

Figure 7   Effect of the aspect ratio on the flexural strength.

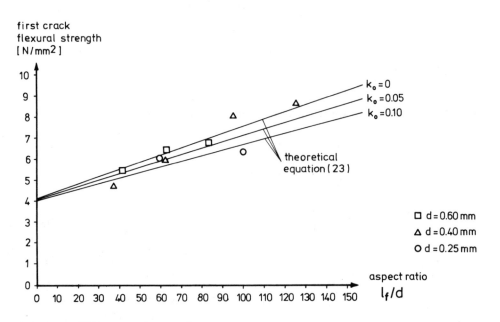

Influence of fibres on the cracks:

   Variation of the maximum size of aggregate

   Specimens as above.

   Maximum size of aggregate: 1, 2, 4, 8, 16 and 32 mm.

   Fibre content: 6% by weight (1.86 vol. %).

   Variation of the shape of aggregate particles

   Specimens as above.

   Maximum size of aggregate: 4, 8 and 12 mm.

   Shape of particles: some fractions are crushed (see Figure 9).

Figure 8 Effect of the top size of aggregate on the strength.

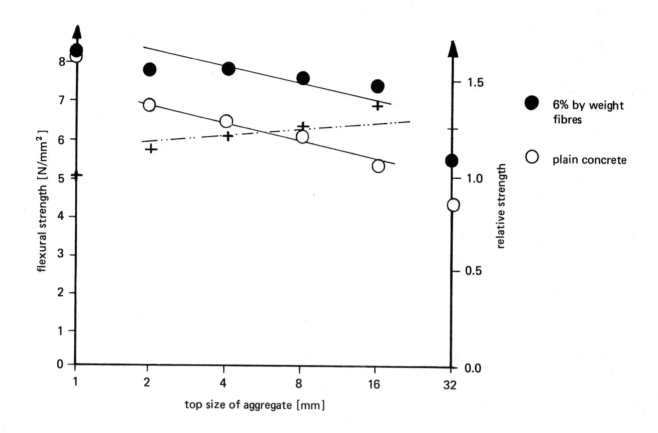

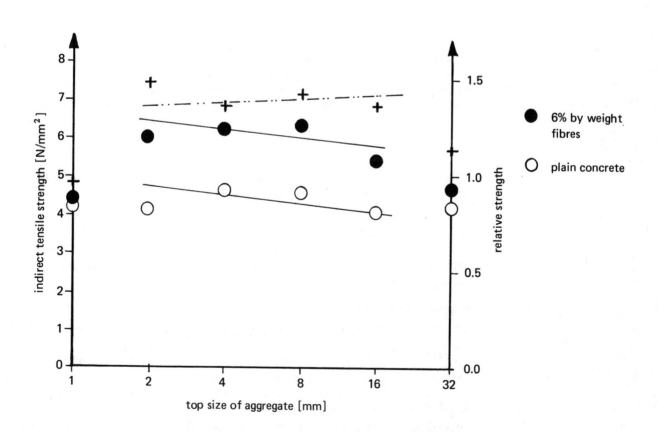

Figure 9 Effect of the form of aggregate particle on the flexural strength and the indirect tensile strength.

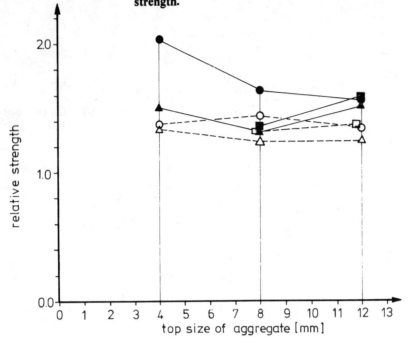

Test results

The results of the investigation are shown in Figures 5 to 9.

Influence of fibre content: As shown in Figure 5 the increase of strength with increasing fibre content corresponds nearly with equation (23). The factor k_1 in Figure 5 takes into consideration the increasing difficulty of workability with increasing value of $V_f \ell_f/d$. The first term in the equations (22) and (23) will be multiplied by the factor

$$\eta_w = 1 - k_o V_f \frac{\ell_f}{d}. \tag{24}$$

A relationship similar to equation (22) could not be obtained in the investigations. That was to be expected on the grounds of the theoretical investigations. The bending tests showed a larger increase of strength than in the equation (23) depending on the age of specimen, but as shown in Figure 6 this increase of strength is dependent on the after-treatment of the specimen. It is a result of the differential shrinkage during curing of concretes with and without fibre reinforcement that gives rise to secondary stresses and to a subsequent apparent increase of strength.

Influence of fibre dimensions: The investigations carried out verify the dependence of strength on the aspect-ratio (ℓ/d) of the fibres as shown in equation (23). However the number of tests is too small to allow a generally valid conclusion.

Influence of the state of cracks: The investigations do not give a clue as to the influence of fibres on the dimensions of cracks in unloaded specimens. The increase of strength is nearly constant with different sizes and shapes of the aggregate as was to be expected from equation (23).

REFERENCES

1 Griffith, A A, The phenomena of rupture and flow in solids. Philosophical transactions, Series A 221, Royal Society, London, 1920, pp 163-198.
2 Irwin, G R, Fracture dynamics. Fracturing of Metals, ASM, Cleveland 1948, pp 147-166.
3 Muskhelishwili, N J, Einige Grundaufgaben zur mathematischen Elastizitätstheorie, Carl Hanser Verlag, München, 1971.
4 Schnütgen, B, Das Festigkeitsverhalten von mit Stahlfasern bewehrtem Beton unter Zugbeanspruchung, dissertation to be published in 1975.

Section 4

Properties and Testing of Steel Fibre Concrete

4.1 Opening Paper: Méthodes d'essais et caractéristiques mécaniques des bétons armés de fibres métalliques

N M Dehouse
Université de Liège, Belgique

Summary

To the extent where the assumptions of the present elementary calculations are not invalidated, it seems that the bending strength of the section reinforced with fibres results from the utilisation of the considerable specific strength of the small diameter wires. To achieve this strength, it is necessary that the fibre should have adequate bonding and have an optimum orientation.

The range of compression—tension diagrams means that a (fictitious) bending strength of up to 3.5 times that of non-reinforced concrete is achieved.

In direct tension, there is only a marginal improvement to be expected with the currently used pecentages (2.5 by volume).

Comparison with reinforced concrete in bending quickly shows that fibre reinforced concrete may not a priori be considered as a substitute.

Résumé

Dans la mesure où les hypothèses de ces calculs élémentaires ne sont pas infirmées, il apparait que la résistance de la section armée de fibres en flexion résulte de la mobilisation de la résistance spécifique considérable des fils de petit diamètre. Pour que cette résistance se manifest, il importe que la fibre trouve un accrochage de qualité et qu'elle soit aussi judicieusement orientée que possible.

Le jeu des diagrammes de compression—tension fait que l'on atteint (fictivement) jusqu'à 3.5 fois la résistance en flexion du béton non armé.

En traction directe, avec les pourcentages couramment admis (2.5 en volume) il n'y a guère que des accroissements marginaux à espérer.

La comparaison avec le béton armé fléchi montre rapidement que le béton de fibres ne peut être à priori considéré comme un substitut au premier.

PRÉAMBULE

En préambule à cet exposé introductif sur les méthodes d'essais et les caractéristiques mécaniques des bétons armés de fibres métalliques, remarquons que notre tâche est à la fois simple et compliquée.

Elle est simple car différents documents de synthèse ont déjà été consacrés à l'état de la question, aux USA et en Grande Bretagne dans les derniers mois notamment.

Elle est compliquée par la multiplicité des aspects qu'évoquent les titres qu'il m'est proposé de traiter.

De nombreux pays ont manifesté un grand intérêt pour les bétons renforcés de fibres et je suis persuadé que différents expérimentateurs ont tenté bien avant la dernière décennie de modifier la résistance du béton par l'adjonction de métal sous diverses formes.

Les scientifiques sont toujours tenaillés par un démon de performance et nombre d'entre eux ont tenté, et nous en sommes au laboratoire du Génie Civil de l'Université de Liège, d'incorporer des déchets métalliques au béton dans ce but.

Ce fut on le sait, pratiquement sans succès avant les travaux de Romualdi, la chose est bien connue.

Depuis, nombre de pays se sont précipités dans la nouvelle voie offerte et en parcourant la littérature nous avons pu constater combien l'engagement était grand non seulement des 2 côtés de l'Atlantique mais aussi au Japon.

Je ne connais rien de ce que l'URSS ait tenté à ce jour mais il serait étonnant que le sujet n'y ait pas été abordé. D'autant que les expériences d'approche sont relativement aisées à réaliser tant que l'on ne se pose pas trop de question: vous permettrez d'ailleurs à un professeur de se réjouir du nombre de travaux d'étudiants de nos universités que les fibres métalliques ont suscités.

Le présent rapport n'a pas la prétention d'être un document de synthèse des travaux exécutés dans le domaine. Je saurais gré au lecteur de le considérer comme un texte introductif à une discussion reposant à la fois sur les travaux des différents auteurs et sur les nôtres évidemment.

Figure 1

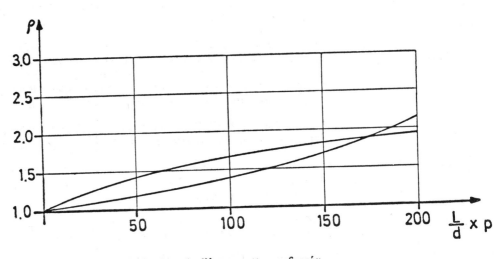

ρ = $\dfrac{\text{résistance en flexion de l'éprouvette renforcé}}{\text{résistance en flexion de l'éprouvette non renforcé}}$

L = longeur de la fibre
d = diamètre de la fibre
p = pourcentage en volume.

RÉSULTATS PRINCIPAUX OBTENUS PAR DIVERS AUTEURS ET FAISANT L'OBJET D'UN LARGE CONSENSUS

Résistance à la rupture en flexion

Cette résistance est fortement accrue par rapport à celle que l'on obtient en flexion pure ou en flexion simple sur des éprouvettes de mortier ou de micro-béton.

Le rapport entre la résistance à la rupture avec fibres et sans fibre dépend de la matrice, de la géométrie de la fibre, du pourcentage de fibres.

La notion d'espacement, soit celle présentée par Romualdi et Mandel, soit une autre, met clairement ce phénomèn en évidence. Pour marquer la tendance on a fait usage de graphiques du genre de ceux de la Figure 1 où les courbes moyennes montrent que les valeurs de ρ atteignent l'ordre de 2 pour des valeurs de $\frac{L}{d}\rho$ l'ordre de 200.

La littérature fait état de différents types d'éprouvettes: prismes fléchis, éprouvettes cylindriques creuses, éprouvettes de formes plus compliquées (Figure 2).

Figure 2

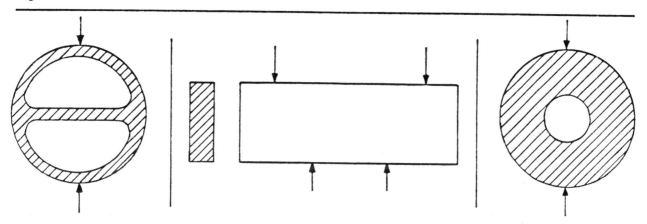

Il est cependant essentiel de remarquer:

1. Que dans cette comparaison, on calcule les contraintes en section homogène soit par la théorie de la résistances des matériaux soit par celle de l'élasticité;

2. Que les courbes moyennes obtenues à la Figure 1 sont tracées au travers d'un nuage de points très dispersés correspondant aux différences de mise en place, types de fibres, orientation des ces dernières... On ne peut les considérer que comme des indications de tendance;

Figure 3

3. A notre connaissance, les essais précités ont été exécutés avec de fibres droites, lisses, empreintées ou ondulées. Nous disposons en Belgique d'une fibre à crochets (Figure 3) pour laquelle il semble que les résultats soient notablement différents (Figure 4).

L'allure du diagramme de la Figure 4 est obtenue aussi bien pour des éprouvettes fléchies classiques que pour des essais de fendage creux de mortier.

Résistance à la traction directe

Assez peu d'expérimentateurs ont emprunté cette voie difficile il est bien connu que la réalisation d'un essai de traction digne de confiance pose de multiples probèmes et donc s'avère d'un coût élevé.

Les résultats obtenus font état en moyenne de majorations de l'ordre de 25% de la résistance à la rupture pour des pourcentages de fibres de l'ordre de 2.5% en volume.

Ce ne sont donc que des améliorations marginales et en tout cas sans commune mesure avec les résultats apparents obtenus en flexion.

Résistance en compression

La résistance ultime n'est guère améliorée, ce qui est aisé à comprendre, le pourcentage de fibres étant géneralement beaucoup trop faible pour espérer une notable majoration.

Figure 4

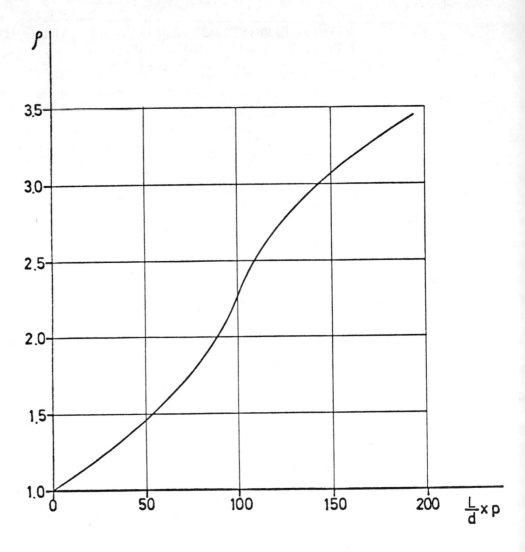

Ici aussi les majorations ne sont que marginales de l'ordre de 25% au plus pour des pourcentages de fibres de l'ordre de 2.5% en volume.

Une telle constatation est d'ailleurs bien conforme aux résultats que l'on pourrait escompter en faisant usage de la loi des mélanges pour le calcul du module d'élasticité moyen:

$$E_{\text{béton renforcé}} = E_{\text{béton}} \times \text{pourcentage en volume de béton}$$
$$+ E_{\text{acier}} \times \text{pourcentage en volume de fibres}$$

soit

$$E_{bf} = E_b \times (1-p) + E_a \times p$$

et en appliquant une relation classique du genre

$$E_{bf} = 21,000 \sqrt{R'_{br}} \quad (\text{kg et cm}^2).$$

Remarquons que si l'on admet la théorie selon laquelle la rupture en compression est due à l'apparition de fissures **parallèlement** à l'effort appliqué, on comprend que l'amélioration escomptée par l'addition de fibres en compression ne pourrait dépasser celle obtenue en traction . . . ce qui est précisément le cas.

Une confirmation supplémentaire a été apportée par la réalisation d'essais de compression localisée sur cylindre où là également on n'a observé que des majorations de même ordre de grandeur (soit donc ± 25% pour des pourcentages de l'ordre de 2.5 en volume).

Consistance

A égalité de composition, un mortier avec fibres est moins fluide qu'un mortier normal.

C'est ainsi que pour 2.5% en volume de fibres, on n'atteint aux essais à la table à secousses que les 65 à 75% des valeurs du mortier normal.

Les résultats obtenus font apparaître les mêmes tendances qu'il s'agisse d'essais au consistomètre VB, à la table à secousses, ou par la mesure du facteur de consistance. Selon le type de fibre, l'effet d'aglomération apparaît entre 2.5 et 4% rendant alors les mélanges inhomogènes. Edgington, Hannant et Williams ont établi une formule précisant ces valeurs limites.

Un facteur d'importance considérable est celui de forme $\left(\dfrac{L}{d}\right)$ dont il semble bien que la valeur limite pour une ouvrabilité raisonnable soit de 100.

L'effet d'adjuvants sur le pourcentage limite admissible de fibres avant agglomération, reste à notre avis à démontrer. Pour rencontrer les difficultés d'insertion et dépasser les pourcentages limites précitées, la firme belge Bekaert a mis au point une technique de collage de 10 fibres par un produit soluble dans l'eau (voir Figure 5).

L'ouvrabilité des micro-bétons renforcés et fabriquées à partir des mortiers de référence est évidemment moins bonne également que celle de micro-bétons non renforcés.

Granulométrie des agrégats et composition

La granulométrie est étroitement liée à la dimension des fibres.

Les plus employées actuellement ayant entre 25 et 35 mm, l'agrégat doit être de dimensions telle qu'un encadrement soit possible (Figure 6). On comprend que dans ces conditions, il soit malaisé d'utiliser une pierraille de dimension moyenne supérieure à 10 mm.

Figure 5

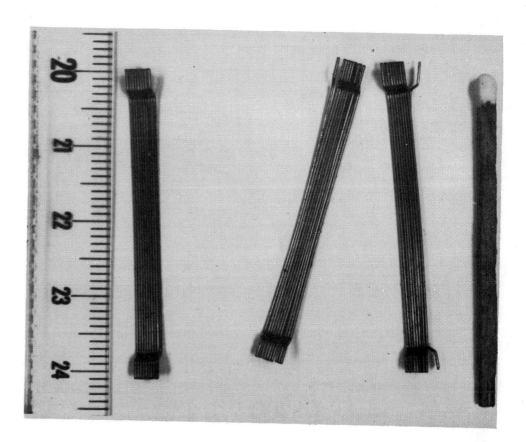

Figure 6

Déformabilité en flexion

Fatigue

Figure 7

Quant à son pourcentage, il convient que le rapport

$$\frac{\text{Pierraille (poids)}}{\text{Pierraille + sable (poids)}}$$ en reste inférieur à 0.5

sous peine de voir décroître sérieusement l'efficacité de l'incorporation. Lorsque l'on parle donc de bétons à base de fibres, il s'agit plutôt de micro-bétons ou de mortiers additionnés de pierrailles de petit calibre. Comme il y a beaucoup de "fines" il y a davantage de ciment ou de pouzzolanes que dans les bétons traditionnels.

Plus le mortier est riche, plus la résistance en flexion est accrue: le mortier ($\frac{\text{ciment}}{\text{sable}} = \frac{2}{3}$ en poids) est légèrement meilleur que le mortier $\frac{1}{2}$ et le mortier $\frac{1}{2}$ est nettement meilleur que le mortier $\frac{1}{3}$ à cet égard.

C'est dans le domaine des déformations que l'amélioration de l'addition de fibres se fait le plus remarquer. Le travail requis en flexion pour extraire les fibres de la matrice de béton augmente considérablement la résilience et transforme en fait le matériau béton à rupture fragile en un matériau quasi plastique (Figure 7).

Si on définit la résilience par l'aire comprise sous la courbe contrainte-déformation en limitant celle-ci à la valeur de la contrainte maximum, on obtient une allure remarquable de l'accroissement de résilience fournie pour les fibres (Figure 8).

Ces résultats sont confirmés par des essais au mouton de Charpy et par des essais qualitatifs de résistance aux explosifs: là où le béton normal est entièrement désintégré, le béton de fibres résiste en assurant l'unité de la structure.

Bien que divers auteurs signalent la résistance en fatigue comme devant être accrue, par l'adjonction de fibres, il semble que peu d'auteurs aient, à ce jour, investigué cette propriété.

Il est vrai que les variables sont très nombreuses. Nous avons réalisé des essais de fatigue en flexion pour des charges comprises entre 25 et 60% de la charge de rupture statique en flexion d'une éprouvette.

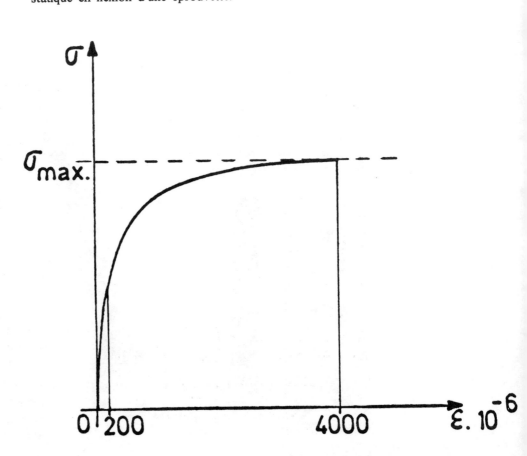

Figure 8

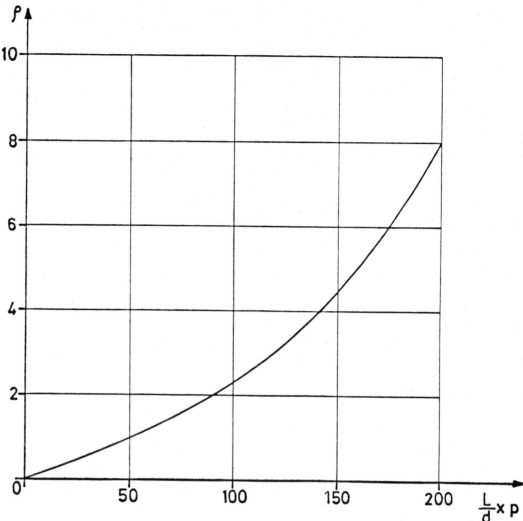

Nous donnons à titre d'information, un extrait des résultats obtenus sur des éprouvettes de mortier ($\frac{c}{s} = \frac{1}{2}$).

Pourcentage de fibres en volume %	Contrainte de rupture en flexion statique N/mm²	Nombre moyen de cycles avant rupture en fatigue pour des cycles compris entre 0.25 et 0.6 des valeurs de la colonne ci-contre
0	5.1	1,843
0.75	9.3	20,080
1.50	11.0	Plus de 30,000
2.25	13.6	Plus de 30,000

Les comparaisons ne pourraient être faites que par l'établissement de courbes du genre de celle de Wöhler et en définissant les cycles de charges. Le travail considérable qu'implique ces essais, n'a pas été fait à ce jour.

Fluage et retrait

D'essais de longue durée effectués pour le fluage (12 mois) et pour le retrait (3 mois), il résulte que les fibres ne modifient pas substantiellement les caractéristiques du béton sous ces 2 aspects, ce qui est aisément concevable compte tenu des faibles proportions en volume d'acier.

Durabilité

Nos observations sont en parfaite concordance avec celles des autres expérimentateurs, en première analyse, la rouille apparaît en surface mais ne progresse nullement en profondeur.

Résistance à l'usure et aux cycles de gel/degel

Peu d'informations sont disponibles quant à ces propriétés bien que l'on cite régulièrement des indices de performances doubles de ceux du béton non armé.

ANALYSE DE L'ORIENTATION DES FIBRES

Il est évident que l'orientation des fibres dans le sens des sollicitations va de pair avec un accroissement de résistance. On connaît plusieurs théories, de caractère géométrique, qui ont été élaborées pour justifier les valeurs d'un facteur d'orientation.

Figure 9

Nous avons voulu apporter à cette question une réponse expérimentale que nous présentons ici aux fins de susciter la discussion.

Désignons par n_f le nombre de fibres par cm² de section transversale quelconque, (Figure 9).

La surface métallique occupée par ces fibres est donc

$$n_f \times \frac{\pi d^2}{4}.$$

Si toutes les fibres sont parfaitement alignées en continu (comme des barres d'acier de béton armé), la section d'acier disponible pour absorber un effort dans le sens des fibres, perpendiculairement à la section est

$$1 \text{ cm}^2 \times \frac{p}{100}$$

si p est le pourcentage en volume des fibres.

Nous désignons par k le rapport entre la surface effective d'acier présente et celle que l'on s'attendrait à trouver dans le cas d'un alignement parfait perpendiculaire à la section:

$$k = \frac{n_f \cdot \frac{\pi d^2}{4}}{\frac{p}{100}} = 25 \, n_f \cdot \frac{\pi d^2}{p}.$$

Appelons-le coefficient d'efficacité.

Nous avons fait 3 déterminations de ce coefficient k.

1. Dans le cas d'une grosse éprouvette où par une aiguille vibrante un effet d'homogénéité a été recherché

De diverses coupes de 50 cm² au disque diamanté, on a déduit

$n_f = 6.24$ fibres

pour $d = 0.35$ mm

et $p = 1.5$ en volume

soit $k = 0.4$.

2. Dans le cas d'une éprouvette de 5 cm d'épaisseur vibrée (Figure 10) par le coffrage

Nous avons obtenu k = 0.62 dans les coupes telles que ABCD avec $n_f = 9.68$.

Figure 10

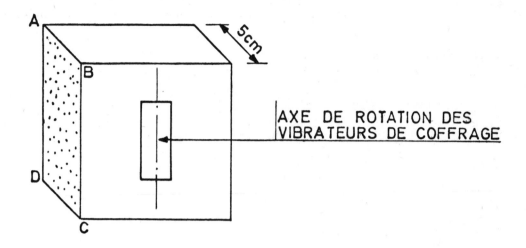

3. *Dans le cas d'une éprouvette de 2.5 cm d'epaisseur vibrée dans les mêmes conditions qu'au 2 ci-avant*

Le résultat obtenu est 0,65, avec $n_f = 10.16$.

On peut donc par des moyens extérieurs opérer pour disposer dans les sections transversales plus de 50% d'acier que celui dont on disposerait dans le cas d'homogénéité maximum. Ces résultats semblent a priori confirmés, ou du moins en accord avec l'article de Hannant et Edgington sur l'anisotropie constatée suite à une mise en vibration d'un moule lors du bétonnage.

Mais une autre expérience nous rend cependant sceptique. C'est celle qui a consisté à mettre en vibration des moules parallélipipèdiques dont les parois étaient transparentes et où le béton était remplacé par des granulats de verre de 5 mm, et le mortier par une solution aqueuse de $ZnCl_2$. Aucune orientation nette ne nous est apparue mais il est manifeste que d'importants moments tourbillonnaires ont été enregistrés malheureusement peu efficaces.

Par contre, deux faits nous sont apparus comme importants:

Figure 11

a. la possibilité d'orientation découlant de la mise en place (par exemple l'extrusion)
b. l'épaisseur des parois; si les parois sont rapprochées, le laminage du béton, lors de sa mise en place dans le coffrage oriente les fibres parallèlement aux parois. (On atteindra dans ce cas les valeurs 0.60 et 0.65 précitées.)

Si l'effort à absorber est lui aussi parallèle aux parois on touchera au maximum d'efficacité.

Un contrôle radiographique est d'ailleurs possible ainsi qu'en témoigne la photo (Figure 11) aimablement envoyée par le Dr Jamrozy de l'Université de Cracovie.

ADHÉRENCE

S'il est un phénomène important dans le béton renforcé de fibres c'est bien celui de l'adhérence ou de l'accrochage au égard à la longueur restreinte des fibres.

Très peu d'essais ont été faits et peu de valeurs ont été citées à ce jour.

Dans un article de Swamy, Mangat et Rao on voit apparaître une valeur de l'ordre de 4 N/mm^2 comme contrainte tangentielle à la surface de contact de la fibre avec le béton, alors que Den Boer dans un "pull out" test cite des valeurs de l'ordre de 0.6 à 1.2 N/mm^2 pour la contrainte de frottement.

Nous avons effectué divers essais à partir d'éprouvettes de Michaelis couramment employées dans les essais de pâte pure de ciment où une zone centrale non adhérente a été ménagée. Les Figures 12 et 13 illustrent le montage employé. Les essais ont porté sur des fibres de 30 mm de longueur, 0.35 mm de diamètre avec crochet ou sans

Figure 12

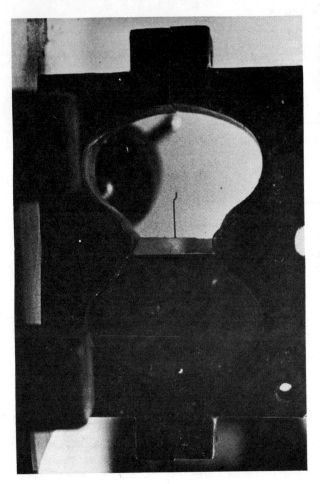

Figure 13

Figure 14

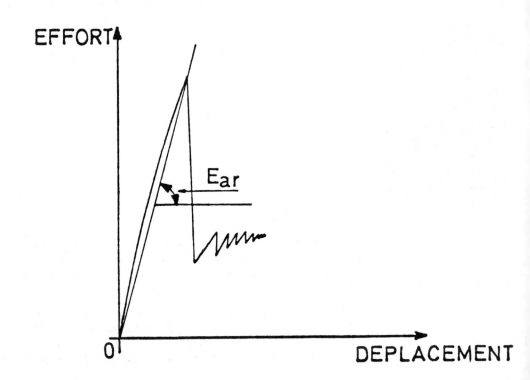

crochet. Les diagrammes effort de traction, déplacement, font apparaître un comportement quasi linéaire suivi d'une rupture soit par extraction de la fibre soit par rupture de la fibre (Figure 14).

Nous donnons ci-après quelques résultats où la contrainte d'adhérence (τ) est calculée en première approximation par la formule

$$\text{Effort} = \pi d \frac{L}{2} \times \tau$$

et où la contrainte de traction est donnée par

$$\text{Effort} = \frac{\pi d^2}{4} \times \sigma_a.$$

Constitution de l'eprouvette		Fibres sans crochet		Fibres avec crochets	
Partie superieure	Partie inferieure	Contrainte de traction (σ_a) N/mm²	Contrainte d'adherence (τ) N/mm²	Contrainte de traction (σ_a) N/mm²	Contrainte d'adherence (τ) N/mm²
Resine	Resine	1480 Rupture de la fibre	8,7	1600 Rupture de la fibre	9,4
Resine	Mortier 2/3	440	2,6	900	5,3
Resine	Mortier 1/3	320	1,84	650	3,85
Resine	Micro-beton	140	0,82	630	3,70

Quant au module d'arrachement des fibres soit E_{ar} (Figure 14) il apparaît comme devant être de l'ordre de 50,000 N/mm² pour le cas de la fibre enrobée dans le mortier 2/3.

Les expériences réalisées à ce jour sont hautement perfectibles de telle sorte que les résultats sont à considérer davantage comme ordre de grandeur plutôt que comme valeur de référence.

ESSAIS SUR TUYAUX

Quelques résultats d'essais sur tuyaux de grand diamètre en béton ont été présentés au récent colloque de l'American Concrete Institute.

Ils correspondaient à des diamètres supérieurs à 1.30 m, c'est-à-dire à des épaisseurs de l'ordre de 12 à 15 cm (au moins) semble-t-il, avec des fibres de 20 mm de longueur.

Une comparaison y est faite avec les tuyaux en béton armé. Notre attention a été attirée parallèlement sur le même sujet mais dans l'optique différente d'une réduction des épaisseurs. Nous avons opéré sur des tuyaux de faible diamètre: 40 cm et d'épaisseurs de 50 mm et 25 mm, en employant du micro-béton (pierraille 2/8 et $\frac{p}{p+s} = 0.5$).

Le béton a été mis en place par des vibrateurs de coffrages. Les caractéristiques des fibres étant les suivantes:

L = 40 mm

d = 0.35 mm

2 crochets d'extrémité

p = 1.5% en volume.

Le tableau ci-après reprend les principaux résultats.

On constate que l'on obtient des résultats parfaitement conformes aux considérations précédentes: l'orientation accroît la valeur de ρ. D'autre part, l'accroissement de déformabilité d'une telle conduite lui confère un caractère nouveau la rendant apte à mobiliser plus rapidement une butée des terres.

Essais d'écrasement avec trou

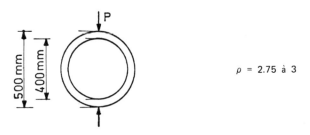

$$\rho = \frac{p^{max}_{\text{béton avec fibres}}}{p^{max}_{\text{béton seul}}} = 1.4 \text{ à } 1.8$$

Pas d'orientation de fibres

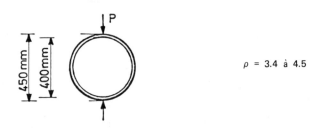

$\rho = 2.75$ à 3

Vibration latérale sur le coffrage

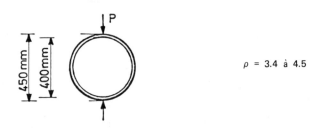

$\rho = 3.4$ à 4.5

Vibration latérale sur le coffrage

TENTATIVE DE JUSTIFICATION APPROXIMATIVE DES RÉSULTATS OBTENUS EN FLEXION ET EN TRACTION POUR LES CHARGES ULTIMES DES MORTIERS ET MICRO-BÉTONS ARMÉS DE FIBRES

Flexion

Plusieurs analyses du gentre ont déjà été proposées (Swamy-Hannant, ...).

Nous avons voulu, in fine de ce rapport, apporter aussi notre contribution à la question en nous servant autant que faire se peut des seules observations expérimentales:

1. Nous remarquons au préalable qu'au moment de la rupture par flexion, les allongements à la fibre tendue sont tels que seules les fibres résistent : il ne peut donc être fait allusion à un matériau homogène, même statistiquement, dans la zone tendue.

2. Le transfert des forces de traction vers les fibres s'effectue d'une manière quasi linéaire (contrainte, déformation) et avec un module d'arrachement E_{ar} de l'ordre de 50,000 N/mm².

3. Les sections d'acier disponibles dans les sections de rupture peuvent être définies par le produit k x p où k est le coefficient d'efficacité et p le pourcentage de fibres en volume.

Sur ces faits expérimentaux nous sommes conduits à nous référer aux schémas de flexion de la Figure 15, où l'on suppose donc que sous la zone comprimée et tendue de béton, il ne subsiste que l'acier dont l'importance est réduite à b.p.k. en largeur. On peut écrire aisément les deux équations de la flexion

$$\sigma'_b = \sqrt{\frac{1}{m}\left[m R_b^2 + (\sigma_{a_2}^2 - m^2 R_b^2)\, kp\right]} \qquad (1)$$

avec $m = \dfrac{E_{ar}}{E_b} = \dfrac{\text{Module fictif d'arrachement des fibres}}{\text{Module d'élesticité du béton}}$

et
$$m^2 \sigma'^3_b + m^2 R_b^3 + m R_b (\sigma_{a_2}^2 - m^2 R_b^2)\, pk$$
$$+ (\sigma_{a_2} - mR_b)^2 (2\sigma_{a_2} + mR_b)\frac{kp}{2} = \frac{3M}{bh^2}(\sigma_{a_2} + m\sigma'_b)^2 \qquad (2)$$

si l'on désigne par M le moment de flexion.

Dans l'hypothèse d'une section homogène on aurait la relation

$$M = \frac{bh^2}{6} \cdot R_{Ff} \qquad (3)$$

ou R_{Ff} serait en quelque sorte la résistance fictive en flexion du béton armé de fibres en section homogène. On peut aisément déduire de 2 et 3, la relation 4 donnant R_{Ff}

$$R_{Ff} = \frac{2m^2(\sigma'^3_b + R_b^3)}{(\sigma_{a_2} + m\sigma'_b)^2} + \frac{3m R_b (\sigma_{a_2}^2 - m^2 R_b^2)}{(\sigma_{a_2} + m\sigma'_b)^2}\, pk$$
$$+ \frac{(\sigma_{a_2} - mR_b)^2 (2\sigma_{a_2} + mR_b)}{(\sigma'_{a_2} + m\sigma'_b)^2}\, pk \qquad (4)$$

Figure 15

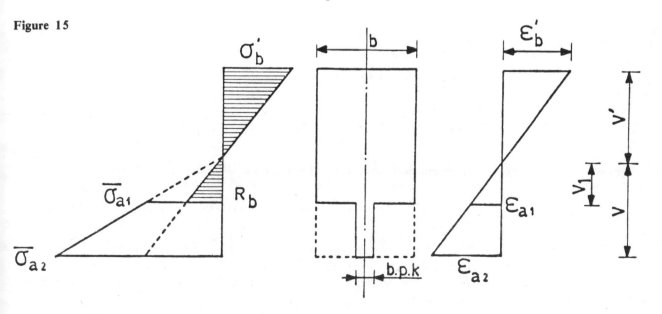

Figure 16

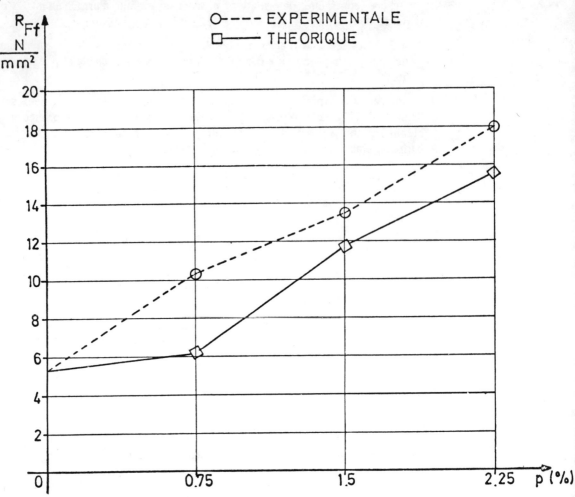

Si donc on connaît R_b résistance du béton en traction), p (pourcentage d'acier), k (coefficient d'efficacité) et $m = \dfrac{E_{ar}}{E_b}$, on peut calculer σ'_b par (1) puis R_{Ff} par (4).

Or nous disposions pour un mortier, des résultats expérimentaux du diagramme de la figure 16 qui correspondent à des essais de flexion sur éprouvettes de 16 cm x 16 cm x 50 cm. Nous connaissions donc R_b = 5.3 N/mm².

La technique employée permettait d'escompter une valeur de k de l'ordre de 0.5.

D'autre part, les essais d'adhérence ont permis de définir ± les valeurs de σ_{a_2} et de E_{ar}.

Au sujet de la valeur de σ_{a_2}, elle est de l'ordre de 900 N/mm², (8.5 N pour une fibre de 0.35 mm de diamètre) ceci grâce aux crochets d'extrémité.

La valeur de E_{ar} est nettement inférieure à 210,000 N/mm² car il s'agit d'un module d'extraction d'une fibre hors d'une gangue de mortier.

Les valeurs relevées sont de l'ordre de 50,000 N/mm² permettant d'identifier le rapport $\dfrac{E_{ar}}{E_b}$ à des valeurs comprises entre 1 et 2.

Nous avons retenu la valeur 1.5.

Les valeurs numériques obtenues ci-après marquent

p	σ'_b N/mm²	R_{bf} N/mm²
0.0075	45.5	5.9
0.0150	64	11.8
0.0225	79.5	16.0

une allure fort semblable à l'allure expérimentale.

On peut d'ailleurs la faire apparaître plus nettement en observant que les relations (1) et (4) peuvent en première approximation être écrites

$$\sigma'_b = \sigma_{a_2} \cdot \sqrt{\frac{kp}{m}} \qquad (5)$$

et

$$R_{Ff} = \sigma_{a_2} \cdot kp \qquad (6)$$

(ce qui est plus logique).

La relation (6) montre bien que pour accroître R_{Ff} il importe en premier chef d'accroître σ_{a_2} k et p.

Il y a cependant une limite supérieure car σ'_b croît aussi et ne peut dépasser la contrainte admissible en compression, la rupture intervenant alors par compression.

Ce phénomène est très bien connu du calcul du béton armé. Il convient donc de ne pas dépasser un certain pourcentage limite défini par

$$R'_b = \sigma_{a_2} \cdot \sqrt{\frac{kp}{m}}.$$

Avec les valeurs que nous avons obtenues, ce pourcentage limite est de l'ordre de 2.25%.

(R'_b = 75 N/mm², σ_{a_2} = 900 N/mm², k = 0.5, m = 1.5.)

Ceci est bien conforme à expérience.

Le diagramme de la Figure 17 reprend ces considérations. On y a ajouté la hauteur de la zone tendue de la poutre fléchie.

En fait elle est tellement importante que l'espoir que l'on caressait d'une économie des fibres ne se trouve pas concrétisé. Ce fait est bien confirmé par l'examen des ruptures en flexion.

La conclusion à tirer de ces considérations est que la résistance en flexion des éprouvettes est accrue:

1. dans la mesure où les fibres ont une résistance à l'arrachement suffisante,
2. sont judicieusement orientées,
3. dans la mesure du pourcentage de fibres employées, avec cependant une valeur limite dépendant de la résistance en compression du béton seul; si le béton est de médiocre qualité, il conviendrait alors de prévoir des formes en té avec table de compression.

Traction directe

Renforcement des fibres en traction: Nous donnons ci-après la valeur de la résistance en traction directe obtenue pour les fibres seules (soit donc après rupture du béton)

$$\Omega \times p \times k \times \sigma_{a_2}$$

si Ω est la section considérée.

La résistance de la même section en béton non armé vaut:

$$\Omega \, R_{br}.$$

Figure 17

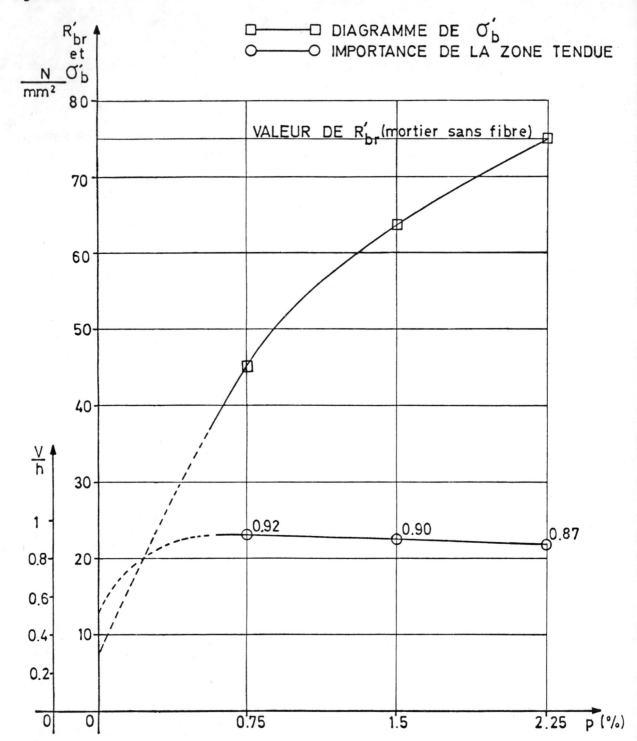

Comparons ces deux expressions:

$$R_{br} \gtreqless p \, k \, \sigma_{a_2}.$$

Avec nos valeurs:

1. $\sigma_{a_2} = 900 \text{ N/mm}^2$

   correspondant à un mortier 2/3 ($R'_{br} = 75 \text{ N/mm}^2$)

2. $k \sim 0.4$ (sans orientation)

cela revient à comparer plus ou moins

$$60 \text{ N/mm}^2 \gtreqless p \times 0.4 \times 900$$

on constate que pour des pourcentages inférieurs à 2%, il n'est pas logique d'escompter une amélioration.

Conclusion   Dans la mesure où les hypothèses de ces calculs élémentaires ne sont pas infirmées, il apparaît que la résistance de la section armée de fibres en flexion résulte de la mobilisation de la résistance spécifique considérable des fils de petit diamètre. Pour que cette résistance se manifeste, il importe que la fibre trouve un accrochage de qualité et qu'elle soit aussi judicieusement orientée que possible.

Le jeu des diagrammes de compression-traction fait que l'on atteint (fictivement) jusqu'à 3.5 fois la résistance en flexion du béton non armé.

En traction directe, avec les pourcentages couramment admis (<2.5% en volume) il n'y a guère que des accroissements marginaux à espérer.

Ajoutons que la comparaison avec le béton armé fléchi montre rapidement que le béton de fibres ne peut être a priori considéré comme substitut au premier.

Ainsi, dans le cas du béton de fibres, nous écrirons le moment de rupture en **première** approximation:

$$M = \frac{bh^2}{6} \cdot R_{Ff} = \frac{bh^2}{6} \cdot \sigma_{a_2} \; k \; p$$

alors qu'en béton armé nous aurions approximativement

$$M = 0.9 \cdot \omega_a \cdot R_{ae} \cdot h.$$

Comparons:

$$\boxed{bhp} \times h \times \left(\frac{k}{6} \cdot \sigma_{a_2}\right) \gtreqless \boxed{\omega_a} \times h \times \left(0.9 \; R_{ac}\right).$$

Si l'on emploit la même quantité d'acier, on est conduit à la comparaison de

$$\frac{k}{6} \sigma_{a_2} \gtreqless 0.9 \; R_{ac}$$

où il apparaît malaisé de faire en sorte que le 1er membre soit supérieur au second.

CONCLUSION

En matière de conclusion, citons quelques sujets qui méritent examen, pour lesquels on dispose déjà d'éléments de réponse mais qui ne sont pas suffisamment étayés selon nous:

1. Contrainte de fissuration, sujet compliqué s'il en est
2. Forme du diagramme des contraintes en flexion ("stress-block")
3. Sollicitations biaxiale et triaxiale pour lesquelles les fibres pourraient s'avérer supérieures aux armatures actuellement employées
4. Fatigue
5. Il semble que l'on ait très peu étudié le comportement au cisaillement et pas du tout à la torsion mais il ne paraît pas raisonnable d'en attendre des améliorations analogues à celle de la flexion.

Ajoutons que pour nous les fibres doivent être regardées comme améliorant très notablement certaines propriétés mécaniques des bétons à petits agrégats et des mortiers, mais qu'il est vain de vouloir les considérer exclusivement comme des substituts aux armatures du béton armé traditionnel. Leur domaine d'application apparaît bien défini: c'est celui de la structure mince où le ferraillage classique est malaisé à placer et où l'effort d'orientation est plus marqué.

Il reste à démontrer si elles peuvent être employées pour les sollicitations tridimensionnelles où le ferraillage classique est toujours un compromis.

4.2 Qualitative testing of fibre reinforced centrifugated concrete

E F P Burnett
University of Waterloo, Waterloo, Ontario, Canada

Summary *When fibre reinforcement is used spinning has a significant effect upon the distribution and orientation of the fibres, the shape of the section and the quality of the concrete.*

In order to evaluate the mechanical properties of the spun fibre reinforced concrete product it is necessary to develop tests that utilize annular, thickwalled, spun, steam-cured, concrete specimens. Since the properties of spun concrete are somewhat different from those of non-spun concrete, it is also necessary to manufacture and test standard non-spun, non-steam-cured specimens in order to provide a qualitative datum. As spun specimens are annular in section it is also necessary to produce and test annular non-spun, non-steam-cured specimens to relate to the solid datum.

This paper documents the experience gained during a fairly extensive developmental research programme to study the feasibility of using fibre reinforcement in spun precast concrete poles. The properties involved are primarily tensile and compressive strength and flexural stiffness. Non-standard test procedures are described and their validity evaluated and discussed.

Résumé *Quand on utilise des fibres, la centrifugation a un effet significatif sur la répartition et l'orientation des fibres, la forme de la section et la qualité du béton.*

Pour évaluer les propriétés mécaniques d'un produit en béton de fibre centrifugé, il faut mettre au point des essais qui utilisent des éprouvettes annulaires, à parois épaisses, centrifugées et traîtées à l'étuve. Comme les caractéristiques du béton centrifugé sont quelque peu différentes de celles d'un béton qui ne l'a pas été, il est nécessaire également de confectionner et de mettre à l'essai des éprouvettes de ce genre pour obtenir des données qualitatives. Puisque les éprouvettes centrifugées ont une forme annulaire, il faudra aussi confectionner et mettre à l'essai des éprouvettes annulaires non-centrifugées et qui n'ont pas passé à l'étuve, pour établir la comparaison.

Cette communication décrit l'expérience acquise au cours d'un programme de recherches destiné à examiner la possibilité d'utiliser des fibres dans des mâts en béton centrifugé. Les caractéristiques en cause sont principalement la résistance à la traction et à la compression et la rigidité transversale. Des procédures d'essai non-normalisées sont décrites au cours de l'exposé et on évalue et discute leur validité.

NOTATION

a	shear span
c	co-efficient of variation (expressed as a percentage)
D	principal sectional dimension
E	modulus of linear elasticity
f_{cc}	compressive strength of solid concrete cylinder
f_{cc}^{a}	compressive strength of annular concrete cylinder
f_{cc}^{as}	compressive strength of annular spun concrete cylinder
f_{ct}^{s}	indirect tensile or splitting strength
f_{ct}^{r}	modulus of rupture strength
f_{ct}^{f}	flexural tensile strength
ℓ	span length
L	specimen length
N	number of test results in sample
s	standard deviation
x	distance between applied loads
$\bar{x}$	mean value
Δ	mid-span deflection

INTRODUCTION

Centrifugation or spinning at relatively high speeds is an integral part of the manufacture of a number of concrete products. Using small proportions of wire, glass or other types of short fibre as reinforcement, the spinning regime can be arranged to distribute and orient fibres advantageously. In order to achieve an economic turnaround of forms, accelerated curing, usually low pressure steam-curing, is necessary. The final product eg pole, pile or pipe, will be annular in section and relatively thick-walled. Clearly the mechanical properties of the spun, steam-cured, fibre reinforced concrete will be different from those of the comparable initial fibre concrete that has not been spun or steam-cured.

A variety of standard tests is required to assess the quality of the fibre concrete both before and after spinning and steam-curing. Of course standardised tests and procedures exist for plain concrete. These tests can readily be performed on the non-spun fibre concrete except that:

(i) the presence of the fibre, which certainly affects the value of the test results, may render the test invalid in that the failure mechanism of the fibre concrete specimen may be different from that of a plain concrete specimen. At the very least the interrelationship that exists between the various plain concrete tests may be altered when these tests are performed on fibre concrete;

(ii) because the fibre concrete is neither spun nor steam-cured the test results may be of little, if any, value in assessing the quality of the spun, steam-cured product.

Spinning causes some segregation of the constituents in the fibre concrete and the resulting non-uniform nature of the wall in the annular product will be increased by accelerated curing. Therefore in order to evaluate the mechanical properties of the spun, steam-cured product, tests should incorporate the whole annular section. While it is simple enough to develop standard tests for the annular spun product it is also necessary to establish the relationship of these results with those results from tests performed on the initial, non-spun fibre concrete.

Over the past four years the author has been involved in a programme to develop spun fibre concrete products. Both fibreglass and wire fibre were utilized and, in addition to reinforcement parameters, the influence of the spinning regime and curing procedures were studied. Some of these results have already been reported (1,2). During the course of the work a number of tests were used to evaluate the mechanical properties of both the spun and non-spun fibre concrete. For comparative purposes these tests were invaluable but, as far as individual parameters were concerned, there was the constant problem of relating test results on spun fibre concrete to those on non-spun fibre concrete and then correlating these results with the customary plain concrete datum. However, sufficient test results are now available to provide a reasonably large statistical base for assessing the value of some of these tests. The objectives of this paper are therefore to outline and discuss these tests and, where possible, to quantify their interrelationship. Tests for the evaluation of compressive and tensile strength and flexural stiffness are of primary concern.

COMPRESSION STRENGTH

The strength of a 152.4 mm diameter by 305 mm solid cylinder loaded in concentric compression i.e. f_{cc} is, in North America, customarily used as the measure of the compressive strength and, to some extent, the quality of the concrete involved. One practical reason for using this simple test is that it provides a means for assessing the relative quality of batching and mixing during production. A comparable test can be performed on the spun concrete except that the test specimen will have been steam-cured and be annular in section. Therefore in order to eliminate this shape effect and provide a means of evaluating the effects of spinning or steam-curing or both, it is necessary to test non-spun, non-steam-cured annular test specimens as well.

Accordingly for each test in the experimental research programme three sets of three compression test specimens were produced. Details are provided in Table 1. The

Table 1 Compression strength tests

Symbol	Shape	Size mm		Production details	Number of specimens per test	Remarks
f_{cc}		D	152.4	Non-spun, fog room cured	3	Standard uniaxial cylinder compression test — ASTM C39
		L	305			
f_{cc}^a		D	152.4	Non-spun, fog room cured	3	Tested in uniaxial compression, as for the standard cylinder test
		D_i	63.5			
		L	305			
f_{cc}^{as}		D	152.4	Spun, steam cured	4	
		$D_i \pm$	63.5			
		L	305			

relationship between the values of f_{cc} and f_{cc}^a is of particular interest since, with the exception of the cross-sectional shape, the materials and procedures involved in each test are identical. An examination of the strength ratio f_{cc}/f_{cc}^a for 25 sets of test results yields the following:

mean value of f_{cc}/f_{cc}^a i.e. $\bar{x} = 1.0102$

standard deviation   i.e. $s = 0.0462$

co-efficient of variation   i.e. $c = 4.57\%$

sample size   i.e. $N = 25$

total number of specimens involved = 150 (i.e. 25 x 3 x 2).

The compressive strength of the specimens ranged from 27.58 N/mm² to 68.95 N/mm² and at the lower end of this range the ratio f_{cc}/f_{cc}^a tended to be slightly less than the mean value. Age of test, 7 or 28 days, did not appear to affect the ratio. Until such time as a larger statistical base is available it may be concluded that for all practical purposes the solid cylinder and the annular thick-walled cylinder provide equal values for compression stress. From a few tests done on cylinders without fibre reinforcement it would appear that the presence of fibres does not have any significant influence on this conclusion.

TENSILE AND FLEXURAL TESTS

Table 2 provides details of the tensile and flexural tests performed during the course of the experimental test programme. Figures 1 and 2 provide additional information regarding the flexural tests on the spun specimens. The following are of concern:

1. the relationship between the splitting strength, f_{ct}^s, and the modulus of rupture, f_{ct}^r, for fibre concrete;

2. the influence on flexural strength, f_{ct}^f, of the length of the spun fibre concrete test specimen;

Table 2 Tension and flexural tests

Symbol	Shape	Size mm		Production details	Number of specimens per test	Test set-up	Remarks
f_{ct}^{s}	circle with D	D	152.4	Non-spun, fog room cured	3		Indirect tensile - standard splitting test - ASTM C496
		L	305				
f_{ct}^{r}	square with D	D	152.4	Non-spun, fog room cured	3		Standard modulus of rupture test - ASTM C78
		L	610				
f_{ct}^{f}	annulus with D_i	D	152.4	Spun, steam-cured	1		Flexural test
		D_i	± 63.5				
		L	1829				
f_{ct}^{f}	annulus	D	152.4	Spun, steam-cured	2		Flexural test (using the two halves of the longer specimen)
		D_i	± 63.5				
		L	± 914				

f^s_{ct}/f^r_{ct}

3. the influence on cross-sectional shape and length on flexural properties.

From studies (3) on plain concrete specimens a value of between 0.7 and 0.8 would be expected for the ratio f^s_{ct}/f^r_{ct}*. The weaker the concrete the greater this value and for f_{cc} values in excess of 41.4 KN/m² values of 0.7 or even less are to be expected. However as the following results suggest, the addition of fibres to concrete does alter the relationship between these two tests.

Comparison of f^s_{ct}/f^r_{ct} values	Specimens with variable fibre content	Specimens with 1% by volume fibre content
$\bar{x}$	0.835	0.899
s	0.1064	0.0768
c	12.8%	8.5%
N	28	16

When fibre content (but not length, size or type of fibre) was kept constant the mean value for f^s_{ct}/f^r_{ct} was 0.899, which is substantially greater than expected. The inclusion of other test results, mostly with lower proportions of fibre, does reduce the mean value. That the fibre reinforcement has relatively more effect on the splitting test than the flexural test is not unexpected if one considers the stress distributions involved in each test. Theoretically the fibre distribution in non-spun specimens is random but with small, differently shaped sections the relative distribution in each case will be affected. Moreover with a reasonably uniform distribution of tensile stress, as in the splitting test, the presence of reinforcement will have relatively more influence than will be the case where the non-uniform stress distribution is a maximum at the essentially unreinforced exterior surfaces.

Flexural tests on spun fibre concrete: length considerations:

To evaluate the flexural tensile strength of spun fibre concrete a 1830 mm test specimen was centrifugally cast and then tested as shown in Figures 1 and 2. Cracking and, hence, damage was localised in such a manner that, after failure, the two broken lengths could be tested in flexure. The same basic test was made in each case except that the span in the latter two tests was one-half that in the initial test. Load and deformation were monitored and the limiting tensile stress and the equivalent elastic modulus for the fibre concrete could be evaluated (see Figure 1). The following comments are relevant.

(i) f^f_{ct}: A comparison of values of flexural tensile strength obtained from the long specimen with the mean value obtained from the two shorter specimens i.e. the ratio f^f_{ct} (1.52) / f^f_{ct} (0.76) yields the following:

$\bar{x} = 1.004$

$s = 0.0835$

$c = 8.32\%$

$N = 22$ (1 x 1.52 m, 2 x 0.76 m).

In spite of the magnitude of the coefficient of variation it may be concluded that, for all practical purposes, the length of these flexural specimens does not influence the value of the flexural tensile strength.

(ii) E: Based on the limiting strength and deflection values and assuming linear elastic response the equivalent flexural stiffness (EI) or elastic modulus (E) may be

---

* See test results in Table 3 (note 3).

Figure 1 Flexural test details.

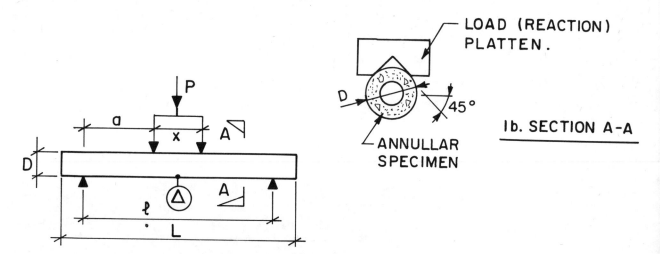

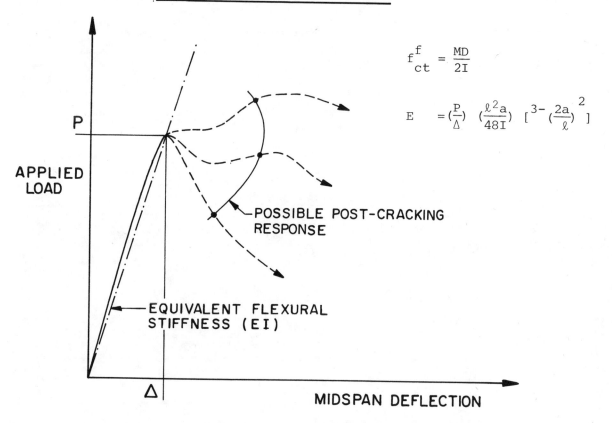

1a FLEXURAL TEST SET-UP

1b. SECTION A-A

1c REPRESENTATIVE LOAD-DEFLECTION RELATIONSHIPS

$$f_{ct}^{f} = \frac{MD}{2I}$$

$$E = \left(\frac{P}{\Delta}\right)\left(\frac{\ell^2 a}{48I}\right)\left[3-\left(\frac{2a}{\ell}\right)^2\right]$$

determined for each test. If the value for the long specimen, E (1.52), is compared with the mean of the two short specimens values i.e. E (0.76) it appears that, whereas E (1.52) is of the correct order of magnitude, the value of E (0.76) is consistently much lower. Over a number of tests it was found that values for the ratio E (1.52)/E (0.76) had the following statistical characteristics:

$\bar{x} = 2.026$

$s = 0.4724$

$c = 23\%$

$N = 23.$

Figure 2   Flexural test set-up.

Table 3 Flexural tests; shape and length considerations

Test specimen		Flexural test set-up			Test results				Comparative results			
Shape	Size $D \times L$ mm	x	ℓ	a/D	f_{ct} N/mm² Mean value -3 tests	c %	Δ Mean of 3 values mm	Equivalent E value[1] N/mm²	$\dfrac{f_{ct}}{f_{cc}}$	$\dfrac{f_{ct}}{f_{ct}^s}$	$\dfrac{f_{ct}}{f_{ct}^m}$	Ratio of equivalent E values
□	152.4 x 610	D	3D	1	4.978	4.1	0.622	2340	0.134	1.385	1.00	0.087
○	152.4 x 610	D	3D	1	3.081	5.9	0.381	4660	0.163	1.690	1.22	0.175
□	152.4 x 1829	2D	10D	4	4.950	2.7	0.650	22820	0.133	1.375	1.005	0.861
○	152.4 x 1829	2D	10D	4	5.481	5.9	0.620	26510	0.147	1.525	1.10	1.000

f_{cc} Mean value - 6 tests: 37.232 N/mm² C.V. = 7.63%

f_{ct}^s Mean value - 6 tests: 3.592 N/mm² C.V. = 17.1%

Notes:
1. Fibre reinforcement was not used in the above tests.
2. See testing details in Figure 1.
3. $f_{ct}^s / f_{ct}^r = 0.72$
4. In computing $f f_{ct}$ neither the self-weight of the platen nor the specimen has been included.

This suggests that the mode of behaviour of the short specimens is non-flexural in nature. That the non-dimensionalised shear span ratio (i.e. a/d) is much less than 3 tends to support this view. It is perhaps fortunate that the computed values for flexural strength agree but clearly any deformational considerations must be restricted to the longer test specimens.

Shape and length considerations

To gain further insight into the problems of shape and length effects in flexural tests, the series of plain concrete tests summarised in Table 3 were performed. The number of tests does not constitute an adequate statistical base but the following comments are pertinent:

(i) For the square section strength does not appear to be significantly affected by specimen length. For the circular section the calculated strength value is not only greater than that for the square specimen but length does not appear to have affected the strength results.

(ii) Cross-sectional shape certainly has an influence on the deformation and, hence, the equivalent stiffness of the concrete. The square section is consistently less stiff than the circular section.

(iii) In tests with low non-dimensionalised shear span ratios (a/D < 3) care should be exercised since there is the possibility of obtaining non-conservative strength predictions and incorrect deformational information. For consistency of results square section specimens are to be preferred.

What additional influences fibre reinforcement would have is difficult to predict but these observations do tend to substantiate previous findings.

CONCLUSIONS

The standardisation and validation of test procedures is vital to the planning and communication of research and the control of production quality. This is particularly important for spun fibre reinforced products where there are many significant variables and their influence is not well understood. This paper attempts to document some experience with various tests and prompts the following conclusions:

(i) The thick-walled annular cylinder and the solid-cylinder tested in concentric compression produce, for all practical purposes, the same value of the maximum compressive stress.

(ii) With fibre concrete the inter-relationship between different test results is not necessarily the same as for plain concrete. For example the splitting strength of fibre concrete is, relative to the modulus of rupture, considerably larger than would be expected. In addition to being a constituent of the mix, the fibre acts as reinforcement and it is essential that the laboratory test does, to some extent, simulate the actual in-service situation.

(iii) Shape and length of test specimens are particularly important parameters in developing a standard test. While the effect of shape or length or both may not significantly affect the computed strength they will have a considerable influence on the behavioural mechanics, particularly the deformation of the specimen. Flexural tests with non-dimensionalised shear span ratios of less than 3 are not recommended.

ACKNOWLEDGEMENTS

Nada Holicki, Peter Cover and Tom Constable have, as research assistants, been involved in the experimental work that provided the data for this study. They have all contributed to aspects of this study and their assistance is gratefully acknowledged. The financial support provided by Stress/Crete Limited, Fibreglas Canada Ltd, and the National Research Council of Canada is greatly appreciated.

REFERENCES

1 Burnett, E F P, Constable, T and Cover P, "Centrifugated wire fiber reinforced

concrete", Publication SP-44 Fiber Reinforced Concrete, American Concrete Institute, Detroit, 1974, pp 455-475.
2   Holicki, Nada, "Glass fiber reinforced spun concrete", Thesis submitted in partial fulfillment of MASc degree, University of Waterloo, January 1971.
3   Walker, S and Bloem, D L, "Effects of aggregate size on properties of concrete", Journal of the American Concrete Institute, Vol. 32, No. 3, September 1960, pp 238-298.

4.3 Reduction of shrinkage cracking in reinforced concrete due to the inclusion of steel fibres

Robert H Elvery
University College London
and Mufid A Samari
University of Baghdad

Summary

Some of the cracking which occurs in reinforced concrete results from the restraint to free shrinkage caused by both the steel reinforcement and the moisture gradient in the concrete.

The tests described in this paper were carried out to examine the influence of steel fibres on the development of shrinkage cracking and formed part of a more extensive programme to study the effect of fibre inclusions upon crack development in reinforced concrete. More than 150 concrete prisms each containing an axially-placed reinforcing bar, and 40 reinforced concrete slabs were tested. The specimens were stored in a dry laboratory atmosphere and examined for the incidence of cracks over periods of time varying from 7 to 520 days.

Those observations showed that, while specimens containing no steel fibres developed numerous cracks, no shrinkage cracks were detectable in any of those containing fibres. Since shrinkage cracking contributes to crack development in normal reinforced concrete the results indicate that fibres included near reinforcement would provide an additional inhibition to cracking of dry concrete and would thus allow an enhanced stress to be used in the reinforcing steel.

Résumé

Une partie du fendillement qui se produit dans le béton armé est le résultat de la contrainte sur la contraction libre occasionnée à la fois par l'armature en acier et la rampe de la moiteur dans le béton.

Les essais décrits dans cette étude ont été faits pour examiner l'influence des fibres d'acier sur le développement du fendillement dû à la contraction, et font partie d'un programme plus étendu pour étudier l'effet des fibres sur le développement des fendilles dans le béton armé.

Plus de 150 prismes de béton, chacun contenant une barre de renforcement sur l'axe, et 40 dalles de béton armé ont été essayés. Les spécimens furent gardés dans une atmosphère sèche, et ils ont été examinés pour déterminer l'occurrence des fendilles pendant des espaces de temps variant entre 7 et 520 jours.

Ces observations ont démontré que, tandis que les spécimens qui ne contenaient pas de fibres d'acier avaient développé beaucoup de fendilles, on ne pouvait pas en découvrir dans aucun des ceux contenant des fibres. Puisque le fendillement dû à la contraction contribue au développement des fendilles dans le béton armé ordinaire, les résultats indiquent que des fibres comprises près de l'armature fourniraient une interdiction supplémentaire contre le fendillement du béton sec, et ainsi elles permettraient l'emploi d'un plus grand effort dans l'acier de l'armature.

INTRODUCTION

The restraints which reinforcing bars embedded in concrete offer to its free shrinkage cause tensile stresses to develop in the concrete when it dries. This effect is more pronounced in the concrete near the surface of a structural member where it is augmented by the effect of similar restaints due to moisture gradients.

This results in the development of surface cracking in the tensile zone of concrete beams when bending moments of modest, or even zero, magnitude are applied to them.

It would therefore be expected that the cracking pattern which subsequently develops in reinforced concrete beams when fully loaded would be influenced by the effects of initial drying shrinkage and the characteristics of the concrete.

Recent experimental work (1) has shown that the inclusion of steel fibres in reinforced concrete modifies the crack development in axially-loaded specimens resulting in the formation of narrower cracks at closer spacing. The practical consequence of this is that higher tensile stresses could be allowed to develop in the steel reinforcing bars without the cracks exceeding the maximum permissible width at the surface.

The essential role of steel fibres included in concrete under tensile load is to maintain a tensile force across a crack immediately after it has formed. This force contributes to the further development of tensile stress in adjacent uncracked concrete as the tensile load is increased and this results in the formation of cracks at more closely spaced intervals than would occur in concrete without fibre inclusions.

It has been found that the tensile force maintained across a crack by steel fibres makes a significant contribution to the tensile resistance in a reinforced concrete member until the crack reaches a width of about 0.1 mm. When the crack width exceeds this value, the tensile force provided by the fibre fails to become almost negligible at crack widths of 0.3 mm.

This paper is concerned with an examination of the influence of drying upon the magnitude of the stress which can be developed in the reinforcing steel corresponding to different surface crack widths in reinforced concrete with and without steel fibres. It also examines the effect of fibre inclusions upon the incidence of initial shrinkage cracking.

DETAILS OF SPECIMENS

The basic concrete materials used were ordinary Portland cement, and Thames Valley sand and gravel with a maximum size of 10 mm. The concrete mixture consisted of 1 part cement to 3.25 parts sand to 2.25 parts aggregate, all by weight, with a water-cement ratio of 0.60.

Two types of steel fibres were used in the present investigation and were obtained from the National Standard Co Ltd. Round fibres 0.15 mm dia x 13 mm long were used in the direct tension specimens and Duoform fibres 0.38 mm dia x 38 mm long in the flexural slabs.

A cold drawn prestressing steel wire 5 mm in diameter was used as the reinforcing bar for the flexural slabs. The reinforcing bar for each of the direct tension specimens was a round ribbed twisted bar with a diameter of 12 mm and a specified characteristic strength of 460 N/mm^2. The specified characteristic strength was 1570 N/mm^2 for the prestressing bar.

All mixing was done in a pan mixer. Materials, except for the fibres were first mixed with water until the mixture had come to uniformity, then the fibres were sieved into the rotating mixer pan to ensure random distribution. The mix was then placed in the moulds which were clamped to a vibrating table. The exposed surface of the concrete in the moulds was covered with polythene sheeting for 24 hours, at which time the specimens were stripped from the moulds and cured. The tension specimens were cured in water while the flexural slabs were placed inside waterproof sheets which were filled with water and sealed for the required time. Some of the specimens were then allowed to air-dry in a laboratory room, with a constant relative humidity of 60% until the time of testing. The rest of the specimens were left in their moist curing condition for the same length of time as the equivalent air dried specimens.

Figure 1   Crack pattern of slabs cured for 7 days under waterproof sheet and then dried at 60% RH for 520 days. ▲ without fibres. ● with fibres.

a) No load

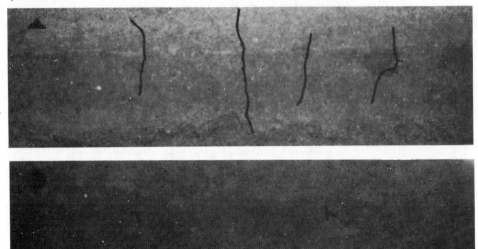

b) At 0.3 mm crack width

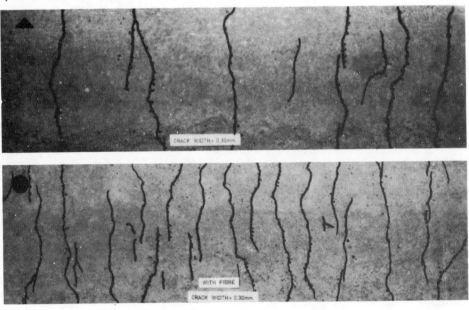

Testing procedure

The tension specimens were 500 mm long with a 75 mm square cross-section, each reinforced with a single bar placed along its axis protruding at each end. The load was applied to the steel and measurements made of crack widths and spacings.

The flexural tests were made on slabs loaded so that the central part of the span had a constant bending moment with the tensile surface uppermost. Measurements were made of the central deflection, the surface stains and crack widths at the tensile face over the constant moment zone. The slabs were 100 mm thick and 1.8 m long being supported at points 0.8 m apart.

Table 1   **Effect of drying on the reinforcing bar stress of specimens with or without steel fibres at various crack widths.**

Designation	Curing period (days)		Fibre content %	Percentage difference in reinforcement stress between dry and wet specimens Crack width, mm			
	Water	Dry Air		0.1	0.2	0.3	0.4
T1	7	7	0	-11	-5	-15	-7
T2	7	21	0	-24	-20	-22	-18
T3	7	82	0	-45	-41	-32	-33
T4	7	173	0	-40	-42	-39	-36
T5	7	353	0	-45	-42	-39	-36
T6	14	14	0	-25	-12	-9	0
T7	14	76	0	-40	-33	-30	-35
T8	14	346	0	-36	-35	-32	-19
T9	28	62	0	-20	-23	-28	-25
T10	7	7	0.8	-9	-7	-9	-6
T11	7	21	0.8	-12	-17	-10	-8
T12	7	82	0.8	-21	-27	-20	-19
T13	7	173	0.8	-18	-16	-15	-20
T14	14	14	0.8	+22	+33	+29	+11
T15	14	76	0.8	-14	-17	-15	-10
T16	28	62	0.8	-5	-9	0	-13
T17	7	7	2.0	+2	-2	0	+1
T18	7	21	2.0	-2	-3	-4	+2
T19	7	82	2.0	-9	-4	-2	-5
T20	7	173	2.0	-1	-8	+6	+4
T21	7	353	2.0	0	+3	+10	0
T22	14	14	2.0	+8	-1	0	+5
T23	14	76	2.0	-7	-8	-6	-4
T24	28	62	2.0	+9	+24	+20	+14
T25	28	346	2.0	+14	+14	+23	+11
S1	7	21	0	-38	-33	-34	-28
S2	7	83	0	-38	-40	-35	-30
S3	7	353	0	-45	-39	-31	-28
S4	28	332	0	-41	-32	-29	-25
S5	0	7	1.0	-17	-21	-19	-11
S6	7	21	1.0	-15	-13	-14	-8
S7	7	83	1.0	-18	-19	-16	-8
S8	7	353	1.0	-18	-16	-18	-8
S9	28	332	1.0	-18	-21	-17	-13

Results and discussion

It has been established that the inclusion of fibres in concrete could be expected to inhibit widening of cracks once they begin to form and this results in the development of a greater number of cracks but of much narrower width (1), (2), (3).

To investigate the influence of steel fibres on the development of shrinkage cracking more than 150 tension specimens and 40 flexural slabs were cast and tested over periods of time varying from 7 to 520 days. Results showing the effect of fibres on the spacing of cracks are presented in Table 1. The results of tests in terms of the nominal stress in the reinforcing bars of the reinforced concrete specimens at a given crack width are presented in Figure 2 and 3 and Table 2.

Table 1 gives the number of cracks formed in the the flexural slabs over the 600 mm constant moment zone at different maximum crack widths. Slabs designated (S) refer to those reinforced concrete slabs without any steel fibres. The rest of the slabs presented in the table, namely those designated (SF), include 1% by volume of Duoform steel fibres incorporated in the region local to the steel bars. Each result is the average of at least two slabs with similar curing conditions.

As is to be expected the slabs containing the fibres, namely SF1 to SF6 exhibited a greater number of cracks than similar slabs without any fibres. The general trend seems to be the development of new cracks with the continuous application of loads, hence the increase in the stresses of the reinforcing bar up to the point where the maximum crack width reached about 0.12 mm. Beyond this point the stresses in the bars between the existing cracks do not seem to reach high enough values to initiate new cracks.

It must also be noted that the periods of water curing and air drying to which slabs were subjected were of little consequence to the final number of cracks developed. Prolonging the drying period from 7 days to 520 days did not even alter the final number of cracks significantly.

Almost all of the slabs without fibre showed surface shrinkage cracks even before loads were applied. The only slabs that did not exhibit any cracks at zero load were those cured for 7 days in water and then dried for 21 days in air. On the other hand, no surface shrinkage cracks were evident in any of the slabs containing fibres even with prolonged periods of drying. This phenomenon could also be seen readily from Figure 1

Table 2   **Number of cracks developed in slabs tested at various crack widths in a 600 mm constant moment zone.**

Designation	Fibre content (%)	Curing period (days) Water	Dry air	At zero load	Number of cracks Crack widths, mm							
					.05	.10	.15	.20	.25	.30	.35	.40
S1	0	7	21	0	1	1	4	6	6	6	6	6
S2	0	7	83	1	2	3	5	6	6	6	6	6
S3	0	7	353	5	5	5	5	5	5	5	5	5
S4	0	7	520	4	5	5	6	8	8	8	8	8
S5	0	14	346	3	3	4	6	7	7	7	7	7
S6	0	28	332	4	4	4	6	6	6	6	6	6
SF1	1.0	0	7	0	3	6	10	10	11	11	11	11
SF2	1.0	7	21	0	2	5	12	12	12	12	12	12
SF3	1.0	7	83	0	2	11	14	16	16	16	16	16
SF4	1.0	7	353	0	3	6	14	14	14	14	14	14
SF5	1.0	7	520	0	2	8	11	14	14	14	14	14
SF6	1.0	28	332	0	1	3	8	11	11	11	11	11

Figure 2   Reinforcing bar stress in slabs tested in flexure after drying for different periods of time.

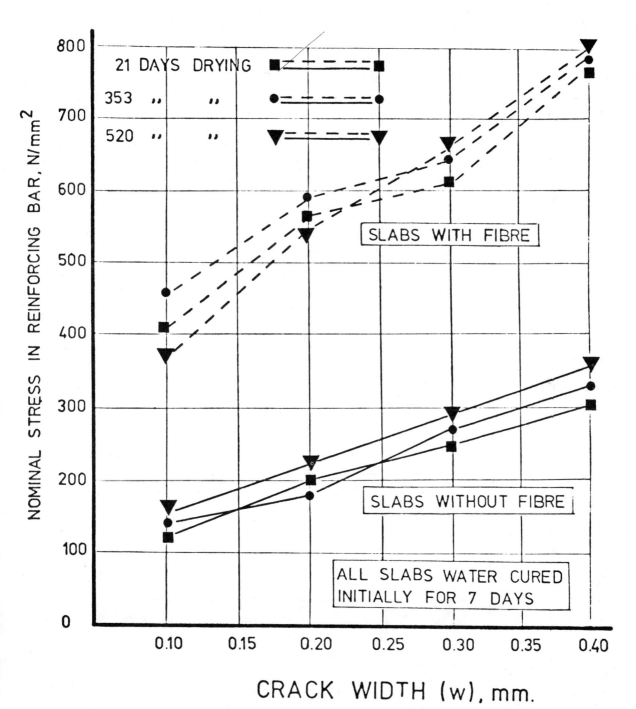

where photographs of slabs with or without fibres dried for 520 days are shown. Before the commencement of loading there were 4 cracks on the surface of slabs without fibres with none on those with fibres. When the final crack pattern was well established, at 0.3 mm crack width, the slabs with fibres not only had greater numbers of cracks but had more numerous branches and sub-divisions.

Figure 2 shows the nominal stress values in the reinforcing bars at different maximum crack widths for slabs dried for 21, 353 and 520 days. The fibres in these slabs were only in the upper 50 mm region around the reinforcing bars. There was a definite increase in the bar stress, at any given crack width, for slabs containing fibres over those without fibres. Increasing the drying period from 21 to 520 days seems to affect the stresses but only slightly. This was true for both sets of slabs as can be seen from the figure. The stresses for slabs without any fibres increased from an average value of 130 N/mm² at a crack width of 0.10 mm to about 320 N/mm² at crack width of 0.4 mm. This is less than half the average stress values for the slabs containing fibres over the same range of crack widths.

The stresses in the reinforcing bars of the tension specimens without any fibres were much reduced when they were air dried as compared to similar specimens continuously left in water. Table 2 gives the results of tests carried out to establish the effect of drying on the nominal stresses in reinforcing bars of tension specimens and flexural slabs as related to similar specimens continuously left in water. The tension specimens were designated with (T) and the flexural slabs with (S). As can be seen from the table the percentage difference in stress between dry and wet specimens (loss of nominal stress) at 0.3 mm crack widths increased from 15% after 7 days drying to about 39% when they were dried for 360 days.

With the inclusion of 0.8 percent fibres by volume the percentage loss in stress between the wet and dry specimens was almost half of those without fibres. This was also the case for the slabs containing 1% fibres as compared to similar ones without any fibres.

Figure 3   Effect of drying for 520 days on the reinforcing bar stress of specimens tested in uniaxial tension with or without steel fibres. Difference scale shows steel stress in dry specimens minus that in wet specimens.

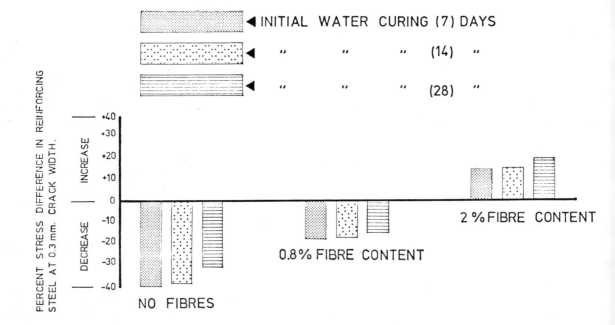

With the inclusion of 2% fibres in the tension specimens the loss in stress was not only reduced to a minimal value but in many instances a gain in stress resulted. The gain in stress, designated by (+) in the table, was greater for specimens with longer initial water curing periods than those with shorter periods. This could readily be seen from Figure 3, where the effect of 520 days of drying on the stress is presented for tension specimens with 0.8% and 2% fibres respectively. It is apparent from the figure that the addition of 0.8% fibres reduced the percent stress difference from that of specimens without fibres. This was true for 7, 14 and 28 days initial water curing periods. Although little improvement was observed when initial water curing periods were increased from 7 to 14 days, the curing of specimens for 28 days gave a more clear improvement. As a matter of fact, all the specimens containing 2 percent fibres showed a small gain in stress when dried for 520 days as compared to those which were water cured for the full period. No explanation could be advanced for this phenomenon.

CONCLUSIONS

1 Reinforced concrete specimens exhibited numerous shrinkage cracks when dried for periods over 21 days while no cracks were detectable in similar specimens, with steel fibres even after 520 days of drying.

2 The length of the period of moist or dry curing to which specimens are subjected had very little effect on the number of cracks formed due to subsequent drying.

3 Fibres included near reinforcing bars of dried specimens inhibited the widening of cracks as compared to those without any fibres, and hence allowed an enhanced stress to be used in the reinforcing steel for a given crack width.

4 The percentage difference in reinforcement stress corresponding to a given crack width between wet and dry specimens was drastically reduced when fibres were incorporated in the specimens.

ACKNOWLEDGEMENTS

The authors wish to thank the Ministry of Higher Education of the Iraqi Government who provided a grant for this work and also Professor K O Kemp, Head of the Department of Civil and Municipal Engineering, University College London, where the work was carried out.

They would also like to acknowledge with thanks the help given to them by the technical staff at University College in the laboratory programme.

REFERENCES

1 Samarai Mufid, A and Elvery Robert, H, "The Influence of Fibres Upon Crack Development in Reinforced Concrete Subject to Uniaxial Tension", Magazine of Concrete Research, Vol 26, Number 89, December 1974, pp 203-211.
2 American Concrete Institute, "Fiber Reinforced Concrete", Publication SP-44, Detroit, 1974, 554 pp.
3 The Concrete Society, "Fibre-reinforced Cement Composites", Technical Report, 51.067, July 1973, 77 pp.

4.4 Durability of steel fibre concrete

D J Hannant and J Edgington
University of Surrey, United Kingdom

Summary

An important aspect of the study of most types of fibre reinforced cement based composites is the durability of the fibres within the cement matrix and the paper gives details of work undertaken to assess the resistance of steel fibres to corrosion in normal weight and lightweight concrete cylinders in the uncracked state. Also some preliminary results are presented on the performance of fibres in normal weight gravel concrete beams exposed after initial loading which induced cracks with widths of 0.1mm to 0.3mm.

The cylinders were placed on three sites covering relatively mild exposure, marine conditions and a polluted industrial atmosphere. Observations over a period of $4\frac{3}{4}$ years suggest that the corrosion of fibres within the concrete is unlikely to cause major problems.

The cracked beams have been exposed to the marine environment for only 11 months but significant carbonation and fibre corrosion has already occurred. The failure loads of specimens tested at this age had not decreased below the initial cracking loads but it is the authors' opinion that this is unlikely to continue to be the case as corrosion increases with time.

Résumé

Un aspect important de l'étude de la plupart des types de composés basés sur le ciment armé par les fibres est la durabilité des fibres dans la matrice de ciment, et la communication donne les détails des travaux entrepris pour évaluer la résistance des fibres d'acier dans les cylindres de béton de poids normal et de poids faible à l'état normal à la corrosion. De plus, quelques résultats préliminaires sont présentés sur le rendement des fibres dans les poutres de béton et de gravier de poids normal après le chargement initial qui a stimulé les fissures d'une largeur de 0.1 mm à 0.3 mm.

Les cylindres étaient placés sur trois emplacements qui comprenaient: l'exposition relativement peu sévère, les conditions marines et une atmosphère industrielle contaminée. Les observations pendant une période de $4\frac{3}{4}$ ans suggérent qu'il est peu probable que la corrosion des fibres dans le béton provoquera des problèmes sérieux.

Les poutres fissurées ont été exposées en milieu marin pendant seulement 11 mois mais la carbonatation significative et la corrosion des fibres a déjà eu lieu. Les charges de rupture des éprouvettes qui ont été essayées à cet âge n'avaient pas diminué au-dessous des charges de fissuration initiales mais d'après les auteurs il est peu probable que cet effet continuerait au fur et à mesure que la corrosion augmente.

INTRODUCTION

Civil engineering codes of practice (1,2) have stringent requirements regarding the concrete cover necessary to prevent rusting of steel reinforcing bars, and it is therefore rather anomalous that the cover to many of the steel wires in fibre concrete is effectively zero. This is especially important since a degree of rusting which might normally be allowable on a large diameter reinforcing bar could completely destroy the small diameter steel fibres.

The acceptability or otherwise of rusting depends upon the use to which the material is subjected. For instance, if the fibre concrete is used for structural applications it may be necessary for the fibres to retain their strength for periods in excess of 50 years. On the other hand, the critical period may be only a few weeks during which, for example, precast units are handled on site. In either case, there may be applications in which the aesthetic effects, rather than structural strength are important and the degree of discolouration by rusting would then be the criterion applied.

It was with these problems in mind that the investigation into the durability of uncracked cylinders was started in 1970 as part of a more general research contract at the University of Surrey into the properties of steel fibre concrete (3,4). During this contact it became apparent that many of the proposed major applications of the material were likely to involve the use of steel fibre concrete in flexural situations where the post-cracking ductility was probably a vital part of the load carrying capacity of the structural system. It therefore follows that if the material is to be effectively utilised it will often have to function in the cracked condition.

A second series of tests was therefore started in 1973 to study the corrosion of the fibres across the cracked zones of beams which are exposed to air borne marine spray. The cracks induced in the beams are between 0.1mm and 0.3mm wide, as permitted by CP 110 1972 (1), and the change with time in the load carrying capacity of the cracked beams is being studied.

FACTORS AFFECTING THE CORROSION OF STEEL IN CONCRETE

The liquid phase and the initial hardened state in concrete made with Portland cement contain large amounts of calcium and other alkaline hydroxides which form a highly alkaline material of pH between 12 and 13 (5). In this environment an insoluble oxide film forms on the surface of the steel which is protected from further corrosion while the film is unbroken.

However, atmospheric carbon dioxide dissolves in the aqueous phase in the concrete to form weak carbonic acid which reacts with the alkalis in the pore water to form carbonates. This reduces the pH to values as low as 8.5 and, because the protective alkalinity of the concrete is then lost, corrosion of the steel can occur if oxygen and water are present.

Factors which accelerate corrosion are cracks which allow accelerated carbonation and access of oxygen and water in critical regions, and permeable concretes made with high w/c ratios and porous aggregates.

The corrosion of the steel can also be accelerated by marine environments (6) where sea water or wind blown salt spray can cause acute differences in the chloride ion concentration in the material which promote electro-chemical action. This type of corrosion could also develop in highway applications in which sodium chloride is used for de-icing during severe winter conditions.

SPECIMEN MANUFACTURE

Uncracked cylinders Two mixes were used, one of which was considered to represent a well made and well controlled durable mix with a hard flint gravel aggregate, a high cement content and a low w/c ratio as required by CP 110 1972 (1) for marine exposure. In view of the likely long-term nature of durability tests on this quality of concrete a relatively permeable, lightweight aggregate concrete was also used in an attempt to obtain an early indication of the possible effects of fibre corrosion in a well carbonated material.

For both mixes, the maximum size of aggregate was 10mm and, of the 96 specimens, half were unreinforced while the remainder were reinforced with 1.22% by volume of 0.38mm diameter by 25mm long Duoform* fibres. This volume represents approximately 4.2% by weight for the gravel mix and 5.3% by weight for the lightweight concrete. Details of the mixes are shown in Table 1.

The cylinders were 152mm diameter by 304mm high and they were cured under wet sacks in the laboratory for seven days before being transferred to one of three Building Research Establishment exposure sites. In order to ensure direct comparibility between plain and fibre concrete specimens, three cylinders of each type were made from the same mixer drum of concrete and one of each type from the same mixer drum was placed on each of the three exposure sites.

Cracked beams

Mix details: Only one type of aggregate (flint river gravel) was used because it was thought that the corrosion of the fibres would be concentrated at the cracked zone and would be relatively unaffected by the permeability of the concrete. However, the surface area of the wire available for rusting was expected to affect the rate of wire damage and thus the loss in load carrying capacity of the beams and two fibre diameters were therefore used (Table 1). The 0.25mm diameter by 25mm long plain round high tensile fibres made by Bekaerts Ltd. had twice the surface area to volume ratio of the 0.5mm diameter by 50mm long wires produced by GKN Ltd. For both fibre types, 1.4% of wire by volume (5% by weight) was included in the mix, details of which are given in Table 1.

The beam size was either 102mm by 102mm by 508mm or 100mm by 100mm x 500mm depending on the availability of moulds and 48 specimens were produced, 24 with each type of wire. The beams were cured for 7 days under wet sacks and loaded in flexure on the 8th day after casting.

Loading to produce cracks and subsequent sealing procedures before exposure: The beams were loaded at 8 days to produce cracks of the desired width. A simple platform, which was adjustable in height, was placed immediately below, but not touching, each beam during loading to 'catch' the beams if they failed unexpectedly, This proved entirely satisfactory, and it was found by trial and error that a gap of about 0.4mm was suitable for the 0.25mm diameter by 25mm long wire beams whereas a gap of 1.5mm was necessary for the crimped fibre beams.

The maximum, minimum and average crack widths were then measured on the lower surface of the beam with a Hilger and Watts TM61-1 crack measuring microscope. The microscope was also used to establish the extent to which the crack had travelled from the bottom of the beam towards the top surface. This was found to average about 80mm which meant that in many cases the neutral axis was within the 10mm of the top surface of the beam.

In order to provide a basis for judging the effects of corrosion on load capacity, some of the beams had their cracks sealed immediately after loading and these were subjected to the same environment as the corroding specimens. However, the results were intentionally slightly biased in favour of higher strengths for the corroding beams by sealing those specimens which had the worst visible damage for a given crack width. Future corrosion effects would therefore probably be conservative.

The sealing procedure consisted of a thin layer of cold cure silicone rubber immediately over the crack followed by a layer of adhesive PVC tape. This was followed by another layer of silicone rubber topped by a thick layer of bitumen applied hot and covered with a final layer of PVC tape.

*Duoform is the trademark of the National Standard Co. Ltd.

Table 1 Details of concrete mixes

Specimen type	Supplier	Wire details Dia. mm	Length mm	Type	Vol. %	Number specs.	Aggregate type	Mix proportions by weight Cement	Water Free	Water Total	Agg.	Cement content kg/m³
Cylinders	National Standard	0.38	25	Duoform	1.22	24	Flint gravel and natural sand	1.0	0.47	0.62	4.62	393
					0	24						
Cylinders	National Standard	0.38	25	Duoform	1.22	24	Lightweight coarse and fines (Aglite)	1.0	0.66	0.75	2.25	420
					0	24						
Beams	Bekaerts	0.25	25	Round	1.47	24	Flint gravel and natural sand	1.0	0.40	0.49	3.40	480
Beams	G.K.N.	0.5	50	Crimped	1.47	24						

Aggregate sieve analysis

	% passing						
	10mm	5mm	2.4mm	1.2mm	600μm	300μm	150μm
Cylinders (Gravel)	100	47	37	28	18	3	1
Cylinders (Lightweight)	95	71	48	34	23	17	11
Beams (Gravel)	100	60	53	45	32	17	5

SPECIMEN LOCATION

Uncracked cylinders

The exposure sites were provided by the co-operation of the Building Research Establishment and cover a range of climatic conditions. They are at Garston, Hertfordshire (mild, temperate, dry with mild air pollution), Hurst Castle, Hampshire (temperature, exposed, coastal), and at a Greater London Council sewerage outfall at Beckton (very high pollution with sulphur compounds).

Each cylinder stood separately on a non-porous surface and the orientation relative to due north was noted.

Cracked beams

For the first $4\frac{1}{2}$ months the cracked beams were exposed on the University site at Guildford after which they were transferred to Hurst Castle, Hampshire.

ASSESSMENT OF DURABILITY

Uncracked cylinders

The main aim of the tests was to determine the depth to which steel fibres are likely to corrode in a gravel concrete and in a lightweight aggregate concrete over a period of many years. Also, information was required regarding the effect, if any, of steel fibres on the rate of carbonation and in particular whether the increase in the volume of the steel on rusting is sufficient to damage the concrete and thus accelerate the process of weathering.

At periods of 7, 16, 36, and 57 months from the date of initial exposure, one specimen of each type was removed from each of the exposure sites and the following procedures were carried out:-

a) A visual and photographic inspection was made of the specimens in order to judge the degree of rusting and spalling that had occurred on the surfaces of the fibre reinforced specimens;

b) The depth of the carbonation front was determined by fracturing the specimens at mid-height and spraying a phenolphthalein indicator onto the newly fractured surface. This coloured the uncarbonated core of the specimen (pH >9.0 to 9.5) pink but left the colour of the carbonated outside layer (pH <9.0 to 9.5) unchanged.(7). The maximum and minimum depths of carbonation were measured and the surfaces were colour photographed. A visual assessment was made of the degree of rusting in the carbonated zone.

Cracked beams

The main aims were to measure the rate of carbonation into the crack and to determine the capacity of the cracked beams to resist loads in the corroded state. Some beams were therefore loaded to failure at one month and others at 11 months and the loads were compared with those of the cracked but sealed beams. Further beams will be tested at future dates. In addition, after failure, the depth of carbonation in the cracked zone was measured as described above and a visual assessment was made of fibre rusting within the cracked section.

DISCUSSION OF RESULTS

Uncracked cylinders

After 7 months exposure the general appearance of both the gravel concrete and the lightweight concrete, reinforced and non-reinforced, was quite good. Slight rust staining was apparent but this had only occurred where a fibre was lying on and parallel to the surface. The surface appearance deteriorated slightly with increasing time of exposure and Figure 1 shows the surface condition of the most corroded set of specimens which was removed from Beckton after 57 months. Also shown on Figure 1, is the penetration of the carbonation front for the fibre reinforced specimens at Beckton, the carbonated zone being of lighter appearance in the photograph.

The maximum distance of the carbonation front at any of the three sites from the periphery of each cylinder is shown in Figure 2 and although it is difficult to measure the position of the front accurately, it is clear that the depth of carbonation for the gravel specimens has not increased significantly after $1\frac{1}{2}$ years, being generally about 2mm at 57 months.

Figure 1 Surface corrosion and carbonation depth of uncracked cylinders after 57 months exposure.

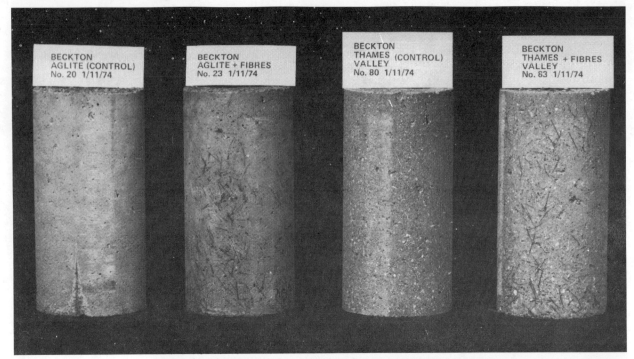

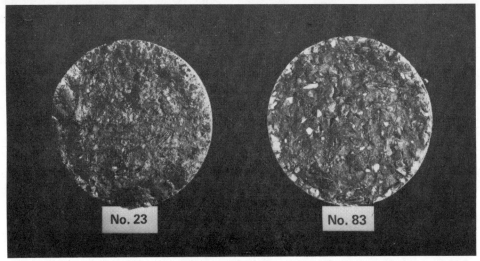

The extent of carbonation was more variable for the lightweight aggregate concrete and depended on the local aggregate concentration as well as on the direction of the prevailing wind. The carbonation depth could vary in one specimen from 20mm on the south facing surface to 4mm on the northern face on the plain lightweight specimens, the variation being rather less in the fibre reinforced cylinders.

In comparison with the variability within single specimens the difference in depth of carbonation between the exposure sites appeared to be small.

The following general observations were made:-

a) The extent of surface staining on fibre reinforced cylinders varied quite considerably from specimen to specimen at each of the exposure sites;

b) The degree of surface rusting was more severe on the fibre reinforced cylinders exposed to the corrosive environments at Hurst Castle and Beckton than at the other sites. Indeed there was very little surface staining on the fibre reinforced specimens situated at Garston;

Figure 2   **Relation between carbonation depth and duration of exposure for uncracked and for cracked sections.**

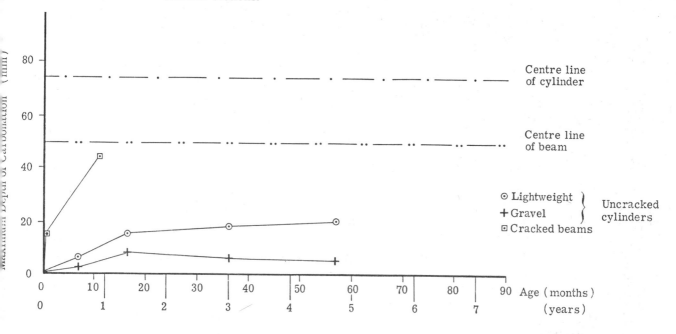

c) Fibre corrosion was more pronounced on the surface of the fibre reinforced lightweight specimens than on the Thames Valley gravel specimens;

d) The severity and depth of rusting on the worst specimen was insufficient to bring about serious surface spalling of the concrete. However, it should be pointed out that the cement paste lying directly over or around the corroded fibres had frequently been dislodged. This was particularly noticeable on the fibre reinforced lightweight specimens taken from Hurst castle, and occasional rusting to a depth of 6mm was observed for these specimens at 57 months;

e) Where a piece of aggregate breached the carbonation front, carbonation was observed to have occurred around the interface of the particle. In the case of the Thames Valley gravel aggregate particles it is considered that this may have been due to the poor bond between cement paste and gravel. In the case of the lightweight particles this may have been due to one, or more, of the following:-

    i. the carbonation path had passed right through the particles due to the porosity of the aggregate being greater than that of the cement paste,

    ii. carbonation had taken place around the interface without penetration of the aggregate,

    iii. salts or sulphates were leached out of the lightweight aggregate and reduced the alkalinity of the matrix in immediate contact with the particle;

f) On all sites, some of the fibres on the cylinder outer surfaces had not completely corroded after nearly 5 years.

Cracked beams

Initial loading: The initial loading tests to induce cracks are worthy of comment as the techniques used may influence the corrosion resistance of the beams.

The initial loads shown in Table 2 are generally not the loads at which cracks were first seen in the beams because of the manner in which the cracks developed to the required width. In the case of the 0.25mm dia by 25mm long round fibres, which were not well bonded at 8 days, fibre slip occurred soon after cracking resulting in a single crack of fairly uniform width in which corrosion could take place.

For the 0.5mm dia by 50mm long crimped fibres, however, the bond was stronger because of the shape of the fibre and a considerable increase in load was necessary

Table 2   Comparison of beam loads, before and after exposure for 34 days and 325 days

Beam No	Wire size mm	Exposure condition	Crack widths mm			Load kN			Exposed initial
			Max	Min	General	Initial 21/12/73	After exposure 34 days	After exposure 325 days	
5B	0.25	Sealed	0.3	0.10	0.25	11.9	10.2	—	0.86
5E	by	Sealed	0.4	0.15	0.30	16.9	12.0	—	0.71
5F	25	Sealed	0.2	0.05	0.15	16.2	15.0	—	0.93
5A	0.25	Rusting	0.2	0.05	0.15	17.0	17.9	—	1.05
5C	by	Rusting	0.3	0.10	0.25	18.9	19.5	—	1.03
5D	25	Rusting	0.4	0.10	0.30	12.8	13.6	—	1.06
8A	0.5	Sealed	0.3	0.05	0.10	17.9	18.6	—	1.04
8C	by	Sealed	0.8	0.20	0.40	16.6	14.3	—	0.86
8F	50	Sealed	1.0	0.10	0.50	16.8	13.9	—	0.83
8B	0.5	Rusting	0.4	0.05	0.15	16.5	15.9	—	0.96
8D	by	Rusting	0.5	0.10	0.30	17.3	17.7	—	1.02
8E	50	Rusting	0.7	0.10	0.40	19.4	20.9	—	1.08
6B	0.25	Sealed	0.4	0.10	0.30	16.2	—	15.0	0.93
6D	by	Sealed	0.2	0.10	0.15	15.5	—	15.0	0.97
6F	25	Sealed	0.1	0.05	0.10	18.0	—	19.4	1.08
6A	0.25	Rusting	0.4	0.10	0.30	15.3	—	15.4	1.01
6C	by	Rusting	0.3	0.10	0.25	16.9	—	17.3	1.02
6E	25	Rusting	0.1	0.05	0.10	16.5	—	21.5	1.34
7A	0.5	Sealed	0.25	0.05	0.15	19.8	—	23.8	1.20
7B	by	Sealed	0.3	0.05	0.15	17.7	—	19.6	1.11
7E	50	Sealed	0.6	0.10	0.40	18.0	—	17.5	0.97
7C	0.5	Rusting	0.3	0.05	0.20	16.4	—	20.7	1.26
7D	by	Rusting	0.8	0.05	0.40	20.2	—	26.6	1.32
7F	50	Rusting	0.1	0.05	0.10	19.1	—	25.4	1.33

Notes: (i) The average initial modulus of rupture for 24 beams with 0.25 by 25mm fibres was 6.19 N/mm^2 with a standard deviation of 0.654 N/mm^2 and coefficient of variation of 10.6%.

(ii) The average initial modulus of rupture of 24 beams with 0.5 by 50mm fibres was 7.23 N/mm^2 with a standard deviation of 0.853 N/mm^2 and a coefficient of variation of 11.8%.

(iii) The average modulus of rupture at 8 days of 6 control specimens of the same mix without fibres was 4.53 N/mm^2 with a standard deviation of 0.244 N/mm^2 and a coefficient of variation of 5.4%.

before the crack would remain open when the load was removed. This caused some difficulty in achieving the correct crack size because many small cracks generally formed and the 0.3mm cracks could only be attained after considerable disruption of the matrix which exposed a large volume of the concrete to attack by corrosive elements.

The crack widths shown in Table 2 indicate clearly that it is difficult to achieve a crack of a uniform size with fibre concrete and a visual assessment was therefore made of the general average crack width and this was used as a basis for the choice of comparably cracked beams for either sealing or leaving to rust.

The footnotes to Table 2 give the values of modulus of rupture of the various composites and of the corresponding plain concrete specimens. These values for the fibre concrete are quoted for comparative purposes only since the concrete is taking no load at the cracked section and the tensile force is therefore concentrated in the fibres at a stress level very much higher than the calculated value (4).

The values for coefficients of variation were 10.6% and 11.8% for the fibre concretes and these are rather higher than one would normally expect from a well controlled laboratory test on plain concrete. Also, the coefficient of variation within a batch was up to 17% and this is attributed to differences in fibre distribution. This may be a critical parameter in practice as the strengths of the composites after cracking will depend entirely on the number of well bonded uncorroded fibres across the cracked zone.

Loading tests after 34 days exposure: The purpose in loading two of the batches of pre-cracked beams to failure after only 34 days exposure to the elements was principally to determine the capacity of cracked beams to resist loads early in their life in the relatively uncorroded state so that the effects of subsequent corrosion on the load carrying capacity could then be judged.

Table 2 shows that the loads at failure on individual beams were 71% to 108% of the maximum loads applied at 8 days. Thus it is apparent that the fibres were well bonded to the matrix even after slip had occurred previously. However, at this age, the sealed specimens retained a smaller proportion of their initial loads than their rusting counterparts, possibly for the reasons stated above.

The sealing technique had effectively prevented any carbonation in the cracked zones but in the unsealed beams the carbonation front had generally penetrated at least 10mm, and in some cases up to 15mm (Figure 2) into the cracks, particularly at the corners of the beams. Fibre corrosion had already started for the 0.25mm dia by 25mm fibres within 5mm from the surface in the carbonated zone and this was in marked contrast with the uncracked cylinders after $4\frac{3}{4}$ years. Also, the 0.25mm diameter fibres had suffered damage where they were wholly exposed on the trowelled surfaces.

Loading tests after 325 days exposure: The general increase in load capacity, shown in Table 2, during the eleven months exposure was rather unexpected and may be attributable to the combined influences of improved fibre-cement bond and autogenous healing of the cracks. The rusting specimens again show more favourable load carrying capacity than do the sealed specimens possibly due to increased bond local to the cracks caused by carbonation shrinkage.

However, it can be noted from Table 2, that for beams 6A and 6C, which showed the greatest carbonation and rusting, the ratio of exposed to initial load is lower than for similar specimens 5D and 5C at 34 days whereas in every other case the relevant ratio at 325 days is higher than at 34 days.

The position of the carbonation front for non-sealed beams 6A, 6C and 6E is shown by the junction of the light and dark coloured areas on the broken sections on Figure 3. The depth of carbonation was dependent on the initial crack width (Table 2), being 45mm for 6A, 30mm for 6C and a local maximum of 15mm for 6E. Severe fibre corrosion was starting at the corners of beams 6A and 6C and some fibres were starting to rust at depths up to 25mm into the cracks. Also beam 6E showed spots of rust up to a depth of 15mm. Thus the effect of carbonation was to greatly increase the local carbonation depth in comparison with the cylinders, as shown on Figure 2.

The effect of a local decrease in alkalinity is shown on Figure 3 where a fibre which was rust free in the matrix shows severe corrosion at the crack interface. However, the majority of the fibres had sufficient reserve of tensile strength after 11 months exposure to exceed the bond forces resisting pull out and the normal fibre strengthening mechanism (4) could still operate.

When the fibre strength is eventually reduced by rusting to below the fibre pull-out load for the majority of the fibres, the failure mechanism will change and the beams will fail in a more brittle manner. Under these conditions, where the mode of failure is expected to change after a critical exposure duration, it is unwise to attempt to extrapolate the load carrying capacity of the beams into the future, because the expected change in slope of the load-time curve cannot necessarily be detected from previous data.

Figure 3   Carbonation depth and fibre corrosion at crack interface for cracked beams after 325 days exposure. (Fibre diameter = 0.25mm)

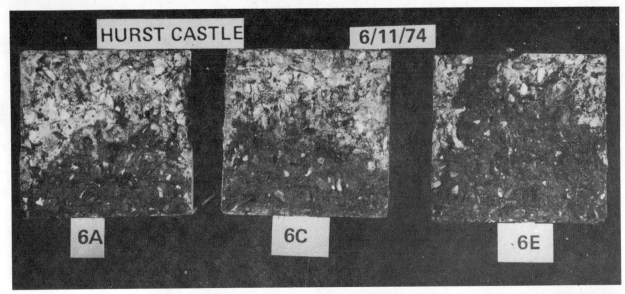

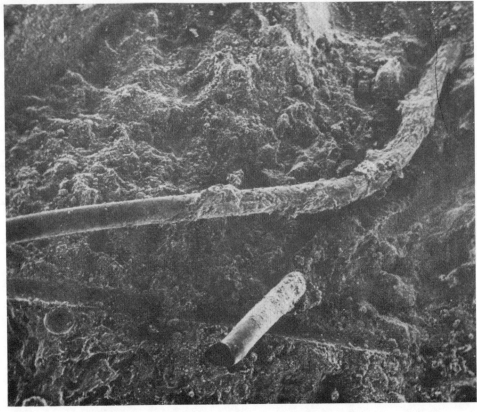

It was not possible to accurately determine the carbonation depth for beams 7C, 7D and 7F because the severe disruption of the matrix during failure removed much of the carbonated material from the crack surface before it was sprayed with phenolphthalein. However, carbonated patches up to depths of 20mm, 10mm and 5mm were observed for 7C, 7D and 7F, respectively, with rust spots at depths of 10mm, 15mm and 3mm.

This minor rusting appeared to have had little effect on the strength of the large diameter fibres and this is substantiated by the considerable increase in the load capacity of the beams as a result of exposure as shown in Table 2.

None of the beams with sealed cracks had any carbonation of the concrete or rusting of the fibres in the cracked zones.

CONCLUSIONS

1. Normal weight concrete made with a cement content and water-cement ratio appropriate to the exposure conditions provides adequate protection in its uncracked state against corrosion of steel fibres for at least 5 years, and probably much longer.

2. Cracked concrete subjected to a marine environment results in a greatly increased rate of carbonation and in fibre corrosion local to the crack when compared with uncracked concrete.

The load capacity of the cracked beams, however, was increased during a period of 11 months exposure but it is the authors' opinion that this is unlikely to remain the case as corrosion progresses. The fibre diameter therefore is likely to be a controlling factor in the rate of loss of strength in the future.

ACKNOWLEDGMENTS

The authors would like to thank the Building Research Establishment of the Department of the Environment for permission to use the exposure sites and for financing the first part of the investigation.

Also they wish to thank Mr R I T Williams, Reader in Construction Materials at the University of Surrey for his advice and encouragement during the work and the laboratory staff of the Construction Materials Research Group at the University of Surrey for their willing help at all times.

REFERENCES

1. International recommendations for the design and construction of concrete structures. Principles and Recommendations, June 1970, CEB-FIP Sixth Congress, Prague.
2. British Standard Code of Practice CP 110: Part 1, 1972, The Structural Use of Concrete.
3. Edgington, J, Steel fibre reinforced concrete, Research report submitted to the Department of the Environment, January 1974. Also available as PhD thesis, University of Surrey, 1974.
4. Edgington, J, Hannant, D J, and Williams, R I T, Steel fibre reinforced concrete, Building Research Establishment, Current Paper 69/74, July 1974.
5. Roberts, N P, The resistance of reinforcement to corrosion. Concrete, October 1970, pp 383-387.
6. Shalom, R and Raphael, M, Influence of sea water on corrosion of reinforcement, Journal American Concrete Institute, Title No 55-76, June 1959, pp 1251-1268.
7. Grimer, F J, The durability of steel embedded in lightweight concrete, Building Research Station, Current Papers, Engineering Series 49.

4.5 Bond studies on oriented and aligned steel fibres

A E Naaman and S P Shah
*Department of Materials Engineering,
University of Illinois at Chicago Circle,
Chicago, Illinois, USA*

Summary   *It is of general agreement that the efficiency of discontinuous fibre reinforcement can be increased by increasing the bond strength at the fibre-matrix interface and/or by aligning the fibres with the loading direction. Bond strength is generally measured from pull-out tests on single fibres. This paper shows first that the pull-out load (or equivalent bond) associated with a randomly oriented steel fibre is not necessarily lower than that of an aligned fibre; and second that the bond associated with a group of randomly oriented fibres decreases drastically when the number of fibres pulling out simultaneously from the same area increases. These results seem to explain why the addition to a concrete matrix of fibres with highly improved bond properties does not lead to an equivalent improvement in composite properties.*

Résumé   *On admet en général que l'efficacité des fibres d'acier dans le béton peut être augmentée en améliorant l'adhérence entre la fibre et le béton ou en alignant les fibres avec la direction de contrainte principale. Généralement la contrainte maximale d'adhérence est mesurée par un essai d'arrachement. Le présent article montre en premier que l'adhérence d'une fibre non alignée n'est pas necessairement plus petite que celle d'une fibre alignée et en second que l'adhérence d'un groupe de fibres non alignées, soumises a un essai d'arrachement, décroit notablement quand le nombre de fibres arrachées simultanément de la même surface augmente. Ces résultats semblent expliquer pourquoi l'addition au béton de fibres aux qualités d'adhérence très améliorées n'améliore pas proportionnellement les propriétés du composite.*

INTRODUCTION

The failure of steel fibre-reinforced concrete composites is generally attributed to the failure of the bond between fibres and matrix. As a result, the strength of the fibres is not fully utilized. Attempts have been made to improve the efficiency of fibre reinforcement by increasing the shear strength of the fibre-matrix interface by chemical or mechanical means (1,2,3) as well as by aligning the fibres in the direction of the principal tensile stress (4). The influence of increased bond strength by chemical or mechanical means is generally measured by a pull-out test on a single fibre (5). The peak load required to pull out a fibre is often expressed as an average bond strength by dividing its value by the embedded surface area.

A substantial amount of analytic work has been reported to evaluate the effect of orientation of fibres in either two or three dimensions (6,7). Almost all of these investigations assume that the matrix is a continuum, that matrix and fibres behave elastically, and that there is no relative slip between fibres and matrix. For fibre-reinforced concrete, these assumptions may be valid up to the stress or strain corresponding to the stage at which the matrix cracks. Because of the relatively small cracking strain of Portland cement concrete or mortar, initial matrix cracking occurs at a relatively early stage and, as a result, the significant contribution of the fibres occurs only after the matrix starts cracking. Following this initial cracking, the slip of the differential strain between fibres and matrix will continue to occur. As a result a purely elastic analysis cannot be valid for calculation of the effect of orientation for fibre-reinforced brittle composites, such as fibre-reinforced concrete.

It has been observed that certain chemical and mechanical treatment can substantially improve the bond strength as measured by a pull-out test on a single fibre (1,3). However, it is also observed that when such treated fibres are used in the concrete (8), the increase in the properties of the composite is much less noticeable when compared with the increase in the pull-out test. This seems to imply that the peak pull-out load of a single fibre is not an accurate measure of bond strength when the fibre is part of a group of fibres pulling out, as is the real case in the post-cracking behaviour of fibre-reinforced concrete.

PURPOSE OF STUDY

In the light of the above inconsistencies and in order to understand the bond efficiency of steel fibre reinforcement in mortar more fully, the following parameters are being investigated at the University of Illinois at Chicago Circle:

1. The angle of orientation of the fibre with the loading direction;

2. The number of fibres being pulled out simultaneously from the same area.

An attempt is made to establish the influence of these two parameters on the peak pull-out load, on the shape of the pull-out load vs pull-out distance curve for a given embedment length, and on the work required to pull out the fibres.

EXPERIMENTAL PROGRAMME

The overall goal of the continuing experimental programme is to investigate the effects of orientation and number of fibres for different lengths and rigidities of fibres as well as for different matrix compositions. In this paper, the results for one type of fibre, one embedment length and one type of matrix are reported.

The fibre length was 25.4 mm and the diameter of the fibres was 0.4 mm. These are commonly used dimensions of steel fibres. The fibres had a smooth surface and were cut from high strength music wires. The embedment length of the fibres in the pull-out test was 12.5 mm.

The matrix consisted of ASTM Type III cement, fine silica Ottawa sand (ASTM C-109), with sand-to-cement and water-to-cement ratios by weight of 2.5 and 0.55, respectively. Standard ASTM half or full briquettes were used for the pull-out tests as

Figure 1 a) Test specimen for inclined fibres
b) Pull out mechanism.

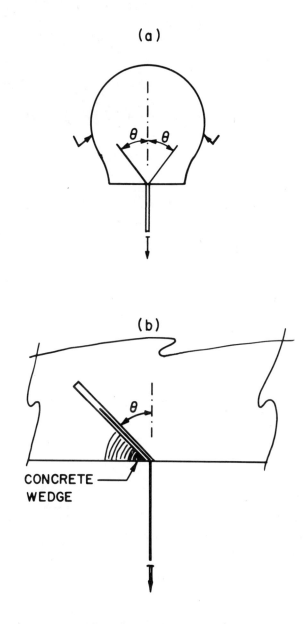

shown in Fig 1-a. The cross-sectional area of the briquette at the centre was 645 mm².

Three series of experiments were performed. The first series included the pull-out tests on two fibres which were symmetrically oriented with respect to the loading direction. The angle of orientation, θ, was varied from 0° to 75° in increments of 15°. The second series consisted of pull-out tests on fibres parallel to the loading direction, the main variable being the number of fibres. From 1 to 36 fibres were uniformly distributed in the available area (Fig 2-a). The third series was similar to the second one, except that the fibres were symmetrically oriented at an angle of 60° (Fig 2-b).

Testing was performed in an Instron universal testing machine with the load-elongation curve automatically recorded on a chart. The results reported here are averages for six to eight specimens.

ANALYSIS OF RESULTS

Some of the results of this investigation are summarized in Figs 3-5. Fig 3-a shows a typical pull-out load vs pull-out distance curve for a fibre parallel to the loading

Figure 2 a) Specimen with 36 parallel fibres
b) Fibres symmetrically oriented at 60°.

(a)

(b)

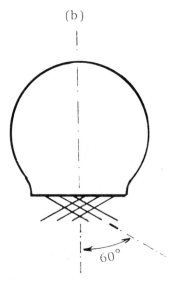

direction ($\theta = 0$). The area under the curve is indicative of the pull-out work. Fig 3-b shows a typical pull-out curve for a fibre oriented at 60° to the loading direction. The following four observations can be made from Fig 3:

1. The peak load observed for $\theta = 60°$ is almost as high as (or higher than) the peak load for $\theta = 0°$.

2. Just prior to complete pull-out, the pull-out load (termed final load) for $\theta = 60°$ is substantial where as it is equal to zero for parallel fibres.

3. The maximum pull-out distance is smaller for $\theta = 60°$ while it is equal to the embedment length for the parallel fibre.

4. The pull-out work is higher for $\theta = 60°$ than for $\theta = 0°$.

The presence of a significant final load for inclined fibres is an important observation. It appears that the final load results from the combination of two bond phenomena: angular friction and dowel action. The higher pull-out work observed for the inclined fibres is the result of the higher final load, provided the total pull-out distance remains almost the same.

Figure 3   **Typical pull-out load vs pull-out distance curves for 1.25 cm (0.5in.) embedded length.**
    a) Straight fibre θ=0°
    b) Inclined fibre θ=60°

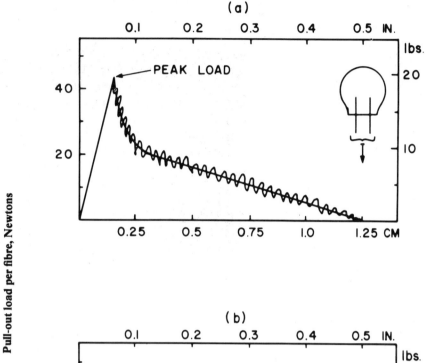

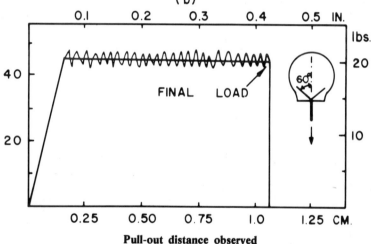

Pull-out distance observed

The smaller pull-out distance observed for the inclined fibre is due to a partial failure of the matrix wedge at the root of the fibre (Fig 1-b) resulting from the dowel action during the pull-out process.

Figure 4 represents in a non-dimensional form variation of observed peak load, final load and pull-out work with the angle of orientation, θ. Note that in these tests two symmetrical fibres were pulling out from the available surface. It can be seen that for θ up to 75°, the peak load for an inclined fibre is of the same order as that of a parallel fibre. However, the final load increases substantially with increasing values of θ. This influences the pull-out work (Fig 4-b). The pull-out work for inclined fibre can reach as much as three times that of a parallel fibre. Note that since the pull-out process for inclined fibres induces a continuous bending of the section of the fibre being pulled out of the matrix, the final load and pull-out work should increase with increasing rigidity of the fibre.

The preceding results would suggest that randomly oriented fibres may be better than aligned fibres for fibre-reinforced concrete. However, these results are based on one or a pair of fibres being pulled out from unit area of matrix. Actually, in the post-cracking behaviour of the composite a large number of fibres pull out from the same area. It is reasonable to assume that the performance of a group of fibres will not be as high as that of one single fibre. This is analogous to a group of foundation piles where it is known that the bearing capacity of a group of piles is less than the sum of that of single piles. Also, since a different mechanism of bond failure is involved for inclined fibres, it is likely that the influence of the number of fibres may be different for inclined fibres compared with that for parallel fibres.

The influence of increasing the number of fibres on pull-out behaviour was studied for both parallel and inclined fibres. It was observed that increasing the number of fibres from 1 to 36 per unit area (645 mm²) of matrix does not influence the pull-out load or the pull-out work for parallel fibres oriented with the loading direction. Note that 36 fibres pulling out from a square inch of matrix correspond to about 3% of fibres by volume of composite if it is assumed that half of the fibres pull out from each of the fractured surfaces.

Figure 4   Typical results for pull-out load and pull-out work vs fibre inclination.

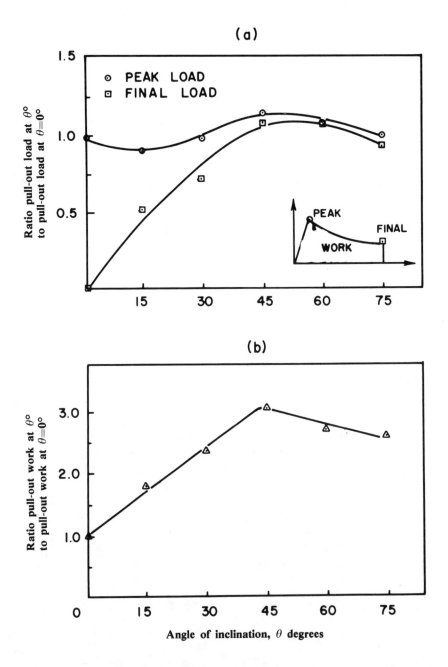

The results of experiments with fibres inclined at 60° are shown in Fig 5. It can be seen that the pull-out work per fibre decreases with the increasing number of fibres pulling out per unit area. Thus it appears that the effect of the increasing number of fibres pulling out from a unit area is more pronounced for inclined fibres than for parallel fibres. This would mean that the previously observed beneficial effects of a single inclined fibre may be negligible for sufficiently large numbers of inclined fibres.

CONCLUSIONS

Based on the results observed in the present investigation, the following conclusions are drawn:

1. Pull-out load tests on single fibres indicate that:

    a. Inclined fibres with angle of inclination of up to 75° show a peak load of the same order as parallel fibres ($\theta = 0°$).

    b. The pull-out work of inclined fibres is higher than that of parellel fibres and seems to reach a maximum for $\theta \simeq 45°$.

Figure 5 Typical results showing the influence of the number of fibres on pull-out load and pull-out work.

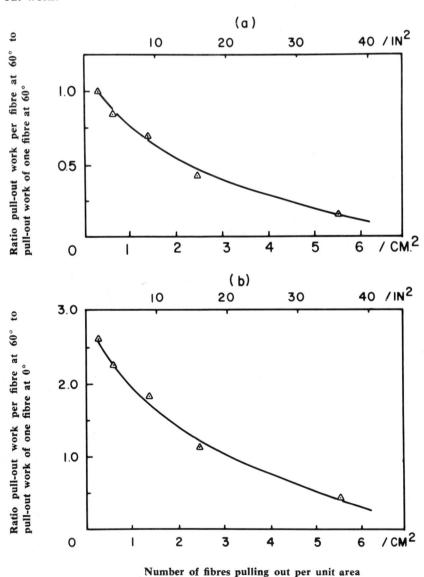

Number of fibres pulling out per unit area

c. The pull-out distance, given the same embedment length, decreases with θ when $\theta \geq 45°$, due to a disruption of the concrete wedge at the root of the fibre.

d. The final load observed just before complete pull-out for $\theta > 0$ depends on the value of θ. This load should also be influenced by the rigidity of the fibres.

2. Pull-out load tests on an increasing number of parallel fibres ($\theta = 0$) has shown that for a number of fibres between 1 and 36 per square inch ($\simeq 600$ mm^2) there is no substantial decrease in peak load and pull-out work per fibre compared with measurements on one single fibre.

3. Pull-out load tests on an increasing number of fibres inclined at 60° have shown that (for the same range described in Conclusion 2), the peak load, the pull-out work, and the pull-out distance per fibre decrease significantly when the number of fibres pulling out from the same area increases.

REFERENCES

1. Tattersall, G H and Urbanowicz, C R, "Bond Strength in Steel Fibre Reinforced Concrete", Magazine of Concrete Research, Vol 26, No 87, June 1974, pp 105-113.
2. "Duoform" Wire, patented by National Standard Co, Ltd.
3. Naaman, A E, McGarry, F J and Sultan, J N, "Developments in Fibre Reinforcement for Concrete", Massachusetts Institute of Technology, Civil Engineering Department, Report No R72-28, May 1972, 67 pp.
4. Hannant, D J and Spring, N, "Steel Fibre Reinforced Mortar: A Technique for Producing Composites with Uniaxial Fibre Alignment", Magazine of Concrete Research, Vol 26, No 86, March 1974, pp 47-48.
5. de Vekey, R C and Majumdar, A J, "Determining Bond Strength in Fibre Reinforced Composites", Magazine of Concrete Research, Vol 20, No 65, December 1968, pp 229-234.
6. Krenchel, H, "Fibre Reinforcement", Akademisk Forlag, Copenhagen, Denmark, 1964, English translation.
7. Pakotiprapha, B, Pama, R P and Lee, S L, "Mechanical Properties of Cement Mortar with Randomly Oriented Short Steel Fibres", Magazine of Concrete Research, Vol 26, No 86, March 1974, pp 3-15.
8. Edgington, J and Hannant, D J, "Steel Fibre Reinforced Concrete", Building Research Establishment Current paper CP 69/74, England, July 1974, p 11.

4.6 Fibre concrete for a folded plate structure

J-P Rammant and M Van Laethem
*Department of Civil Engineering, Katholieke Universiteit te Leuven
(University of Louvain), Belgium.*

Summary *A model study of a folded plate structure in steel fibre reinforced concrete is investigated. The model has an omega shaped cross-section and is made out of parts which are post-tensioned together. The results of the model testing are compared with an original finite element theory including the material properties. The paper describes the comportment of the structure till cracking. The effect of post-tensioning on longitudinal stress as well as on transverse bending moments is discussed.*

Résumé *Un modèle de voile plissé a été construit en béton armé de fibres métalliques. Le modèle a une section en omega, est composé de 4 parties, assemblées au moyen de fils de précontrainte. Les résultats de l'essai sont comparés aux valeurs calculées par éléments finis, comprenant les propriétés du matériau jusqu'à la fissuration. L'effet de la précontrainte sur les tensions longitudinales et les moments transversaux est examiné.*

INTRODUCTION

As part of a doctoral work the behaviour of structures in fibre reinforced concrete is being studied. Since 1968 Belgium has taken a great interest in the quality of fibre reinforcement; a special committee was set up to coordinate the research (1). A special shape of fibre* was considered to give the best results. The length of the fibre is 40 mm, the diameter is 0.35 mm; the fibre has a hook at each end to improve the adherence to the matrix. Reference (1) gives full details about handling, static and dynamic behaviour, adherence, durability etc.

The choice of a structural element was influenced by a special application in shell design, namely a folded plate structure. Rühle (2) shows the importance of the different kinds of prefabricated folded plates like VT-18. The placing of the reinforcing network is very time consuming (about 50% of the worktime). The replacement by steel fibres increases the total percentage of reinforcement (eg 0.26% in transversal direction becomes 0.62% in the overall direction for the special case of VT-18). This allows a reduction in the quantity of prestressing cables. The higher price of the fibre (approximately 0.35 £/kg in Belgium) can be offset by savings in the handwork-price in industrialised countries, so that a good saving of total cost can be reached, even for highly prefabricated shell structures.

CONSTRUCTION

The dimensions of the model are given in Fig 1. The model has a scale factor of 1/3 with respect to the prototype discussed in (3). Four parts of equal length (1.5 m) were cast in a vertically standing double framework. The mixture and material properties are listed below.

Mixture: Portland cement P-500   635 kg/m³

Aggregates 2/5   330 kg/m³

Rhinesand 0/2   990 kg/m³

Water   300 ℓ/m³

Steel fibres $\ell_f/d = 114.3$ 120 kg/m³

Material properties (age 28 days):

Cube (100 mm) crushing strength $f'_{cu} = 55$ N/mm²

Cylinder (150 x 300 mm) splitting strength $f_{ct} = 8.9$ N/mm²

Flexural strength in tension $f_{cb} = 16.8$ N/mm².

These properties are excellent, compared with theoretical and experimental formulae (4,5). The flexural strength of the FR concrete is more than 2.5 times higher than that of the same concrete without fibres.

Figure 1   **Dimensions of the model of a folded-plate structure.**

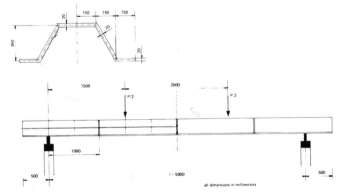

*Licence by Bekaert N V, Belgium

Figure 2   The folded plate structure model.

The fibres are delivered in a special form. About ten fibres are glued together like staples and when mixed in the concrete they become separated. No problems like the balling-up effect were encountered for the applied volume ratio of fibres to the mortar of 2%. No preferential distribution of fibres was intended, the concrete was only slightly vibrated into the formwork (with thickness of 20 mm). Plastic tubes were incorporated for the prestressing wires, to facilitate the assembly of the four parts which were glued together with Araldite. The forces in the prestressing wires of diameter 2.43 mm were measured with a strain-gauge system designed for laboratory use (6). More details can be seen in Fig 2. The shell was lifted and placed on two diaphragms made of the same fibre concrete. The free span is 5 m.

LOADING ARRANGEMENT AND MEASUREMENTS

Two point loads (Fig 1) are transmitted by means of dynamometers sustained by a frame carrying counterweights. A screw on the vertical standing bars of the frame makes it possible to increase the load step by step. The measuring equipment consists of dial gauges for horizontal and vertical displacements and electrical resistance strain gauges (Tokyo Sokki Kenkyujo, 120 Ω, 60 mm in length) to register the deformations at the shell surface. Each loading cycle lasted about 20 minutes, the increment of total load being 1 kN.

Before starting to load, the post-tensioning forces were altered in order to register the prestressing effect on the shell by means of the strain gauges. The location of the prestressing wires (Fig 2) was fixed beforehand to facilitate casting and assembling of the four parts. The wire forces after assembling were chosen (see Calculation) as follows: in the lower plates 3.29 kN, in the inclined plates 5.44 kN and in the upper plate 1 kN.

The test was stopped after the appearance of the first crack, the load being gradually removed from the structure. The crack appeared at an early stage (total load 7.5 kN = 2.3 times dead load), possibly due to an insufficiency of fibres near the middle section. This section was at the top of the formwork during casting. The strain in the lower side of the horizontal plate was 98×10^{-6}, the crack developed suddenly to the neutral axis. Near to the point where the load was applied a strain of 250×10^{-6} in the transverse direction was measured without causing any damage, which would not be the case in fibreless concrete. In the near future the crack will be repaired and the test will be repeated to complete failure of the structure.

Fig 3 gives the load versus displacement curve. Till the appearance of the first crack the behaviour of the structure is nearly elastic. Fig 4 gives the diagram of the measured longitudinal strains in the middle section for a total load of 6 kN (ten points).

Figure 3 Load versus displacement curve.

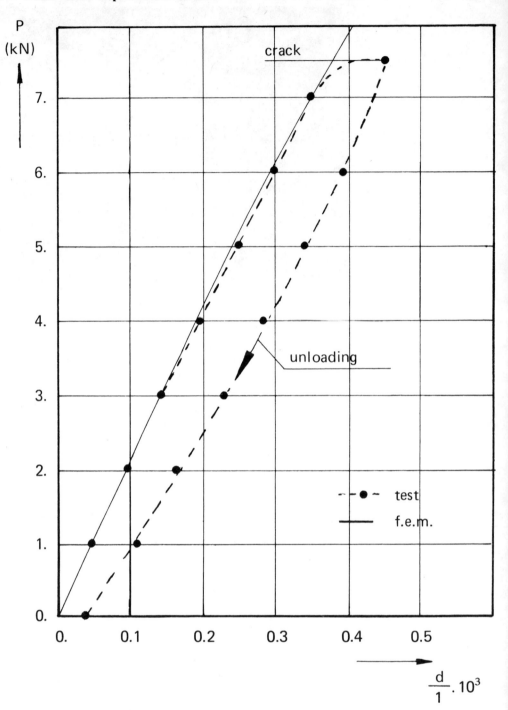

CALCULATION

A finite element theory was developed, including the elasto-plastic properties of the material (non linear σ-ε diagram in compression and tension). A membrane element of Argyris (7) known as the Rem-4 element was modified for in-plane bending problems (8,9) and a plate element (10) was superposed following the Kirchhoff-hypothesis. The subdivision in elements (Fig 1) was taken as small as possible for the computer memory whilst being sufficient for a good convergence. No separation between membrane and bending stiffness was allowed. Each element was separated in a finite number of layers in which a biaxial stress state can be adopted, following the laws of elasto-plasticity. In each layer the relation between stress- and strain increment is given by the elasto-plastic material matrix.

$$D^{ep} = D^e - \frac{D^e \frac{\partial F}{\partial \sigma} \frac{\partial F^T}{\partial \sigma} D^e}{-\frac{\partial F}{\partial K} \sigma^T \frac{\partial F}{\partial \sigma} + \frac{\partial F^T}{\partial \sigma} D^e \frac{\partial F}{\partial \sigma}}$$

in which D^e is the elastic relationship between stress and strain, and $F(\sigma, K) = 0$ the plasticity criterion with K a hardening parameter. A plasticity criterion for concrete (11) was modified taking account of the plastic behaviour of fibre concrete in tension. As can be seen in Fig 3 there is a good agreement between theory and experiment up to cracking. The structure behaves beamwise at the beginning of the loading. Near cracking there is an important distortion of the cross-section. Other boundary conditions, difficult to realize in experiment (12), are easily incorporated in the finite element method. The finite element model could not take into account the local loss of strength due to a possible unequal distribution of fibres. Fig 4 gives the comparison of the computed membrane strains with the measured surface strains at the middle section.

Figure 4   **Measured longitudinal strains in the middle section for a total load of 6 kN.**

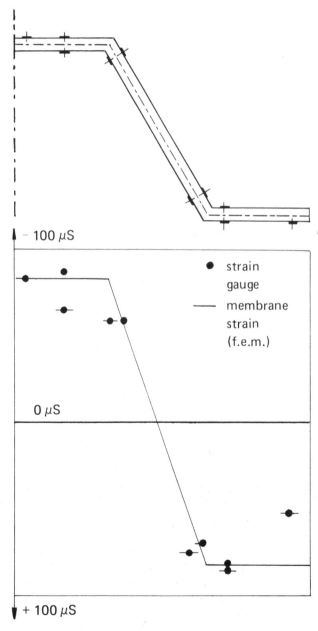

Figure 5 Influence lines for symmetric prestressing forces.

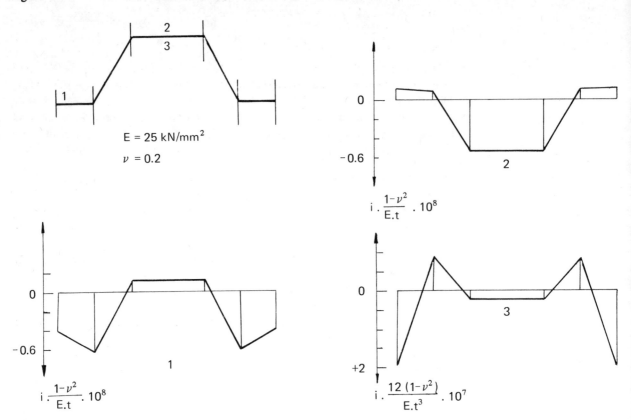

Fig 5 gives influence lines for symmetric prestressing forces. These lines were computed by the finite element method and show the stress distribution following a prestressing force. Fig 5.1 shows the influence line for the longitudinal membrane force at the middle section in the lower plates, eg two forces of 3.3 kN in the lower plates give a theoretical stress −0.88 N/mm² (measured stress −0.74 N/mm²). Fig 5.2 gives the influence line for the longitudinal membrane force at the middle section in the upper plates, eg two forces of 3 kN in the inclined plates gives a theoretical stress −0.36 N/mm² (measured stress −0.36 N/mm²). Fig 5.3 gives the influence line for symmetric prestressing forces at the end section for the transverse bending moment at the middle section at the upper plate. As can be seen, the influence of the prestressing forces in the lower plates is to increase the positive bending moment, already existing from self-weight. For this reason the wire in the inclined plate was more highly tensioned than the others. The experiment showed that no cracking arose by the high prestressing forces on the thin (thickness 20 mm) shell planes.

CONCLUSION

A completely isotropic distribution of fibres was not achieved so that early cracking occurred in the weakest section. Further tests to complete failure of the structure will show the importance of the stratification of the fibres. Other processing techniques like shotcreting, vacuum handling etc. should be investigated.

The use of FR concrete together with prestressing wires seems very promising.

The saving in expenses by replacement of the classical reinforcement in shell structures by steel fibres can be very important.

ACKNOWLEDGEMENTS

The authors wish to sincerely thank Mr R Dutron, chairman of OCCN, Mr Moens and Mr Nemegeer (N V Bekaert) for their stimulating ideas. Assistance was capably rendered by Mr Jang, Mr de Coen, Mr Rogier and the staff of the Construction Department. Technical aid was kindly provided by Mr Van Gramberen.

REFERENCES

1. Dehousse, N, "Les mortiers et bétons renforcés de fibres métalliques", Séance de la Commission belge du béton armé du 28 mars 1973, publication ABEM No 401.
2. Rühle, H, "Räumliche Dachtragwerke—Konstruktion und Ausführung", Band 1, VEB Verlag für Bauwesen, Berlin, 1969, p 63 and p 280.
3. Harry, WC, "Precast folded plates become standard products", Journal of ACI, Oct 1963.
4. Naaman, A E, Moavenzadeh, F, McGarry, F J, "Probalistic analysis of fibre reinforced concrete", Journal of ASCE, EM2, April 1974, p 397.
5. Swamy, R N, Mangat, P S, Rao, CVSK, "The mechanics of fibre reinforcement of cement matrices", presented at the International Symposium on fibre reinforced concrete, Ottawa, Oct 1973.
6. Van Gemert, D, "Load distribution in skew anisotropic girder bridges", Doctor's thesis, Univ. of Louvain (not yet published).
7. Argyris, J H, "Energy theorems and structural analysis", Butterworth Scientific Publications, London, 1960.
8. Wilson, E L, Taylor, R L, "Incompatible displacement models", Report Structural Engineering Laboratory, University of California, Berkeley, 1971.
9. Blaauwendraad, J, Kok, A W M, "Elementenmethode voor constructeurs", Parts 1 and 2, Agon Elsevier, Amsterdam-Brussels, 1973.
10. Zienkiewicz, O C, "The finite element method in engineering science", McGraw Hill, 1971.
11. Kupfer, H, Hilsdorf, H K, Rüsch, H, "Behaviour of concrete under biaxial stresses", ACI Journal, Proceedings, Vol 60, No 8, Aug 1969.
12. Stroeven, P, "Six folded plate constructions of micro-concrete under uniformly distributed load and test-loaded to failure", IASS Symposium on folded plates and prismatic structures, Vol 1, Vienna, Sept-Oct 1970.

4.7 Flexural behaviour of fibre concrete with conventional steel reinforcement

R N Swamy and K A Al-Noori
*Department of Civil and Structural Engineering,
University of Sheffield, United Kingdom*

Summary

Fibre reinforcement in the form of discrete fibres cannot be used as direct replacement of conventional steel in reinforced and prestressed structural members. The superior resistance of fibre concrete to cracking and crack propagation may, however, be utilised to improve the resistance of structural members to cracking, deflection and other serviceability conditions. Tests are reported on the flexural behaviour of reinforced concrete beams with fibre concrete in the tension or compression zone or as a tensile skin. It is shown that fibre concrete in the compression zone can develop a higher degree of compressibility and plastic deformations at failure. Fibre concrete in the tension zone enables high strength steels to be used in practice: with characteristic strengths of 700 N/mm^2, both crack width and deflection are controlled to within acceptable limits, and the beam is able to develop plastic deformation characteristics at failure. The use of a single layer tensile skin of fibre concrete transforms a conventional over-reinforced beam to behave in a ductile manner. Fibre concrete can thus enable higher steel percentages to be used in practice without fear of brittle type failure.

Résumé

Une armature de fibres sous forme de fibres distinctes ne peut pas être utilisée en remplacement direct d'une armature d'acier traditionnelle dans les pièces en béton armé et précontraint. On peut cependant mettre à profit la meilleure résistance du béton de fibres à la fissuration et à la propagation des fissures pour améliorer la résistance des pièces à la fissuration, déflexion et autres conditions d'exploitation. On décrit des essais sur le comportement en flexion de poutres de béton armé avec du béton de fibres dans la zone tendue ou comprimée ou comme une couche extérieure tendue. On montre que dans la zone comprimée, le béton de fibre peut développer un plus grand degré de compressibilité et de déformations plastiques à la rupture. Dans la zone tendue, le béton de fibre permet d'utiliser des aciers à haute résistance en pratique. Avec des résistances caractéristiques de 700 N/mm^2, la largeur des fissures et la flèche sont maintenues dans des limites acceptables et la poutre peut montrer des caractéristiques de déformation plastique à la rupture. L'emploi d'une couche extérieure tendue en béton de fibre porte un béton sur-armé à la manière traditionnelle à se comporter d'une manière ductile. Le béton de fibres peut donc permettre d'utiliser des pourcentages d'acier plus élevés en pratique sans danger de rupture fragile.

INTRODUCTION

It is now well established that one of the important properties of steel fibre concrete is its superior resistance to cracking and crack propagation. As a result of this ability to arrest cracks, fibre composites possess increased extensibility and tensile strength, both at first crack and at ultimate, particularly under flexural loading; and the fibres are able to hold the matrix together even after extensive cracking. The net result of all these is to impart to the fibre composite pronounced post-cracking ductility which is unheard of in ordinary concrete. The transformation from a brittle to a ductile type of material would increase substantially the energy absorption characteristics of the fibre composite and its ability to withstand repeatedly applied, shock or impact loading.

When the fibre reinforcement is in the form of short discrete fibres, they act effectively as rigid inclusions in the concrete matrix. Physically, they have thus the same order of magnitude as aggregate inclusions; steel fibre reinforcement cannot therefore be regarded as a direct replacement of longitudinal reinforcement in reinforced and prestressed structural members. However, because of the inherent material properties of fibre concrete, the presence of fibres in the body of the concrete or the provision of a tensile skin of fibre concrete can be expected to improve the resistance of conventionally reinforced structural members to cracking, deflection and other serviceability conditions.

The design of concrete structural members is primarily governed by considerations of deflection and local damage. The limitations of cracking and deformation, in turn, restrict the full exploitation of the constituent materials. Since the presence of fibres enhances directly the resistance to cracking and deformation, the use of fibre concrete in conventionally reinforced structural members should not only prove beneficial to the overall behaviour of the member but lead to a better usage of the constituent materials. Further, the resistance to cracking of the tensile skin should also enable higher steel stresses to be carried under service load conditions, and there is thus the distinct possibility of being able to use steel reinforcement with characteristic strengths of 500–700 N/mm^2.

The fibre reinforcement may be used in the form of three-dimensionally randomly distributed fibres throughout the structural member when the added advantages of the fibre to shear resistance and crack control can be further utilised. On the other hand, the fibre concrete may also be used as a tensile skin to cover the steel reinforcement when a more efficient two-dimensional orientation of the fibres could be obtained.

The results presented here form part of an extensive study into the use of steel fibre reinforcement in reinforced and prestressed structural members and only typical results are reported. In the first series of tests the effectiveness of fibre concrete in the tension and/or compression zones of conventional reinforced members has been studied in relation to strength, cracking and deflection. In the second series of tests, the influence of a tensile skin of fibre concrete on the use of high tensile steel reinforcement and the behaviour of "conventionally over-reinforced" beams is investigated. It is shown that the use of fibre concrete in conventional structural members can lead to substantial improvements of their serviceability performance, and better utilisation of steel and concrete.

FIRST SERIES: EXPERIMENTAL DETAILS

The first series of tests was designed to study the flexural behaviour of conventionally reinforced beams with fibre concrete in the tension or compression zones or throughout the depth of the beam. Three different tensile steel percentages were used and for each steel percentage, an additional reinforced beam without fibres was also tested.

Ordinary Portland cement was used in all the tests. The fine aggregate was a graded natural sand and the coarse aggregate, a graded crushed gravel with 19 mm maximum size. The steel fibres used were made of low carbon, straight and round, 0.4 mm × 25 mm. After extensive mix design studies a mix of 1:2.5:2.5 (cement:fine:coarse) with

0.55 total water-cement ratio and a fibre content of 7% by weight of the total mix was used throughout. The longitudinal steel in the beams was hot rolled high yield steel with a characteristic strength of 410 N/mm². All the beams were provided with web reinforcement in the shear regions and tested over a span of 2286 mm under two point loads.

All the beams were extensively instrumented to study the strain, cracking, deflection and rotation characteristics over the entire range of loading up to failure. Each beam was loaded in two or three cycles. From the large amount of data obtained only typical results are presented here but the data presented are representative of the general behaviour of all the beams.

TEST RESULTS

Load at first crack

The load at first flexural crack was determined by visual inspection, and all the beams with fibre concrete in the tension zone showed consistently higher first cracking load. The ratio of the moment at first crack to the design moment varied from 32% to 69% for beams with ordinary concrete (OC) in the tension zone compared to 45% to 103% for beams with fibre concrete (FC) in the tension zone. The ratio of the moment at first crack to the ultimate moment varied from 14% to 23% for beams with ordinary concrete in the tension half of the beam compared to 20% to 35% for the fibre concrete beams.

Deflection characteristics

Typical load-mid-span deflection characteristics for two beams with ordinary and fibre concrete in the tension zone are shown in Figs 1 and 2. A comparison of the two figures shows that at a given load the beam with fibre concrete in the tension zone shows less deflection than the other beam, and this reduction in deflection becomes more pronounced in the post-cracking stage. A study of all the deflection data showed that the fibre concrete in the tension zone contributed much more significantly to the flexural stiffness of the beams than those with ordinary concrete in the tension zone. The final deflection shown in the figures was measured prior to failure and is necessarily conservative since large changes in deformation occur quite rapidly at this stage prior to the total collapse of the beam.

The span-deflection ratios at design loads were 10% to 20% higher for the fibre concrete beams than for those with ordinary concrete. This compares with the maximum increase of about 10% generally observed in the flexural elastic modulus of fibre concrete over plain concrete. These results thus show that the flexural stiffness of fibre concrete beams is higher than that indicated by the increase in flexural modulus thus emphasizing that the crack control characteristics of fibre concrete enables a much higher proportion of the tensile concrete to be mobilised to resist deformation and contribute to the overall stiffness of the member.

Cracking characteristics

All the crack characteristics such as crack width, crack spacing and crack height were measured in all the beams. The crack widths were measured both at the tension face and at the tensile steel level. The measured crack widths at the steel level for two typical beams are shown in Figs 1 and 2.

Beams with fibre concrete in the tension zone showed consistently smaller crack widths for the same loads, than beams with ordinary concrete in the tension zone. This superior resistance to crack propagation was observed for a large part of the loading range, although near to failure loads these differences in crack widths tended to decrease. Nevertheless, the crack arrest properties of fibre concrete observed in conventional modulus of rupture tests were fully reproduced in the larger structural members. There is, however, a fundamental difference in the stress redistribution due to flexural cracking between plain concrete and fibre concrete. The fibres in the latter bridging across the crack are still able to transmit stresses, and the bond stress at a crack is therefore no longer zero as in ordinary concrete. The bond strength of steel reinforcement in fibre concrete will therefore be higher as has been shown elsewhere (1).

Since crack width is directly related to the steel stress, it is obvious that for a given

Figure 1 Deflection and cracking characteristics of reinforced beam with fibre concrete in the compression zone and ordinary concrete in the tension zone.

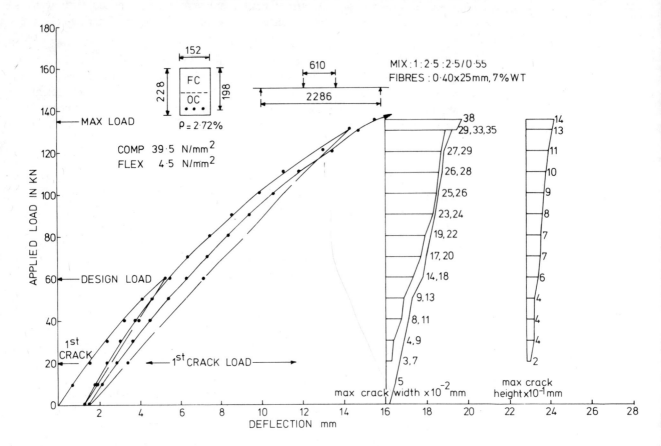

crack width a higher steel stress could be permitted in the longitudinal reinforcement, and the use of fibre concrete would therefore permit the use of steels with higher characteristic strengths than at present.

The presence of fibres in the tension zone also controls the propagation of the cracks over the depth of the beam and all the results consistently showed that the flexural cracks penetrated less upwards into the compression zone in fibre concrete than in ordinary concrete (Figs 1 and 2).

Behaviour of fibre concrete in compression zone

It is generally thought that the inclusion of fibres in the compression zone of a structural member is not beneficial, since they are primarily added to improve the tensile properties of the matrix. The results of this investigation, however, showed that fibre concrete in the compression zone could effectively improve the performance of the structural members. From the tests, the following were observed:

(1) the presence of fibres in the compression zone prevented disintegration of the compression concrete; unlike ordinary concrete, there was no breaking up of the compression concrete, and no falling off of debris, and the integrity of the beam was preserved after failure,

(2) the fibre concrete in the compression zone tended to show a higher degree of compressibility, and the concrete strain prior to failure was higher in fibre concrete in the compression zone than in ordinary concrete,

(3) fibre concrete in the compression zone tended to show greater ability for rotation and plastic deformation than ordinary concrete.

Figure 2   Deflection and cracking behaviour of reinforced beam with ordinary concrete in the compression zone and fibre concrete in the tension zone.

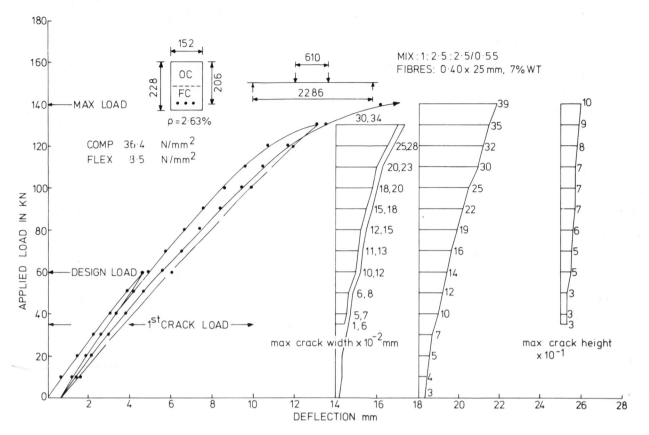

SECOND SERIES: EXPERIMENTAL DETAILS

The second series of tests was designed to study the influence of fibre reinforcement on the use of higher steel stresses and on the behaviour of conventionally over-reinforced beams. The experimental details were largely similar to those of the first series except for the following:

All the beams were provided with a fibre concrete skin to a depth of 50 mm or the depth of the longitudinal reinforcement. The concrete mix proportions were the same as before but the fibre content was reduced to 3.5% by total weight of the mix; further, crimped fibres 0.5 mm × 50 mm were used.

In addition to hot rolled high yield reinforcement, two beams were reinforced with steels of yield stress 670–690 N/mm², namely BRI-strand (beam 3BFC2/3.5) and Bi-steel (3HFC2/3.5). These beams were designed for a steel stress, under serviceability conditions, of 345 N/mm². Beams 4CFC2/3.5 and 81CFC2/3.5 were over-reinforced (with tensile steel of characteristic strength 410 N/mm²) in the conventional sense—58% and 81% over-reinforcement respectively compared to the balanced conditions. These two beams were also confined in the compression zone with 6 mm diameter high yield steel (410 N/mm²) 75 mm deep stirrups at 50 mm centres. Companion beams with ordinary concrete were also tested for comparison purposes.

The tests reported here are only part of the main study and only representative test details are reported (Table 1).

TEST RESULTS

In the following discussion only those benefits of incorporating fibre concrete in structural members other than those already discussed are presented.

Table 1   Details of test beams, series two

Beam no	Type of concrete	Type of fibre mm	Concrete cube strength N/mm^2 FC	Concrete cube strength N/mm^2 OC	Concrete flexural strength N/mm^2 FC	Concrete flexural strength N/mm^2 OC	Overall dimensions $b \times d$ mm	Effective depth d_1 mm	Reinforcement No.	Reinforcement size mm	Area of steel reinforcement A_s mm^2	Steel ratio A_s/bd_1 %
1FC2/3.5	▢	wavy 0.5 × 50	39.1	33.80	7.73	4.50	152 × 228	200.0	1 2	10+ 12	304.7	0.99
3FC2/3.5	▢	wavy 0.5 × 50	39.1	33.80	7.73	4.50	152 × 228	200.0	1 2	16+ 20	829.5	2.70
3BFC2/3.5	▢	wavy 0.5 × 50	39.1	33.80	7.73	4.50	152 × 240	205.0	1 2	6.9+ 11.3	475.0	1.50
3HFC2/3.5	▢	wavy 0.5 × 50	38.8	33.50	7.73	4.32	152 × 240	205.0	5	14	515.0	1.60
4CFC2/3.5	▢	wavy 0.5 × 50	38.8	33.50	7.73	4.32	152 × 228	200.0	1 2	20+ 25	1296.0	4.23
81CFC2/3.5	▣	wavy 0.5 × 50	38.8	33.50	7.73	4.32	152 × 228	201.0	3	25	1472.7	4.94

Notes:
1. Number 2/3.5 indicates 50 mm thick fibre concrete on the tension side of the beam and 3.5% of steel fibre by total weight of the mix.
2. Letter B indicates Bi-steel, H indicates BRI-strand, FC indicates fibre concrete and the letter C indicates confined concrete at the compression zone of the beam.
3. ▣ indicates concrete beam with 50 mm skin fibre concrete.

Fibre concrete as tensile skin

Comparison of the first cracking load, deflection and cracking characteristics of comparable beams made with ordinary concrete, fibre concrete in the tensile half of the beam and fibre concrete tensile skin showed that there was no noticeable difference in the behaviour of the beams made with fibre concrete in the tensile half of the beam or as a tensile skin, provided that the fibre concrete was a workable mix and produced a uniform distribution of the fibres throughout the beam. The tensile skin had the added advantage that some preferential orientation of the fibres was obtained through compaction of the reduced depth of concrete. The results also showed that crimped fibres were superior to straight round fibres in the control of cracking and associated properties.

Use of high yield steel

The use of high yield steel reinforcement in reinforced concrete members is largely limited by the allowable crack width without the danger of corrosion of the reinforcement. The higher the steel stress under serviceability conditions, the higher the crack width and the associated deflection. With the currently available materials, characteristic strengths of 420–460 N/mm² in the tension steel appear to be the limiting stresses to reduce cracking and deflection to acceptable limits under service load conditions. If cracking and deflection can, however, be controlled as shown earlier by the use of fibre concrete, there is then the distinct possibility of being able to utilise the higher strength steels currently available.

In the tests reported in this series, the deflection and cracking characteristics at design loads of beams with Bi-steel and BRI-strand reinforcement were comparable to those with deformed bars. The BRI-strand reinforcement, however, tended to give slightly higher deflection and crack width at design loads than the Bi-steel reinforcement (Fig 3). With these high yield steels, at steel strains of 1600 to 21μs at design loads the maximum crack width at steel level varied between 0.10mm and 0.19mm whereas with hot rolled deformed bars, at steel strains of 1100 to 1200μs, the maximum crack width at design loads varied from 0.06mm to 0.10mm.

Figure 3   **Deflection and cracking characteristics of beam with high strength steel (700 N/mm²) and fibre concrete tensile skin.**

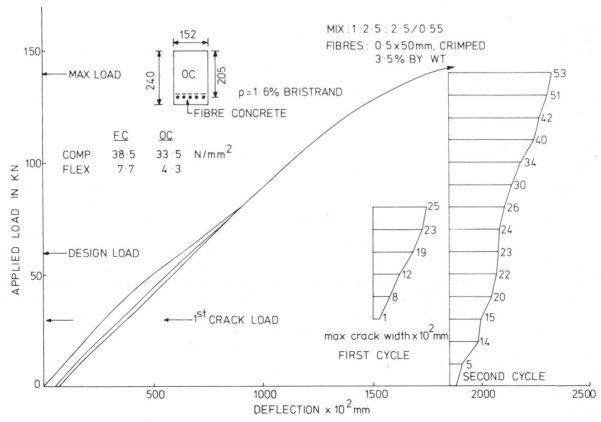

The increase in deflection due to the higher steel stress at design loads was only marginal even with the lower steel ratio due to the higher steel stress. The decrease in span-deflection ratios under short-term service loading in beams with 700 N/mm^2 tension steel was about 15% compared to that in beams with 410 N/mm^2 steel reinforcement. Since the fibres become much more effective in the post-cracking stage, it is unlikely that the deflection under sustained loading will be more than 10% in excess of that of beams with deformed bars. The results of cracking and deflection obtained from these tests clearly show that steel stresses $350–400 \text{ N/mm}^2$ are quite feasible at service loads without exceeding the serviceability limit states provided a fibre concrete tensile skin is used to control deformation.

Mode of failure

The beams with high yield steel (700 N/mm^2) failed by yielding of the tension reinforcement and showed the same degree of ductility as beams with deformed bars (410 N/mm^2). The presence of fibres in the tension zone, in addition, enabled the beams to develop much higher rotation characteristics towards failure. The ability of the fibre concrete for increased extensibility and to bridge the cracks even after extensive cracking enabled the beams to undergo much higher plastic rotation than that normally obtained with ordinary concrete. The results of this study show that with fibre concrete it is possible to attain, at collapse, conditions closer to the fully plastic state in steel than with plain concrete, the increased contribution to ductility arising solely from the fibre reinforcement of the concrete matrix. The results also show that the more ductile behaviour of fibre concrete structural members would enable simpler limit state theorems to be applied to predict their performance.

Effect of over-reinforcement

Beams 4CFC2/3.5 and 81CFC2/3.5 were provided with 4.23% and 4.94% respectively of tension steel (deformed bars, 410 N/mm^2) corresponding to 58% and 81% over-reinforcement compared to the balanced steel ratio. Both these beams had their compression zone confined by the provision of stirrups as detailed earlier. It is in the nature of the over-reinforced beams that cracking and deflection are well within acceptable limits during most of the loading range. Indeed the only disadvantage of over-reinforcement is the lack of ductility at failure due to the primary compression failure, and the consequent lack of adequate cracking and deflection which would serve as ample warning of impending collapse. Hence only the failure characteristics of these beams are examined here.

Failure characteristics

Both the conventionally over-reinforced fibre concrete beams failed by the yielding of the tension steel and showed considerable ductility at failure. The comparable beam OC4 made of ordinary concrete with 4.05% tension steel failed in a brittle manner due to concrete crushing although it was able to carry a slightly higher load than beam 4CFC2/3.5. The provision of a thin layer of fibre concrete in the form of a stiff tensile skin thus enables the beam to develop full ductile behaviour, with extensive cracking, deflection, and plastic rotation characteristics. The fibre concrete skin thus transforms a conventionally over-reinforced beam to behave like an under-reinforced beam with ductile characteristics.

Table 2 gives details of the experimental ultimate moment of resistance of the beams, and the theoretical ultimate moment based on Whitney's (2) theory. These moments are also compared to the ultimate moment computed from CP110 (3). The results show that all the fibre concrete beams developed adequate overall load factors varying from 1.50 to 2.65. The conventionally over-reinforced beams had load factors of 2.20 and 2.65 respectively for the 58% and 81% over-reinforcement: it is thus clear that because of their ductile behaviour, these beams could be designed to carry much higher service loads as if they were under-reinforced beams. The provision of a fibre concrete skin thus enables a much higher proportion of the tensile reinforcement to be used to resist bending forces than is currently allowed.

Table 2   **Test results of beams of series two**

Beam no	Design moment M(CP110) KN-m	Experimental moment M exp KN-m	Ultimate moment Whitney MU KN-m	Ultimate moment CP110 Mui KN-m	$\frac{M\,exp}{M}$	$\frac{M\,exp}{Mu}$	$\frac{M\,exp}{Mui}$	mode of failure
1FC2/3.5	12.10	33.97	23.20	18.85	2.81	1.46	1.78	steel yielding
3FC2/3.5	23.60	57.87	55.00	37.60	2.45	1.06	1.54	steel yielding
3BFC2/3.5	24.50	58.71	56.50	39.20	2.40	1.04	1.49	steel yielding
3HFC2/3.5	25.00	58.71	57.70	40.10	2.35	1.02	1.46	steel yielding
4CFC2/3.5	18.70	67.10	67.80	30.60	3.60	0.99	2.20	steel yielding
81CFC2/3.5	18.80	81.36	68.00	30.80	4.34	1.20	2.65	steel yielding
OC1*	12.20	45.93	23.33	19.05	3.76	1.79	2.18	steel yielding
OC3*	26.04	58.46	58.12	41.27	2.33	1.01	1.42	steel yielding
OC4*	22.29	70.99	81.17	36.52	3.19	0.87	1.94	concrete failure

* Companion beams in ordinary concrete corresponding respectively to beams 1FC2/3.5, 3FC2/3.5 and 4CFC2/3.5

CONCLUSIONS

From the study reported here, the following conclusions are drawn.

1. Beams with fibre concrete in the tension zone showed a consistently higher load when the first flexural crack became visible compared to conventional concrete Beams. The ratio of the first crack moment to the design and ultimate moment was about 50% higher for the beams with fibre concrete in the tension zone.
2. Beams with fibre concrete in the tension zone showed less deflection than conventional concrete beams particularly in the post-cracking stage. The data showed that the fibre concrete in the tension zone contributes much more effectively to the flexural stiffness of the member. This increased stiffness arises from the ability of the fibre concrete to arrest cracks and of the fibres bridging the cracks to carry loads.
3. Beams with fibre concrete in the tension zone showed consistently smaller crack widths for the same loads compared with conventional beams. For a given crack width, therefore, a higher steel stress could be permitted in the longitudinal reinforcement compared to ordinary reinforced beams.
4. Tensile cracks in the pure bending region penetrated less into the compression zone in fibre concrete beams than in ordinary concrete beams.
5. Fibre concrete in the compression zone is shown to effectively improve the performance of structural members. It prevents disintegration and preserves the integrity of the compression zone, it shows a higher degree of compressibility and enables the beam to develop plastic deformations at failure.
6. The provision of fibre concrete in the form of a tensile skin is just as beneficial as providing fibre concrete in the whole of the tension zone. The tensile skin has the added advantage of obtaining some preferential orientation of the fibres.
7. The ability of fibre concrete to control cracking and deflection enables higher strength steels to be used in practice. With steels of characteristic strength 700 N/mm^2, the crack width at design loads (at steel strains of 1600 to 2100 μs), varied from 0.10 to 0.19 mm.
8. The increase in deflection due to the higher steel stress at design loads was only marginal even with a lower steel ratio due to higher strengths. The increase in span-deflection ratios under short-term loading due to the high strength steel was about 15%.
9. Beams with high strength steels and a tensile skin of fibre concrete were able to develop plastic deformation characteristics at failure similar to that in steel structures.
10. The provision of a tensile skin of fibre concrete transforms a conventionally over-reinforced beam to behave like an under-reinforced beam with ductile characteristics. These beams could therefore be designed to carry a much higher service load than is currently possible because of their brittle behaviour.

ACKNOWLEDGEMENT

The work reported here was made possible through financial support to the junior author by the Ministry of Oil, Iraq. The authors would like to record their grateful thanks for this support.

REFERENCES

1. Swamy, R N, and Al-Noori, K A, Bond strength of steel fibre reinforced concrete. Concrete, Vol 8, No 8, Aug 1974, pp 36–37.
2. Whitney, C S, Plastic theory of reinforced concrete design. Transactions, American Society of Civil Engineers, Vol 107, pp 251–282, Discussion, pp 283–326.
3. British Standards Institution. The structural use of concrete: Part 1. Design, materials and workmanship, CP110, British Standards House, London, 1972, pp 154.

4.8 Some properties of high workability steel fibre concrete

R N Swamy and H Stavrides
*Department of Civil and Structural Engineering,
University of Sheffield, United Kingdom.*

Summary Tests are reported on the mix design and properties of high workability steel fibre concrete. It is shown that slumps of the order of 100 mm and Vebe times of 2-3 seconds could be obtained and the quality, strength and ductility of fibre concrete still maintained by using pulverised fuel ash as a direct replacement of 30% by weight of the cement content and a water-reducing agent. The effects of the size and type of aggregate, the size of the specimen and the method of casting and compaction are reported. The results show that fibre concrete gives a better performance if vibrated externally. Vertical casting leads to the creation of weak zones and unfavourable fibre orientation which results in considerable decrease in flexural strength. Smaller sizes of aggregates and smoother surfaces give fibres more scope in their role as crack arrestors. There is a progressive reduction both in the first crack and ultimate flexural strength of fibre concrete with increase in the length and depth of the test specimen. It is shown that fly ash fibre concrete has adequate properties for structural use.

Résumé On présente un rapport sur des essais sur le proportionnement du mélange et les caractéristiques d'un béton à fibres d'acier de grande ouvrabilité. On montre que des étalèments de l'ordre de 100 mm et des temps Vebe de 2 à 3 secondes pouvaient être obtenus et que la qualite, la résistance et la ductilité d'un béton de fibres pouvaient cependant se maintenir en employant des cendres volantes en remplacement direct de 30% par poids de la teneur en ciment et en utilisant un agent de réduction de la teneur en eau. On décrit les effets de la granulométrie et du type de granulats, des dimensions des éprouvettes et de la méthode de moulage et de damage. Les résultats montrent que le béton de fibres se comporte mieux avec vibration extérieure. Le moulage vertical produit des zones faibles et une orientation défavorable des fibres conduisant à une perte considérable de résistance à la flexion. Des granulats de moindres dimensions et des surfaces plus lisses permettent aux fibres de mieux jouer leur rôle d'empècher la fissuration. Il y a une réduction progressive dans la première fissure et la résistance à la flexion à la rupture du béton de fibres quand il y a augmentation de la longueur et de la profondeur de l'éprouvette. On montre que le béton de fibres aux cendres volantes possède des caractéristiques adéquates pour son emploi dans les structures porteuses.

INTRODUCTION

One of the practical difficulties in the application of steel fibre concrete is its lack of adequate workability. The difficulties of placing and compacting fibre concrete arise from the particle interference between the fibres and the coarse aggregates. Recent tests (1,2,3,) show that it is not only the fibre geometry and fibre volume that influence the rheological properties of the fresh fibre concrete but also the size, shape and volume fraction of the coarse aggregate. To ensure adequate compactibility of the fresh fibre concrete, and to achieve uniform distribution without bundling or curling up of fibres, it is, therefore, necessary to control the relative fibre-aggregate volume in addition to the fibre and aggregate geometry. Conventional concrete mixes cannot therefore be used with steel fibre reinforced concrete.

Experience with steel fibre concrete shows that if conventional procedures and equipment are used for the handling and placing of steel fibre concrete, a higher proportion of fine material is then generally found necessary for fibre concrete than in plain concrete. The difficulties of compacting fibre reinforced concrete can be appreciated from the fact that even with specially designed mixes, to maintain the same degree of compactibility for comparable plain and fibre concrete mixes, about 40% more water content is required for the fibre mix with a 1% volume of fibre of aspect ratio 76. The increased water content would obviously reduce the strength and elasticity properties of the fibre concrete. The need to improve the workability properties of fibre concrete without necessarily increasing the free water added to the mix is thus obvious.

Recent tests (4,5) have shown that slumps of the order of 75-100 mm could be obtained and the quality, strength and ductility of fibre concrete still maintained by using pulverised fuel ash as a direct replacement of part of the cement content in conjunction with water-reducing and air-entraining agents. Fly ash has been used for many years in concrete, partly on economic grounds and partly on account of its influence on workability and heat evolution. Apart from the added attraction of utilising a waste product in concrete construction, fly ash has also the advantage of gaining strength with time due to its pozzolanic nature.

The ability of fly ash to enhance workability is a decisive attraction for its use to ease the inherent compaction problems of fibre concrete. Further, in fibre concrete more fine material is generally required to coat the large surface areas of the fibres and render them as efficient crack arrestors, fully embedded in the mortar matrix. Fly ash has the advantage of being available in the same degree of or higher fineness as ordinary Portland cement.

FLY ASH FIBRE CONCRETE

Fly ash can be used in a wide variety of combinations with cement and sand; in the tests reported here it has been used as a straightforward substitution of cement by weight. In the first series of tests a mix suitable for use with steel fibres was chosen and the optimum cement-fly ash combination determined. In the second series of tests, a water-reducing agent was introduced to restore the compressive strength properties of the cement-fly ash mix to those of the all-cement mix. The final mix of cement-fly ash combination with a water-reducing agent has been used for all subsequent studies. The parameters investigated include the effects of the size and type of aggregate, the size of the test specimen and the method of casting and compaction. It is shown that the fibre concrete mix with fly ash has adequate properties of strength and elasticity for structural applications.

Mix design considerations

The aim in the mix design tests was to establish whether a given amount of the Portland cement could be substituted by fly ash, with no detrimental effects on the quality of the fibre concrete, in both the fresh and hardened state, and to ensure that the fly ash remained as an active material and contributed to strength at later ages.

Ordinary Portland cement was used in all the mixes. The fly ash used was obtained from the Ferry Bridge Power Station and contained typically 50.5% of silica, 26.9% of alumina and 9.6% of iron oxide. It had a specific surface of 399 m²/kg and a density of 2.17 gm/cm³. The fine aggregate consisted of washed and dried river sand conforming to zone 2 (6) with a fineness modulus of 2.66. The coarse aggregate was graded crushed gravel with 10 mm maximum size and a fineness modulus of 6.20. The steel fibres were cold drawn and straight and round, 0.5 mm x 38.1 mm.

To establish the optimum fly ash content, several mixes with fly ash alone varying from 0 to 50% and replacing equal amounts of cement by weight were cast and their strength, elasticity and damping properties determined at various ages. From all these results, the mix with 30% fly ash content gave the optimum strength and elasticity properties. Indeed at six months the 30% fly ash mix had nearly the same flexural strength as the all cement mix; however, the reduction in compressive strength compared to the all cement mix was about 30%.

Table 1A   Typical properties of fly ash fibre concrete mixes

Mix proportions*	Fibre content	Fibre geometry mm x mm	P.f.a.*** content	Slump mm	Cube† strength N/mm^2	Flexural† strength N/mm^2	Dynamic† modulus kN/mm^2	Damping † capacity $x\ 10^2$
1: 2.0 : 2.5/0.54	1.5%	38.1 x 0.406	30%	50	27	6.4	30.69	5.13
1: 1.80: 2.25/0.42**	1.0%	38.1 x 0.5	30%	90	44.3	7.25	35.85	4.85

Notes:
* cement: p.f.a.: sand: c agg/w/(cement + p.f.a.) 0.7 : 0.3 : 2.0 : 2.5 /0.54 and 0.7 : 0.3 : 1.8 : 2.25/0.42.
** mix includes a water-reducing agent, at 280 gm per 50 kg of (cement + p.f.a.).
*** p.f.a. content by weight of cement.
† values at 28 days.

Table 1B   Details of mixes used

Mix	Type of aggregate	Mix proportions*	28 day strength† Flexural N/mm^2	28 day strength† Compressive N/mm^2	Slump mm Fibre mix	Slump mm Plain** mix	V.B. time secs. Fibre mix	V.B. time secs. Plain mix
D	10 mm Gravel	1 : 1.8 : 2.25/0.42	7.25	44.30	90	175	4	1
F	20 mm Gravel	1 : 1.8 : 2.25/0.40	5.93	42.20	100	180	3	1
G	20 mm Gravel***	1 : 1.8 : 2.25/0.40	5.50	40.00	110	185	3	1
H	10 mm Limestone***	1 : 1.8 : 2.25/0.44	6.21	37.00	100	180	3	1
I	20 mm Limestone***	1 : 1.8 : 2.25/0.40	5.37	41.15	130	210	2	0.5
J	20 mm Granite***	1 : 1.8 : 2.25/0.40	7.00	48.10	40	80	8	5
K	25 mm Granite***	1 : 1.8 : 2.25/0.40	6.50	44.20	45	90	8	4
L	38 mm Granite***	1 : 1.8 : 2.25/0.40	5.20	42.30	50	100	7	3
M	10 mm Lightweight (Lytag)	1 : 1.05 : 2.45/0.50	4.78	30.40	100	180	3•	1

Notes:
* All mixes include 30% p.f.a. by weight of cement and 280 gm of water-reducer per 50 kg of (cement + p.f.a.).
** The plain concrete mixes had exactly the same mix proportions as the fibre concrete mixes.
*** Single size grading.
† Fibre concrete strength

To compensate for the loss in compressive strength further mix design tests were carried out with a water-reducing agent and reduced water content and minor variations in aggregate content. The resulting mix proportions were 1:1.80:2.25 with a water- (cement and fly ash) ratio of 0.42. With 1% fibre volume, this mix had a slump of 90 mm and a Vebe time of 4 seconds, and average 28 day compressive and flexural strengths of 44.3 N/mm^2 and 7.25 N/mm^2 respectively. Typical properties of both fly ash mixes without and with water-reducing agent are shown in Table 1A.

Effect of aggregate type and size

To study the effects of aggregate type and size on the properties of the fly ash fibre concrete, four different types of aggregates were used with their corresponding mixes in plain concrete without fibres. The details of the mixes are shown in Table 1B. The water content for each mix was adjusted to give approximately similar workability properties, although the granite concrete mixes were consistently less workable. All the mixes had the same fibre volume of 1% of 0.5 x 38.1 mm straight round fibres.

The crushed gravel aggregate was used both in the graded form and as a single-size. The limestone and granite aggregates were used as single sizes. The lightweight aggregate was also single-sized with 10 mm maximum. All the test specimens were cast in steel moulds and compacted by table vibration. They were demoulded at 24 hours and subsequently cured under controlled conditions of temperature of 60 $\pm$ 2°F and relative humidity of 50 $\pm$ 2%. Only data concerning flexural and compressive strengths are reported here.

Effect on workability

All the fly ash fibre concrete mixes, except those made with granite aggregate, showed excellent compactibility properties. Their slump varied between 100 and 130 mm, and the Vebe time between 2 and 4 seconds. The granite concrete mixes were comparatively less workable with an average slump of about 45 mm and 7 to 8 seconds Vebe time (Table 1B); it was however, still possible to compact these without undue difficulty.

Effect on flexural strength

The flexural strength results of the fibre concrete mixes are shown in Table 1B and in Fig 1. The results show that the larger the size of the aggregate, the lower the ultimate flexural strength. Fig 1 also shows that the increase in flexural strength due to the presence of fibres is more pronounced in mixes with smaller maximum aggregate sizes. With the 20 mm granite mix, for example, the increase in flexural strength of the fibre concrete over the corresponding unreinforced mix was about 46%, whereas when the aggregate size was increased to 38 mm, the increase in strength was reduced to 30%.

However, aggregate geometry and surface texture also influence the effect of the fibre presence on flexural strength. With both 10 mm and 20 mm aggregates, the rougher the aggregate surface and shape, the less efficient the fibres appeared to be in their role as crack arrestors although the 20 mm granite concrete gave better results than the crushed gravel or limestone concrete. Whereas in plain concrete the crushed aggregates lead to higher flexural strength, in fibre concrete more evenly shaped aggregates were generally found to render the fibres more effective.

Although limited, the results also showed some effect of aggregate grading on flexural strength. Whereas in plain concrete higher strengths could be achieved through gap grading, particularly for stiff mixes (7), the direct contact of aggregates achieved with gap grading appeared to limit the effectiveness of the fibres. In the tests reported here, the single size gradation gave marginal increases in flexural strength for the plain concrete mixes; however, the situation was reversed with the corresponding fibre concrete mixes. Single size aggregates in fibre concrete showed a distinct reduction in flexural strength, and it appears that in fibre concrete large single sized aggregates cannot produce a better performance than that obtained with continuous grading of aggregates.

Effect on mode of failure

Although all the fibre concrete mixes showed first cracking load distinct from the ultimate strength, the smaller maximum aggregate sizes and the smoother aggregates showed better post-cracking behaviour. With larger aggregates, the bond area is smaller and the bond failure at the aggregate-matrix interface becomes more critical

Figure 1 The influence of size and type of aggregate on the flexural strength of fly ash fibre concrete.

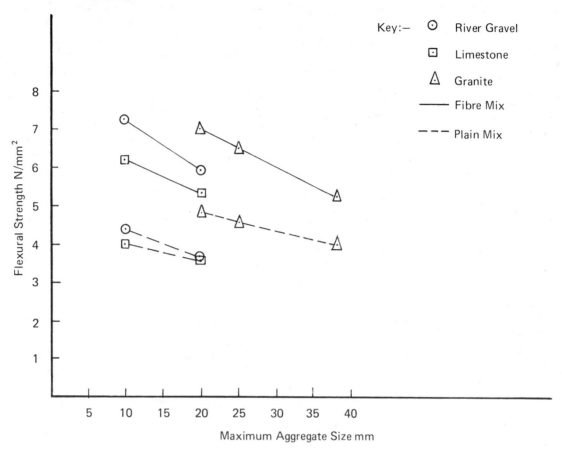

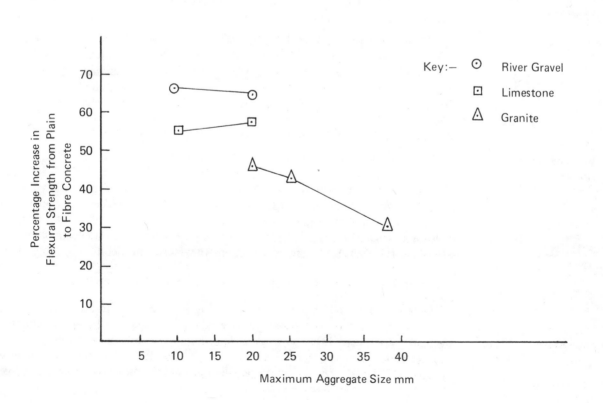

(8). The fibre-aggregate interparticle friction appears to be more pronounced with irregularly shaped aggregates, and the fibres become distorted and their efficiency as long and straight crack arrestors reduced. As a result, a large proportion of the fibres is reduced to act as mere solid inclusions than as effective crack arrestors. The smaller and smoother aggregates give more scope to the fibres to embed themselves undistorted in the matrix, and hence act more efficiently in arresting the propagation of cracks. The fibres are thus more efficiently oriented and more effectively embedded in the mortar matrix to continue to carry the load after the formation of the first crack.

Visual and microscopic examination of the fractured surfaces confirmed the above observations. Aggregate-matrix bond depends primarily on the surface texture of the aggregate, and has an important influence on the strength of concrete, particularly the flexural strength (8); however, size is also an important factor, especially in fibre concrete. With larger aggregate sizes, the fibres become less effective due to the aggregate-fibre interaction; when the aggregate size then reaches that of the fibre length, the propagation of flexural cracks is influenced more by the distribution of aggregate particles than by fibre distribution. X-ray studies have confirmed the aggregate-fibre interaction; the tests also showed that with aggregates larger than 20 mm, the cracks invariably propagated in a more erratic manner, indicating that with larger aggregates, it is the aggregate size which is the crucial factor in crack propagation and not the fibre distribution.

Effect on compressive strength

Compressive strength is influenced in the same way as flexural strength by the size and type of the aggregate, but to a much smaller scale. The effect on compressive strength of the 1% fibre content in all the mixes used was practically negligible. All the fibre concrete mixes showed an increase in compressive strength over the corresponding plain concrete mixes, varying from 0 to 3% for the natural aggregates, and 7% for the lightweight concrete. With the exception of the 10 mm limestone and the lightweight aggregate, the average compressive strength at 28 days varied between 40 and 48 N/mm^2; the lightweight concrete gave a strength of 30 N/mm^2. These tests show that with adequately workable mixes, the presence of fibres need cause no reduction in compressive strength (1), and indeed give a modest increase.

Effect of specimen size on flexural strength

These tests were designed to examine the effect of the depth and length of the test specimen on the flexural strength of steel fibre concrete and on the reproducibility of test results. The results are compared with those predicted by the weakest link theory and the composite mechanics approach.

The same mix proportions, 1:1.80:2.25/0.42 were used with 30% fly ash replacing an equal weight of cement, the fibre content being 1% by volume of 0.5 x 38.1 mm straight fibres. All the beams were tested under third point loading which is known to give less variable results than central loading (9). Companion plain concrete beams were also tested for comparison purposes. For fibre concrete, twelve specimens were tested in each group; only three were tested for plain concrete. The results of the tests are shown in Table 2.

Variability of results

The results showed that there was a progressive reduction in both the first crack and ultimate flexural strength of fibre concrete with increase in the length and depth of the test specimen (Table 2). In the case of beams of different lengths, the variability of the results, however, had no direct relationship with the length of the beam although beams longer than 710 mm showed a progressive reduction in the coefficient of variation with increase in length. Beams of different depths, on the other hand, showed more variable results i.e. a higher coefficient of variation with increasing depth.

COMPARISON OF TEST RESULTS WITH THEORY

The results given in Table 2 are shown in Fig 2 compared with the predicted values of the flexural strength according to the weakest link theory (10) and the composite mechanics approach (11). The weakest link theory (12) was primarily developed for plain concrete, and more factors influence the strength of fibre concrete mixes. Nevertheless, if all the known factors which may influence the appearance of the mix as

Table 2 Influence of beam length and depth on flexural strength

Beam	Dimensions $b \times d \times \ell$ mm	First crack M.o.R.** N/mm^2	Ultimate M.o.R. N/mm^2	Percentage decrease in first crack M.o.R. with respect to RB_1	Percentage decrease in ultimate M.o.R. with respect to RB_1	Plain concrete M.o.R. N/mm^2
RB_1	150 x 150 x 500	6.30	6.38	–	–	4.20
RB_2	150 x 150 x 710	5.90	6.00	6.34	5.95	3.90
RB_3	150 x 150 x 1220	5.45	5.55	13.40	13.00	3.60
RB_4	150 x 150 x 1500	4.80	5.10	23.80	20.00	3.20

Beam	Dimensions $b \times d \times \ell$ mm	First crack M.o.R.** N/mm^2	Ultimate M.o.R. N/mm^2	Percentage decrease in first crack M.o.R. with respect to RC_1	Percentage decrease in ultimate M.o.R. with respect to RC_1	Plain concrete M.o.R. N/mm^2
RC_1	100 x 100 x 500	7.10	7.20	–	–	4.45
RC_2	100 x 150 x 500	6.30	6.84	11.26	5.00	4.30
RC_3	100 x 200 x 500	5.80	6.44	18.30	10.55	4.00
RC_4	100 x 250 x 500	5.00	5.80	29.57	19.44	3.65

** First crack determined by visual inspection

cast in the mould, and hence affecting the flexural strength, are kept carefully and reasonably constant, this theory could be applied to fibre concrete beams as well. The composite mechanics theory applies directly to fibre concrete and has shown excellent correlation with results from a wide range of sources (11).

Fig 2 shows that the weakest link theory always overestimates the flexural strength whereas the composite mechanics approach sometimes overestimates and sometimes underestimates the flexural strength. Nevertheless, the predictions based on the composite mechanics theory are closer to the actual results. It must be noted that since the first crack was visually detected, the first crack flexural strength is almost certainly overestimated; the differences between the composite mechanics theory and the test results are therefore less than those shown in Fig 2 since it is in the case of the first crack strength that this theory underestimates the test results. The weakest link theory predicts a very small reduction in strength both with increasing beam length and beam depth; the composite mechanics theory prediction, however, follows the pattern of the actual test results, and provides a better approximation to reality.

Effect of method of casting

The orientation of the fibres relative to the stress trajectories is a major factor in their efficiency. When short discrete fibres are mixed in concrete, the orientation tends to be largely random three-dimensional, at least just after the concrete has been mixed. The only stage during which some kind of redistribution and reorientation of the fibres can occur is during the casting-compacting stage. Under table vibration, steel fibres tend to align in planes at right angles to the direction of vibration (13). In this test series the

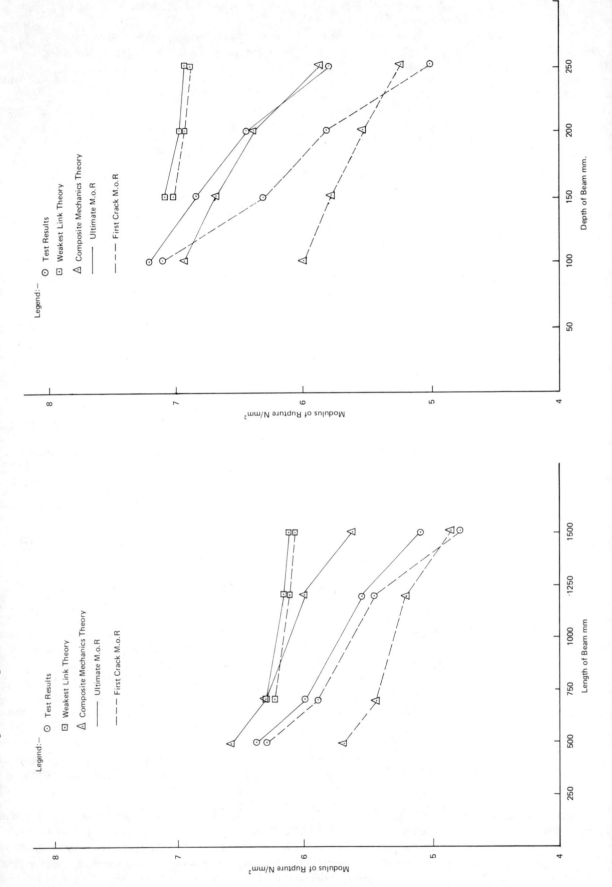

Figure 2 The effect of depth and length of test specimen on the flexural strength of fly ash fibre concrete.

effect of the method of casting, horizontal or vertical, in connection with two different methods of vibration, external and internal, has been studied. Although the problem has been studied for different types of aggregates and different properties, only data concerning 10 mm crushed gravel and flexural strength and dynamic modulus are reported here.

The mix proportions used for the fly ash fibre concrete were the same as before with fibre volume varying from 0 to 1.25%. All the beams were demoulded at 24 hours and cured under constant temperature and humidity conditions for 27 days, and tested dry. Four types of beams were tested, namely,

Type A1—beams cast horizontally and vibrated externally

Type A2—beams cast vertically and vibrated externally

Type B1—beams cast horizontally and vibrated internally

Type B2—beams cast vertically and vibrated internally.

For the beams cast vertically and vibrated externally, special moulds with one side made of thick perspex sheet were used. This enabled a continuous assessment of the movement of the fresh fibre concrete within the mould; and in the case of external vibration, the concrete was vibrated in four layers, each for 30 seconds, to ensure that the fibres were distributed uniformly, leaving no planes "fibre free".

Effect on flexural strength

The influence of the method of casting and type of vibration on the flexural strength of fly ash fibre concrete for various fibre contents is shown in Fig 3a. The results show that when fibre concrete specimens are cast horizontally external vibration gives the optimum strength whereas if the specimens are cast vertically, internal vibration is more effective than external vibration. The results confirm that horizontal casting and external vibration are the most efficient method of casting and compacting fibre concrete and that fibre orientation occurs during the casting-compacting stage.

When fibre concrete is cast horizontally, the reduction in strength due to internal vibration is only marginal, the maximum being 5%. The effect of vertical casting on flexural strength is, however, substantial, the reduction in strength, compared to external vibration of horizontally cast beams, varying up to 32% under internal vibration and 35% under external vibration. Internal vibration consistently produced higher flexural strength with vertical casting, the increase compared to external vibration varying from 3% to 8%. The results give strong evidence of fibre alignment in vertically cast sections under internal and external compaction methods.

It is interesting to note that the flexural strength of specimens with 0.5% and 0.75% fibre content, cast vertically, were even smaller than that of the plain concrete specimens cast horizontally and externally vibrated. The specimens with 1% fibre content cast vertically were only marginally stronger than the plain concrete specimens cast horizontally. The fact that the flexural strength of fibre concrete could be smaller than that of plain concrete cast in a different way cannot be explained only in terms of favourable and unfavourable fibre alignment. The full explanation to this must lie in the creation of weak bond characteristics and imperfections (14,15), which not only weaken the fibre concrete but reduce the strength to such an extent that fibres no longer act as agents of increasing flexural strength.

It must also be added that the combination of the fibre length (38 mm) and the dimensions of the specimen (100 x 100 x 500 mm) may have contributed to the creation of such imperfections. It is probable that in a specimen of larger cross-sectional area, the effect of vertical casting may not be so severe.

Examination of fractured surfaces showed that fibre orientation and fibre distribution were clearly affected by the method of casting and the type of vibration. In a horizontally cast specimen, a large proportion of the fibres will be acting directly along the stress trajectories, whereas in vertically cast specimens, the fibres will be in a direction at right angles to the direction of stress, thus contributing very little to the

Figure 3 The influence of method of casting and type of vibration on a) the flexural strength and b) dynamic modulus of fly ash fibre concrete.

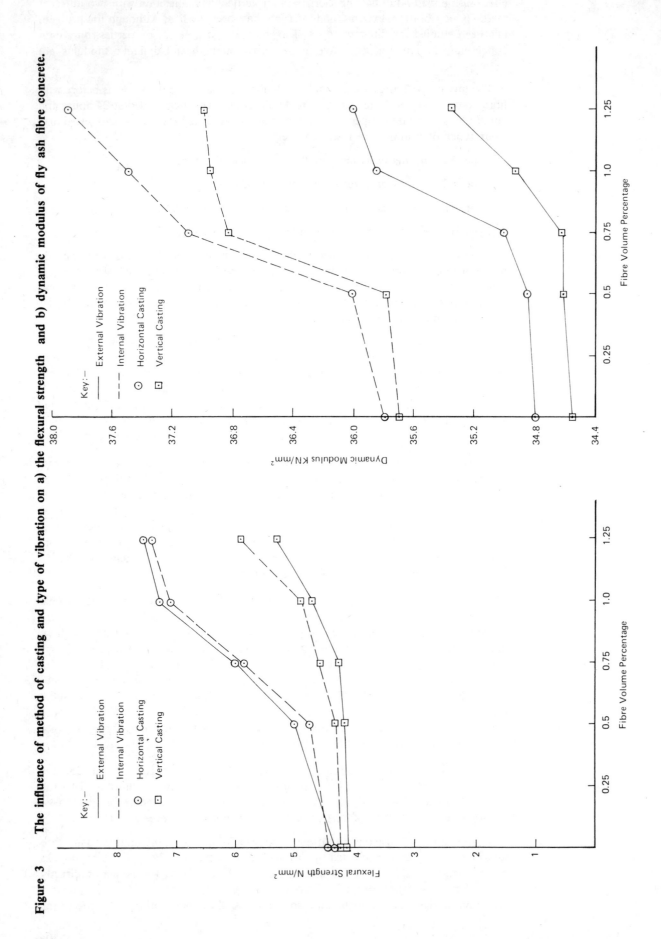

Effect on dynamic modulus

flexural strength. At the initial stages of external vibration in the vertically cast specimens, fibres could be seen to rotate and occupy horizontal directions.

The influence of the method of casting and type of vibration on the dynamic modulus for various fibre contents is shown in Fig 3b. The results show two important conclusions—whatever be the method of casting, internal vibration consistently gave higher modulus than external vibration. Secondly, for a given type of vibration, horizontally cast sections gave consistently higher modulus than vertically cast specimens. However, the variations in the dynamic modulus due to casting or compaction were only marginal and within 5%. In general the reduction in dynamic modulus was larger in the externally vibrated beams.

CONCLUSIONS

From the results obtained in this study the following conclusions may be drawn:

1. The use of pulverised fuel ash as a partial replacement of cement together with a water-reducing agent directly enhanced workability, thus easing the inherent compaction problems of fibre concrete. It also made possible the use of large aggregate sizes.

2. A fly ash replacement of cement, 30% by weight, was found to be the optimum amount. This fly ash mix showed a higher rate of increase in flexural strength with age than the all-cement mix.

3. The smaller and smoother the coarse aggregate, the higher the flexural strength and the better the post-cracking behaviour of the fibre concrete.

4. With large aggregate sizes, the fibres become less effective due to aggregate-fibre interaction and this interparticle friction becomes more pronounced with irregularly shaped aggregates. When the aggregate size reaches that of the fibre length, the propagation of flexural cracks is influenced more by the distribution of aggregate particles than by fibre distribution.

5. With adequately workable mixes, the presence of fibres causes no reduction in compressive strength.

6. Increasing the depth of the test specimen from 100 mm to 250 mm reduced the first crack and ultimate flexural strength by about 30% and 20% respectively. An increase in the length of the specimen from 500 mm to 1500 mm reduced the first crack strength by about 24% and the ultimate by about 20%.

7. Increasing the depth of the specimen caused a small increase in the standard deviation and coefficient of variation.

8. The weakest link theory consistently overestimated the flexural strength of fibre concrete. The composite mechanics approach, on the other hand, provided a better approximation to the test results.

9. Both the method of casting and the type of vibration affected the flexural strength and, to a lesser extent, the elastic properties of the fibre concrete.

10. When fibre concrete is cast horizontally, the reduction in strength due to internal vibration was less than 5%. The effect of vertical casting on flexural strength was more substantial, the reduction in strength, compared to externally vibrated horizontally cast sections, was 30-35% irrespective of the mode of vibration.

11. The results showed that vertical casting not only produced unfavourable fibre alignment but also created weak bond characteristics and internal imperfections.

12. The variations in the dynamic modulus due to variations in casting or compaction were only marginal and within 5%.

ACKNOWLEDGEMENTS

The work reported here was supported by Johnson and Nephew (Steel) Ltd., to whom the authors are most grateful.

REFERENCES

1. Swamy, R N and Mangat, P S, Influence of fibre geometry on the properties of steel fibre reinforced concrete. Cement and Concrete Research, Vol 4 and No 3, May 1974, pp 451-465.
2. Edgington, J, Hannant, D J and Williams, R J T, Steel fibre reinforced concrete. Current Paper CP69/74, Building Research Establishment, Department of the Environment, July 1974, p 17.
3. Swamy, R N and Mangat, P S, Influence of fibre-aggregate interaction on some properties of steel fibre reinforced concrete. RILEM Materials and Structures, Vol 7, No 41, Sept–Oct 1974, pp 307-314.
4. Kesler, C E, Mix design considerations. Proc. Conf. Fibrous Concrete—construction material for the seventies, May 1972, M-28, Construction Engineering Research Laboratory, Champaign, Illinois, Dec 1972, pp 29-37.
5. Kesler, C E and Schwarz, A W, Steel fibre reinforced concrete—mix design considerations. Highway Focus, Vol 4, Oct 1972, pp 22-35.
6. British Standards Institution, Specification for aggregates from natural sources for concrete. BS882 and 1201, British Standards House, 1965, p 22.
7. Ramakrishnan, V, Contribution of gap-grading to the development of high strength concrete. Paper presented at the Canadian Capital Chapter Seminar, Ottawa, Canada, Oct 1973, p 14.
8. Swamy, R N, Aggregate-matrix interaction in concrete systems. Proc. Civil Engineering Materials Conference, University of Southampton, 1969, Wiley, Interscience, Part 1, 1971, pp 301-315.
9. Wright, P J F, The effect of the method of test on the flexural strength of concrete. Magazine of Concrete Research, No 11, Oct 1952, pp 67-72.
10. Tucker, J Jr, Statistical theory of the effect of dimensions and of the method of loading upon the modulus of rupture of beams. Proc. American Society for Testing Materials, Vol 41, 1941, pp 1072-1088.
11. Swamy, R N and Mangat, P S, A theory for the flexural strength of steel fibre reinforced concrete. Cement and Concrete Research, Vol 4, No 2, March 1974, pp 313-325.
12. Weibull, W, A statistical theory of the strength of materials. Proc. Royal Swedish Institute of Engineering Research, No 151, 1939.
13. Edgington, J and Hannant, D J, Steel fibre reinforced concrete. The effect of fibre orientation of compaction by vibration, RILEM Materials and Structures, Vol 5, No 25, Jan-Feb 1972, pp 41-44.
14. Williamson, G R, Fibrous reinforcement for Portland cement. Technical Report No 2-40, US Army Engineer Division, Ohio river, Cincinatti, May 1965, p 29.
15. Hughes, B P and Ash, J E, Water gain and its effects on concrete. Concrete, Vol 3, No 12, Dec 1969, pp 494-496.

4.9 Full scale fibre concrete beam tests

G R Williamson and L I Knab
Construction Engineering Research Laboratory, Champaign, Illinois, USA

Summary   *One major application of fibre concrete to be developed is in the area of structures. Laboratory studies have shown that the addition of steel fibres increases the shear and flexural strengths and the ductility of concrete. However, full scale tests of fibre concrete dolosse have indicated that the failure mode of full scale fibre concrete elements may differ from that of laboratory specimens.*

To determine the effectiveness of steel fibres in full scale structures, four beams 305 mm x 546 mm x 7.01 m were designed in accordance with ACI Code 318-71 to resist a 494 kN four point loading. One beam was fabricated without shear reinforcement, one beam contained U stirrups as shear reinforcement, and two beams were made with 1.5 volume per cent steel fibre concrete with the fibres as the shear reinforcement. All four beams were tested to destruction and load deformation data obtained. It was found that the fibres increased the shear strength of the concrete by 39 per cent, and the load carrying capacity by 42 per cent over a beam without shear reinforcement. Although the fibre beams exceeded the design load by 6 per cent, a shear failure occurred. The load deformation characteristics of all the beams were similar.

Résumé   *C'est dans le domaine des structures qu'une des applications principales du béton fibreux peut encore se faire. Des études de laboratoires ont montré que l'ajoute de fibres d'acier augmente la puissance de courbure et de flexibilité, ainsi que la ductilité du béton. Toutefois, des contrôles à grande échelle sur du béton fibreux "dolosse" ont indiqué que la méthode par manque des éléments du béton fibreux à grande échelle peut être différente de celle des spécimens de laboratoires.*

Afin de déterminer l'efficacité des fibres d'acier dans des structures à grande échelle, on a pris en considération quatre poutres de 30.5 cms × 7.01 m × 54.6 cms en accord avec le Code ACI 318-71 pouvant offrir une résistance à un chargement de 50 Mg en quatre points. Une poutre fut réalisée sans renforcement de courbure, une poutre contenait des étriers en U en guise de renforcement et deux poutres furent construites avec un volume de 1.5 pour cent d'acier de béton fibreux, les fibres représentant le renforcement de courbure. Les quatre poutres furent contrôlées du point-de-vue de leur destruction et les données concernant la déformation de charge furent obtenues. On a remarqué que les fibres augmentaient la puissance de courbure du béton de 39 pour cent, et la capacité de charge de 42 pour cent, comparativement avec une poutre sans renforcement de courbure. Bien que les poutres de fibres aient dépassé de 6 pour cent la charge projetée, une défaillance de courbure s'est produite. Les caractéristiques de déformation de charge de toutes les poutres étaient similaires.

INTRODUCTION

Background: Small scale beam tests have shown that steel fibres increase the shear and flexural strength of concrete. Work by Batson (1) has shown that fibre volume percentages as low as 0.44 will produce moment failure in tensile reinforced beams under four point loading. However, tests of 38.1 tonne steel fibre concrete dolosse have indicated that the failure mechanism of full scale structures may differ from that of laboratory size specimens. The areas of primary concern are shear resistance and ductility, and it is toward a better understanding of these parameters that this study was directed.

Objective: The objective of this investigation was to study the use of fibre concrete as a structural material. More specifically, it was to determine the effect of steel fibres upon the shear and flexural strength of full scale reinforced concrete beams.

Approach: Four tensile reinforced concrete beams, 305 mm x 546 mm x 7.01 m were designed in accordance with the American Concrete Institute (ACI) Code 318-71 to resist a total design load of 494 kN in flexure. One beam contained no shear reinforcement, one beam contained shear reinforcement as specified in ACI Code 318-71, and two beams were made with 1.5 volume percent of steel fibre concrete and no shear reinforcement. Each of these beams was tested to destruction.

Procedure: Two beams were fabricated at the same time. The two conventional (plain) concrete beams were poured first, followed by the two fibre concrete beams six days later. The concrete was designed for a 28-day strength of 31-34.5 N/mm². The design mix and 28-day strengths are shown in Table 1. The variation in water content was due to inadequate aggregate moisture control at the batch plant. Low variability of the cylinder compressive strengths, indicated by low coefficient of variation values of 3-4%, were observed in both mixes.

Table 1 **Design mix, per m³**

Cement	308 kg
Sand	800 kg
9.5 mm Aggregate	800 kg
Water - Plain Concrete, 230 mm Slump	W/C = 0.62 190 kg
Water - Fibre Concrete, 125 mm Slump	W/C = 0.69 212 kg
Fibres - Fibre Concrete Only	118 kg
Air - Darex, Mfg's Rec	4-6%
Plain Concrete	
Compression, 29 da., Avg of 9	32.1 N/mm² ; C.O.V.* = 0.037
Flexural, 34 da., Avg of 3	3.65 N/mm²
Splitting, 30 da., Avg of 3	3.36 N/mm²
Fibre Concrete	
Compression, 35 da., Avg of 9	28.5 N/mm² ; C.O.V. = 0.046
Flexural, 35 da., Avg of 3	4.78 N/mm²
Splitting, 35 da., Avg of 3	3.44 N/mm²

*Coefficient of variation.

The compression and splitting tensile specimens were 152 mm x 305 mm cylinders, and the flexural specimens were 152 mm x 152 mm x 457 mm beams loaded at the third points.

The full-scale beam cross-sections and loading arrangements are shown in Figure 1. The main loading was applied at the third points by two 222 kN actuators. The

Figure 1   Beam cross-sections and leading arrangement.

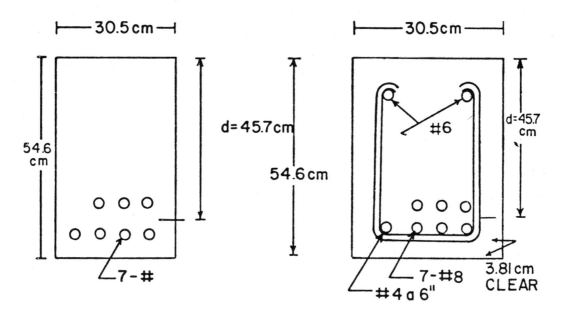

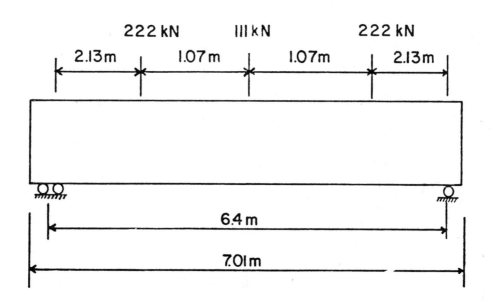

actuators were part of a closed loop servo-hydraulic testing system. For three of the beams, the 445 kN load was insufficient to produce failure. Therefore, when the capacity of the 222 kN actuators was reached, a 111 kN actuator was applied at the centerline, and loading continued. The time required to load the beams to ultimate varied from 30 to 60 minutes.

RESULTS AND DISCUSSION

The results of the four beam tests are shown in Table 2.

Beam No 1 (no shear reinforcement), as was expected, failed in shear at a total load of 369 kN and a shear stress of 1.36 N/mm². This is 75 percent of the 494 kN ultimate flexural design load. At a load of 289 kN, the centerline deflection was 18.5 mm. Because of the anticipated catastrophic failure, measurements were discontinued beyond the 289 kN load.

Beam No 2 (stirrups for shear reinforcement) supported a total load of 600 kN without failing. This exceeded the flexural design load of 494 kN by 22 percent. The shear stress developed at this load was 2.15 N/mm², or 58 percent greater than Beam No 1 with no shear reinforcement. The centerline deflection at 289 kN was 19.0 mm, and at 529 kN, it was 42.0 mm.

Beams No 3 and 4 (fibre concrete) failed in shear at a total load of 529 and 520 kN, respectively. Using the average of 525 kN, this is 6 percent in excess of the 494 kN ultimate flexural design load. The average shear stress of 1.88 N/mm² is 39 percent over that developed in the non-shear reinforced beam. The 39 percent increase is probably low since the cylinder compressive strength of Beam No 1 was, on the average, 3.65 N/mm² more than Beams No 3 and 4. The average centerline deflection at the 289 kN load was 16.0 mm, and at 525 kN it was 40.6 mm. Failure of both fibre concrete beams was classified as catastrophic.

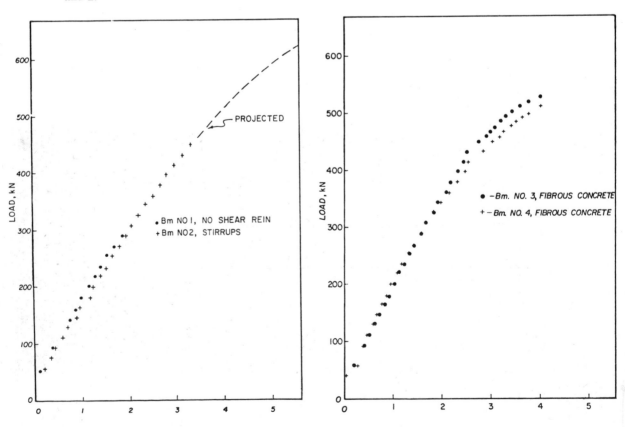

Figure 2 Centerline deflection, mm Beams No 1 and 2.

Figure 3 Centerline deflection, mm Beams No 3 and 4.

Table 2 Results of beam tests

Beam No	Reinforcement	Reinforcement	Cylinder compressive strength N/mm^2	Ultimate load kN	Ultimate moment kNm	Ultimate [+] shear stress N/mm^2
1	7 - #8	None	32.1	369	405	1.36
2	7 - #8	#4 @ 15.2 cm	32.1	600*	705	2.15
3	7 - #8	Steel Fibres	28.5	529	591	1.9
4	7 - #8	Steel Fibres	28.5	520	576	1.87

* Total failure not achieved.

[+] Average ultimate shear stress, V_u/bd where V_u = shear force at ultimate;

b = beam width;

d = depth of tensile reinforcement.

The fact that the fibre concrete beams failed in excess of the design load and the beam without shear reinforcement did not, shows quite clearly that the steel fibres are effective in shear and that conventional stirrups can be replaced by the fibres and the design moment achieved. However, the ratio of the ultimate load to the design load for the fibre concrete beams was 1.06 compared with a ratio value of over 1.22 for the shear reinforced beam. This indicates a reduced margin of safety for the fibre beams as compared to the shear reinforced beam. Further, and probably even more important, the failure mechanism of the fibre beams was shear, and thus catastrophic. The stress in the steel at the 525 kN average failure load had just reached the yield stress of the steel, 413.7 N/mm²; therefore, the desired yielding of the flexural steel that is characteristic of beams with stirrups did not occur.

Figure 4 A view of the shear and main reinforcing in Beam No 2.

Figure 5 Shear failure of beam made with fibrous concrete.

The load-deflection diagrams of Figures 2 and 3 show clearly that the deformation characteristics of all four beams are similar. The energy absorbed at a load of 449 kN by beams No 2, 3 and 4, was 8250, 7120, and 8250 Nm, respectively.

CONCLUSIONS AND RECOMMENDATIONS

1. Steel fibres are effective in increasing the shear strength of the concrete to the extent that the full flexural design capacity of the beam is attained. However, since there was no yielding of the reinforcement steel in the fibre beams, they failed at a considerably lower load than the beam with shear reinforcement.

2. Steel fibres are not effective in preventing catastrophic shear failures in full scale beams.

3. Load-deflection characteristics of steel fibre concrete beams are similar to conventional beams with shear reinforcement.

Because of the significant increase in the shear, moment, and energy capacity of the fibre concrete beams over that of beams without shear reinforcement, research should be conducted in possible applications for fibre concrete structural members, including earthquake resistant structures.

ACKNOWLEDGMENT

The research was performed at the Construction Engineering Research Laboratory (CERL), Champaign, Illinois, USA.

REFERENCE

1 Batson, G, Jenkins, E and Spatney, R, "Steel Fibers as Shear Reinforcement in Beams", ACI Journal, Proceedings Vol 69, No 10, Oct 1972, pp 640-644.

The views of the authors of this paper do not purport to reflect the position of the Department of the Army or the Department of Defense.

Section 5

Properties and Testing of Concrete containing Fibres other than Steel

5.1 Opening Paper: Properties and testing of concrete containing fibres other than steel

J J Zonsveld
*Visiting Reader in the Department of Civil Engineering,
University of Surrey, United Kingdom*

Summary

Two categories of fibres other than steel are considered. The fibres with a high modulus of elasticity (high-E fibres): glass, asbestos and carbon fibres are less suitable for random dispersion in concrete. Glass and carbon because of their vulnerability to damage during mixing and compaction, asbestos because of high cost and health hazards. If not randomly dispersed, glass and carbon, aligned and moulded into resin in the form of bars, can be used in concrete in a similar way as steel reinforcement, but only in special applications.

Low-E fibres are the most important for use in concrete as they can be cheaply dispersed in a variety of mixes. They do not confer a satisfactory increase in tensile strength upon the hardened concrete, but the improvement in impact resistance is remarkable.

Polypropylene is the most suitable material for the low-E fibres. The fibres are commercially available for many other uses as monofilaments, as well as in the form of fibrillated yarns, and both types of fibre produce finished concrete articles with desirable properties. The monofilaments are added to an aerated concrete to achieve a mix with thixotropic properties. After placing, decorative designs can be created on the fresh concrete surface without using expensive moulding techniques.

Fibrillated fibres are well bonded into the concrete as a result of their open net structure, although their hydrophobic surface does not allow adhesion in the physico-chemical sense. Because of the high impact resistance of this type of fibre concrete it has found full industrial application in piling shells which withstand repeated hammer blows considerably better than the previously used steel reinforced shells. The fibre content in this use is 0.5% by volume.

Some work is reported on combinations of high-E and low-E fibres and on the use of fibres in concrete that has been modified with other polymers.

Résumé

Deux catégories de fibres sont considerées. Celles avec un module d'elasticité élevé (fibres haut-E): les brins de verre, d'amiante et de carbone ne sont pas adaptés à une dispersion au hasard dans le béton. Les brins de verre et de carbone peuvent être endommagés aisément pendant le mélange ou le compactage, les brins d'amiante sont hasardeux pour la santé. Le verre et le carbone comme fibres continues se laissent mouler dans une résine sous forme de tige pour l'armature du béton. On l'utilise alors d'une manière semblable à l'acier, mais seulement dans les cas spéciaux.

Les fibres bas-E sont les plus importantes parce qu'ils ne sont pas couteuses et se laissant disperser simplement dans le mélange. Ils ne procurent pas au béton durci une plus grande résistance à la traction, mais la résistance aux chocs est très améliorée. Les fibres de polypropylène sont au marché comme monofilaments et comme ficelles. Les brins des monofilaments dispersés dans un béton aéré le donnent des propriétés thixotropiques. Dans la surface du béton frais on peut créer dessins decoratifs sans l'emploi des techniques de moulage couteuses.

La structure ouverte des fibrilles dans les ficelles de polypropylène a pour résultat une bonne adhérence mécanique dans le béton. La forte résistance aux chocs de cette type de béton a abouti à une application entièrement commerciale dans la fabrication des pieux. Les éléments tubulaires en béton fibreux résistent les chutes du mouton beaucoup mieux que les gaines en béton armé qu'on avait employées pendant des dizaines d'années.

Des recherches, qui s'occupent des combinaisons de fibres haut-E et bas-E sont discutées, et finalement on fait mention des utilisations interessantes de fibres dans des bétons modifiés dans la matrice avec d'autres polymères.

INTRODUCTION

Fibres other than steel can be classified in two categories:

1. Those with a high modulus of elasticity: glass, asbestos and carbon fibres;
2. Those with a low modulus of elasticity: natural and synthetic, organic fibres.

One can truly speak of fibre-reinforced concrete if the fibre/concrete composite shows higher strength characteristics than the control, and this can only be expected from fibres in the first category. The low-E fibres impart other desirable properties to a concrete but the increase in tensile or compressive strength is never impressive. It is therefore suggested that a concrete containing organic fibres should be called a *fibre concrete* and not a *fibre-reinforced concrete*.

Table 1 **Fibres other than steel for concrete**

Type	Specific gravity	E kN/mm^2	Tensile strength N/mm^2	Cost £/kg	% added by wt.	% added by vol.	Cost of composite £/m^3
High-E							
Asbestos	2.6	170	3,000	0.10	10	10	31
Alkali-res. glass	2.7	80	2,500	0.90	3	3	73
Steel	7.8	200	1,100	0.30	5	1.5	45
Carbon	2.0	400	2,000	60	2.5	3	3,600
Low-E							
Polypropylyne monofilament	0.9	5	400	0.60	0.1	0.20	11 (aerated)
fibrillated	0.9	8	400	0.60	0.2	0.5	13
Nylon monofilament	1.14	4	900	0.90	3	6	73
Concrete control	2.4	30	4				10

Table 1 summarises some relevant numerical data of the fibres and the cost of composites to be discussed below. Accurate ranges of properties in this context would be imaginary, so typical values are shown. Steel—as it is discussed by others in this symposium—is included for comparison only. The base material for the fibrous composites is assumed to be an average good-quality concrete costing £10/m³. The percentages added are the ones that have been proposed for actual application but some composites listed in this table are largely hypothetical as explained in the text.

HIGH-E FIBRES

Glass

Glass (like asbestos) is being used in cement where the fabrication is carried out in a factory, first to an intermediary sheet form, then to moulded products. In fresh concrete, however, glass fibres will suffer severe damage and loss of strength since—during mixing as well as during compaction—considerable abrasion and impact forces are generated by the movement of aggregates. To manufacturers of glass-fibre-reinforced plastics such abrasion has long been known. Glass filaments and rovings are handled with great care by guiding them smoothly over pulleys and through eyes while the continous fibres are positioned in a resin matrix. It appears that the weakening of the fibres originates in the surface of the glass fibres and it is obvious that the rough sand and aggregate particles in a concrete mix will play havoc with the tender glass. The Cem-FIL fibres in cement are much less at risk due to the adopted technique of careful placing and the low level of abrasion in a cement slurry compared with a concrete mix. The vulnerability of glass fibres to becoming damaged in fresh concrete rules out the successful development of glass-fibre-reinforced concrete with random fibres.

A development which—so far—does not appear to have been widely used in practice was reported by Klink (1), and consists of using E-glass fibres which have been coated with an epoxy resin compound. The glass is thus protected from alkaline attack by Portland cement and the coating provides a certain degree of mechanical protection to the fibres.

Another way of using glass to reinforce concrete has been reported by Rehm (2,3). He suggested bars of 60% by vol of aligned glass fibres embedded in 40% of resin, which have a tensile strength of 1500 N/mm^2 and $E = 50 \text{ kN/mm}^2$. It is emphasised that the bars cannot be sharply bent and that anchoring in the concrete beam presents problems. Compared with conventional steel reinforcement, the lower E of the glass fibre rods causes increased deformation and wider cracks under load. In prestressed beams, however, the modulus of elasticity of these bars, which is higher than that of concrete but about a quarter of the modulus of steel, offers advantages in case the concrete shrinks or creeps, and losses in pre-stress would be lower. The research so far has consisted of preliminary tests, and has not yet developed into a practical application.

Asbestos

Asbestos fibres in a concrete mix are less prone to being affected by rough plant handling but their short length (around 5 mm) and probably their low length-to-diameter ratio reduces the reinforcing action considerably. The long but expensive, textile-quality asbestos fibres might well reinforce concrete, but would be unattractive for economic reasons. Another factor to cause increased cost could be the high asbestos content needed and the extensive safety measures now required to protect personnel in the plant against health hazards. In asbestos/cement products the content ranges from 10-70%, and it would also be expected to be high in asbestos/concrete. No literature on asbestos/concrete tests seems to be published and the author regrets that he cannot present here any quantitative data on a subject that could show interesting aspects of fibre/concrete composites.

Carbon

Carbon fibres form the most recent and also the most spectacular addition to high-E fibres. They are very expensive but have strength and stiffness characteristics superior to steel. However, they are even more vulnerable than glass fibres to surface damage and subsequent weakening, and must be used in the clumped form, i.e. embedded in, or sized with a resin coating. This excludes the random addition of short fibres in a mix. The high cost makes it obligatory to apply the carbon fibres in the most efficient design, which means their use would be similar to that of steel bars.

Pomeroy (4) calculated that the favourable ratio of the modulus of elasticity of carbon fibre to that of the cement matrix (ie 13) should result in a considerable increase of E in the composite. With carbon fibre aligned parallel to the direction of stress, the increase would be nearly 50% at a volume loading of 3% of fibre. It would be interesting to learn whether improvements of that order have actually been observed in research that may not have been published so far, or whose publication has escaped our attention.

A fairly recent new semi-manufactured product on the market is the pultruded carbon fibre/epoxy resin profile. Carbon fibres carefully aligned in the resin matrix are drawn through a die in the so-called pultrusion process, and cured to a solid rod. With the advent of such profiles long enough to be used like steel bars as a continuous reinforcement in structural members, it is now perfectly feasible, however costly, to apply such carbon fibre reinforcement in concrete under special circumstances.

LOW-E FIBRES

Vegetable Fibres

Vegetable fibres in bygone days were used in mortars and occasionally in cement renderings in a way similar to the use of jute as a scrim in plaster of Paris. As recent as 1969 the Building Research Station in the UK undertook an investigation of sisal reinforcement in concrete at the request of the United Nations Hard Fibres Study Group who were concerned about serious overproduction of sisal in African countries. The results were disappointing as water absorption by the fibres required a high overall water/cement ratio for a workable mix, while the setting was retarded by impurities leaked out of the chopped sisal fibres, and no additional strength was obtained by the addition of up to 5% by weight.

Low-E fibres of vegetable or animal origin can be ignored in the present paper as fully synthetic materials of high purity are now available in forms precisely adapted to the end use in view, free from corrosion by the cement constituents, and of a consistent product quality.

Nylon

Nylon, in the form of cut monofilaments, was the first of these fibres to be recommended in 1965 by Goldfein (5) for the construction of blast-resistant buildings for the US Corps of Engineers. Nylon fibres, being more expensive than polypropylene fibres and not available in other than monofilament form, have not been utilised in the more recent developments. However, these are the only reasons for not exploiting nylon any further, since strength, chemical inertness, durability and elevated temperature resistance are no less in polypropylene.

Polypropylene

Shortly after 1965 Shell Chemical started tests with fibrillated polypropylene twine (6), which in the first place is cheaper than nylon or any other synthetic fibre, but has also a unique texture for bonding into concrete. The net structure of fibrillated fibres is obtained by extrusion and stretching of polypropylene tape, which is then twisted to form a yarn. In the concrete mix the fibres open up sufficiently to allow the cement matrix to penetrate between the mesh and to form a continuous phase in which the fibre is held firmly, together with sand and aggregates. Figure 1 shows the texture of a fibrillated fibre opened up to a width of about 20 mm.

Most authors considering the theoretical aspects of the behaviour of fibres in a cement matrix, stress the benefits of having good adhesion between fibre and matrix. Their reasoning is that after the failure strain of the matrix has been exceeded, incipient cracking results in stress transfer to the fibres and subsequent creation of interfacial bond stresses. The quality of the bond is greatly affected by surface polarity, and a hydrophilic fibre surface would promote a good bond to resist these stresses. However, many of the organic synthetic fibres are poorly wetted by water and a pure hydrocarbon like polypropylene is the worst of all. The adhesion between hydrated cement and the polypropylene surface is such that sheets of this plastic can be used for easy release from shuttering. All emphasis must therefore be placed on mechanical bonding. Monofilaments with regular constrictions or of sinusoid form (crimped fibres) have improved mechanical hold, but the best mechanical bonding is offered in fibrillated tape, either in the open form or twined into string. The latter are the fibrillated fibres that have been used in the composite developed by Shell under the designation "Caricrete", which was successfully commercialised in the piling shells of West's (see below) and a number of other applications.

Summarising this aspect of bond between fibre and matrix, it is postulated that a lack of interfacial adhesion can be replaced satisfactorily by suitable mechanical bonding into the continuum of the cement matrix.

The concrete/polypropylene fibre composite has little added strength compared with the same concrete without fibre. Although the tensile strength of polypropylene is much higher than that of concrete, the modulus of elasticity is so much lower, that under increasing stress the concrete will reach its ultimate deformation and will start cracking

Figure 1   Polypropylene fibrillated fibre yarn opened up to about 20 mm wide.

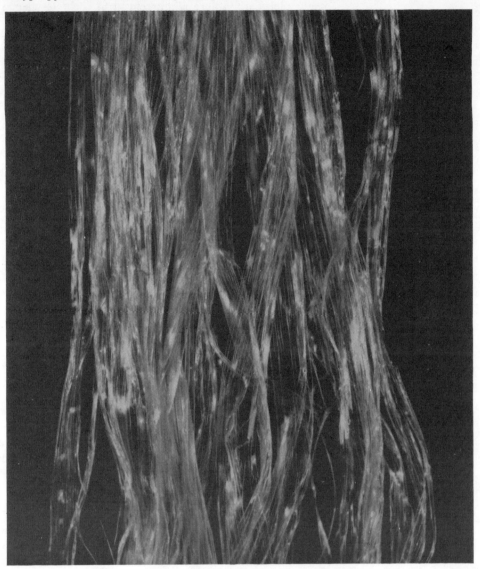

well before the fibre can develop a constraining stress. Besides, stresses in plastics are time-dependent and the strength of a structural member could not be relied upon, if short term strength tests had provided the data for its design. ⌉

In case of accidental damage incorporation of fibres has the obvious advantage of keeping the fractured pieces together. Concrete elements like cladding panels or manhole covers which have conventionally been steel-reinforced, but are never expected to be subjected to their breaking loads once they are in position, should be tested by dropping or banging them the way they suffer in practice. This sounds crude but standard test methods exist which describe how such tests have to be done in a civilised manner. It is therefore strongly recommended that existing standard methods of testing concrete be applied with caution, and that tests of final products be given preference over tests which characterise the starting material only. Manufactured products whose impact strength is of prime importance are the best examples of this principle.

IMPACT PROPERTIES

Concrete/polypropylene fibre composites demonstrate in most laboratory and practical field tests a surprising improvement in impact strength. In contrast to the less desirable example of time-dependent behaviour of plastic materials under long-term stress that was mentioned above, the time dimension in stress-strain relations can be beneficial under conditions of high-speed deformation. When stress-strain diagrams of polypropylene fibres on their own are recorded at various rates of strain one observes a stiffening at higher rates. When plotting the rates of strain against the moduli of elasticity derived from these diagrams, it is possible to extrapolate and estimate the value of E at extreme rates of strain. Under conditions of impact testing where these high rates of strain apply, E appears to be double or treble the standard value and equals roughly the E value of concrete. These observations, far from quantitatively explaining impact results, indicate in which direction concrete products containing polypropylene fibres can be used with confidence.

West's Piling and Construction Co Ltd, have since 1969 developed the so far largest single commercial application of any concrete/fibre composite. West's use 40 mm long pieces of string consisting of fibrillated polypropylene twine, of which about 0.5% by volume is added to the mix for their piling shells. A representative shell is a 915 mm long pipe of 380 mm diameter and approx 50 mm wall thickness. The production amounts to half a million shells per annum. Impact strength in West's plant is assessed by dropping a hammer of 3 ton on a shell taken from the production line. The end of a shell without packing is subjected to a sequence of hammer blows from increasing heights until failure is observed. Fairweather (7) has described the details of these tests, and reported 40% improvement in comparison with the steel-mesh-reinforced shells which have been used prior to 1970.

Notwithstanding the present emphasis on practical impact tests done on manufactured products, the need remains to measure impact resistance as a fundamental property of the material, and to put a figure on the dissipation of impact energy in the composite. A reliable test for impact resistance of the composite material might in future establish a better correlation with the practical impact testing of manufactured elements. This is the aim of a research programme at the University of Surrey, sponsored by the Transport and Road Research Laboratory, basically utilising the Charpy machine with refinements added to achieve a more complete energy account.

HANDLING PROPERTIES OF FRESH COMPOSITES

Workability

The workability or the rheology of the mix is measured in standardised tests like the slump test, the V-B consistometer test and the compacting factor, which are all very relevant to fresh ordinary concrete, but are less so to a mix containing fibres. For instance, the slump of a mix with a low fibre content can be zero, although the mix flows all right when kept moving, and responds well to vibration. Edgington et al (8) recommend the V-B test for these reasons and show that a clear differentiation in V-B times can be obtained for mortars and concretes containing steel fibres (a) for varying amounts of fibre, and (b) to discern between varying aspect ratios of fibres used in the mixes. Systematic work as done on steel fibre composites has not been carried out for mixes with low-E fibres, but the experience in laboratory and plant is certainly in agreement insofar as a workability test should tolerate conditions of flow or vibration.

In colloid-chemical systems like drilling muds, viscosity seldom follows the pure laws of viscous oils. The viscosity measurement gives different results depending on the rate of shear in the liquid. A vigorously stirred dispersion is seemingly less viscous than the slow-moving liquid, and the mud stiffens when stirring stops. This is termed thixotropy. The reverse, which is called dilatancy or rheopoxy, also occurs. These properties of suspensions and gels can all be encountered in fresh fibre/concrete composites, and can be used to advantage in developing new applications, or to streamline the production in the factory. If a mix needs an addition of plasticiser to enhance workability, the admixture may prove to have altered the rheological picture in a surprising direction and the mix composition or the operational procedure may have to be adjusted.

The way of handling fresh composites in the plant is further dependent on the equipment available, and on the daily routine which was followed prior to the

Figure 2   "Cheese" of polypropylene monofilament and fibres chopped to 20 mm length.

introduction of fibrous mixes. Alterations to the operational routine for normal concrete will no doubt eliminate all sorts of complications. Examples of changes implemented in practice have been: a wider opening for emptying the mixer drum, a conveyor belt in place of wheelbarrows and spades, a different external vibrator or mould shuttering, and other similar adaptations in the plant. Curiously enough, mixes with low-E fibres respond well to the conventional vibrating tables or pokers, and presses.

Faircrete

An interesting rheological benefit was discovered by John Laing Research and Development Ltd, who developed an aerated concrete composite with only 0.1 to 0.2% by volume of chopped polypropylene monfilaments added to a mix, designed with a view to creating a decorative sculptured finish, mainly for cladding panels.

Variation of the amount of air entrainment or of the content of synthetic lightweight aggregates allows of a choice of ultimate densities from 700 up to about 2000 kg/m^3. For the latter density ordinary gravel or limestone is used in the mix, which is then only lightly aerated. The polypropylene monofilaments have diameters ranging from 0.1 to 0.2 mm and are cut to lengths of 10 to 20 mm. Figure 2 shows the spool or "cheese" of fibre as purchased and the chopped material. Hobbs (9) likened the action of the fibres in the mix to that of a three-dimensional sieve, stopping the air to pass up through the sieve and holding the aggregate so that it cannot pass down. The resulting properties of the mix, particularly when assisted by very light vibration, are easy flow out of hopper outlets, into restricted areas, and against mould faces. The thixotropic properties enable the concrete after placing to be formed into various shapes and patterns that would not be possible with ordinary concretes. The imprint does not slump back and remains exactly as formed on the hardened concrete.

Full details of the mechanical, thermal and other properties of Faircrete are reported in the leaflets obtainable from John Laing and Son Ltd.

CONCLUDING REMARKS ON LOW-E FIBRE CONCRETES

The sections above have described properties of concretes containing low-E fibres because these are the only non-steel fibres that have attained significance in practical applications. There is considerable experience with polypropylene fibres over periods up to 5 years to support the following conclusions in respect of the beneficial effects that have been observed. In summary, they are:

1. The thixotropic behaviour of the mix as described for Faircrete in particular;
2. The inertness of the material to boiling water temperatures, to alkalinity, or to any reagent that would not by itself destroy the concrete;
3. The limitation of the propagation of cracks and the high ultimate elongation under increasing stress;
4. The much improved impact resistance;
5. A residual strength after cracking;
6. The ability of an article to remain virtually in one piece after being broken.

COMBINATIONS OF DIFFERENT FIBRES IN CONCRETES

High-E and low-E fibres can be combined in random dispersion to achieve increased strength plus increased impact resistance. Work in France on asbestos/polypropylene combinations resulted in a composite with cement, that had much improved impact resistance compared with the control asbestos/cement. However, after production trials in the factory had failed, further research was stopped. From the academic point of view the result was very positive, but the sophisticated machinery in the asbestos/cement plant would have to be adjusted at great expense. At this stage the cost of new machine development and plant alterations could not be justified by the improved properties.

In future, when the Cem-FIL development of alkali-resistant glass fibres in cement will have reached a more established phase, it is hoped that a combination of polypropylene and glass, not only in cement but also in mortars or concrete, would be considered as it might offer extra benefits.

FIBRES IN POLYMER-MODIFIED CONCRETE

The first publication on this interesting aspect of fibre concretes was by Kubota and Sakane for the RILEM Symposium in Paris in 1967 (10). The present author referred to it briefly in 1970 (6) but the combination of two organic polymers, one in the matrix and one in fibre form, deserves some attention.

Kubota and Sakane used polyethyl acrylate emulsion up to a volume concentration of 15% in a 3:1 sand/cement mortar. The fibres used were monofilaments of either polypropylene or Saran (vinyl-vinylidene chloride copolymer). As the polypropylene fibres gave the better results, the Saran fibres will be ignored here. Fibre contents up to 9% by volume were tested. The most notable results were obtained in bending according to the Japanese standard testing method for flexural strength.

A.	Mortar control	3.0 N/mm²
B.	15% by volume of polyethyl acrylate added	4.3
C.	9% by volume of polypropylene fibres added	4.0
D.	Additions B and C combined	6.6

The synergistic effect under D is striking as the effects of the additions under B and C do not add up to the combined effect. The polymer added into the matrix (under B) had halved the dynamic modulus of elasticity of the mortar. This is a factor to be considered more closely in future when judging what value should be attached to the combined effect of polymer plus fibre on the flexural strength. Impact strength was discussed in Section 4 above. In the light of those comments, the values obtained by the Japanese researchers on small prisms in Izod tests should be judged critically, but their results clearly point towards a similar synergistic effect as reported for flexural strength.

Pomeroy and Brown (11) have recently described the use of steel fibres to restore ductility and to remedy brittle fracture of high strength concretes. The polymer used in this case was polymethyl methacrylate added to the mix as an emulsion up to a polymer content of 6% by volume. Steel fibres were added up to 1.5% by volume to plain concrete as well as to polymer modified concrete and the authors observed greater benefits from fibre addition in the polymer modified concrete. They believe that this is due to an improved fibre-matrix bond, resulting from the polymer present in the matrix. The better ductility showed the higher differentials at the greater beam deflections, that is, past the elastic stage in the deformation.

Many interesting aspects still await discovery and it would appear that the combination of fibre addition and polymer modifications of concrete is a worthwhile field for further research.

REFERENCES

1. Klink, S A, Fibro-cement composites. UNIDO report of Expert Working Group Meeting, Vienna, 20-24 October 1969.
2. Rehm, G, GRP rods as reinforcement. Betonwerk + Fertigteil-Technik, Heft 9/1973, pp 631-634.
3. Rehm, G, Fibre-reinforced concrete types and their problems. Ibid, pp 638-641.
4. Pomeroy, C D, Fibre-reinforced concrete—Its properties and applications. Departmental note of the Cement & Concrete Associaton, DN/4023, October 1973.
5. Goldfein, S, Fibrous reinforcement for Portland cement. Modern Plastics, April 1965, pp 156-159.
6. Zonsveld, J J, The marriage of concrete and plastics. Plastica Vol 23, October 1970, pp 474-484.
7. Fairweather, A D, The use of polypropylene fibrillated fibres to increase impact resistance of concrete in prospects for fibre reinforced construction materials, Proceedings of a London conference, November 1971, pp 41-44.
8. Edgington, J, Hannant, D J and Williams, R I T, Steel fibre reinforced concrete. Building Research Establishment Current Paper 69/74, July 1974.
9. Hobbs, C, Faircrete: An application of fibrous concrete. Ibid, pp 59-67.
10. Kubota, H and Sakane, K, A study on the improvement of cement mortar by admixing polymer emulsion and synthetic fibre. RILEM Symposium Paris, September 1967.
11. Brown, J H and Pomeroy, C D, Mechanical properties of some micro-concretes and mortars modified by polymethylacrylate. Cement & Concrete Association, Technical Report 42.507, March 1975, 24 pp.

5.2 Contribution à l'étude du comportement mécanique des bétons renforcés avec des fibres de polypropylène

J Dardare
*Centre d'Études et de Recherches de l'Industrie
du Béton Manufacturé, Epernon, France*

Summary   *The present study deals with the reinforcement of concrete by means of polypropylene fibres. Experiments were carried out with concretes of different fibre contents (0.2%, 0.6%, 1%, 2%) and fibre length (35 mm, 45 mm, 75 mm).*

The preparation of such concrete raised some problems with regard to the mixing process. Besides, the amount of water required could be regarded as being responsible for the decrease in strength and the increase of shrinkage, especially with high fibre contents.

Experiments showed that the bond between fibres and concrete was satisfactory and that the value of the modulus of elasticity of fibre-reinforced concrete was very low. The morphology of fracture, under compression or tension, was found to be different from that of plain concrete.

The use of propylene fibre may be considered for the reinforcement of fragile elements, which are not subject to heavy loads but are exposed to shocks.

Résumé   *Cette étude a pour sujet le renforcement du béton avec des fibres de polypropylène. On a fait des essais avec des bétons à différentes teneurs en fibres (0.2%, 0.6%, 1% et 2%) et avec différentes longueurs de fibres (35 mm, 45 mm, 75 mm).*

La confection de ces bétons a donné lieu à certains problèmes en ce qui concerne le malaxage. De plus on peut considérer que la quantité d'eau requise est responsable de la réduction de la résistance et de l'augmentation du retrait, particulièrement pour de hautes teneurs en fibres.

Les essais ont montré que l'adhérence entre fibres et béton était satisfaisante et que la valeur du module d'élasticité du béton armé de fibres était très bas. On a constaté que la morphologie de la rupture, en tension ou en compression, était différente de celle du béton ordinaire.

On peut envisager l'emploi de fibres de propylène pour le renforcement de pièces fragiles qui ne sont pas soumises à des charges éleveés mais sont exposées à des chocs.

INTRODUCTION

L'industrie du béton manufacturé demande de la part des bétons certaines propriétés mécaniques différentes de celles qui peuvent être exigées d'un béton traditionnel.

En outre, s'il est possible dans la plupart des cas, de pallier le manque de résistance du béton par l'adjonction d'armatures (béton armé, béton précontraint), il n'en est pas toujours de même pour certains produits élaborés par l'Industrie du béton.

Il en est ainsi, notamment, pour certains éléments qui, par suite de leurs faibles dimensions, impliquent de placer l'armature dans des zones où son efficacité est faible. Aussi, certains auteurs ont-ils pensé à reconstituer un matériau aussi homogène que possible en renforçant le béton par des fibres.

Objectif de la recherche

L'objet de la recherche, effectuée au CERIB, était de déterminer le comportement, vis-à-vis des sollicitations mécaniques, d'un béton renforcé par des fibres de polypropylène.

Les essais réalisés avaient donc pour but de vérifier la conservation des principales propriétés mécaniques d'un béton (homogénéité, retrait, résistance) et d'analyser les améliorations apportées par ce type de renforcement.

LA FIBRE DE POLYPROPYLÈNE

Le polypropylène est un polymère cristallisable de la famille des polyoléfines. Découvert en 1954, ce matériau de synthèse a connu depuis une extension croissante dans l'industrie du textile où il apporte les avantages suivants :
— bonne résistance,
— déformabilité élevée,
— imputrescibilité.

La fibre est obtenue suivant le processus classique d'extrusion étirage qui confère une orientation prépondérante aux molécules et qui engendre des propriétés mécaniques élevées. En effet, la contrainte de rupture en traction se situe aux environs de 7.000 bars.

Du point de vue de la durabilité, le polypropylène est sensible aux ultra violets, son exposition prolongée au rayonnement solaire provoque en effet une oxydation se traduisant par :
— une transformation au niveau des molécules,
— une diminution de la masse moléculaire,
— une augmentation de la plasticité,
— une fissuration de surface se propageant dans la masse.

Cependant, noyées dans le béton, ces fibres sont protégées de l'oxydation due aux ultra violets.

Les fibres retenues pour cette recherche étaient obtenues à partir de ficelle dont le titrage est de 900 m au kg (Figure 1).

MISE AU POINT DU BÉTON DE POLYPROPYLÈNE

Choix de la composition du béton

Etant donné le rôle d'armature joué par la fibre, il était impératif d'étudier un béton de plasticité suffisante pour assurer un enrobage correct des fibres et améliorer ainsi l'adhérence.

A la suite d'essais préliminaires, on a choisi la composition suivante :
pour 1 m³
— 750 kg de sable 0/5 (module de finesse 2.32),
— 1.025 kg de gravillons 5/15 roulés,
— 400 kg de ciment CPA 400,
— 210 litres d'eau.

Une telle composition donne une faible compacité, un rapport $\frac{E}{C}$ égal à 0.525 et un affaissement au cône d'Abrams de 60 mm environ.

Figure 1 Fibres de polypropylène utilisées.

Choix du pourcentage et de la longueur des fibres

Quelques essais préliminaires ont montré que l'ouvrabilité du béton était fortement influencée par le pourcentage et la longueur des fibres. Sur le plan d'une mise en place correcte, ces essais ont montré que l'on devait limiter le pourcentage à 2% en valeur maximale et que l'on devait utiliser des fibres de longueur comprise entre 20 mm et 80 mm. En effet, dans le cas d'une fabrication par malaxage classique, des pourcentages plus élevés ou des longueurs de fibres plus grandes conduisent à la formation de "balles" et à une mauvaise homogénéité du mélange.

Dans le cadre de cette étude, on a sélectionné:
3 longueurs de fibres 35 mm, 45 mm et 75 mm,
4 pourcentages en poids 0.2%, 0.6%, 1% et 2%, la masse volumique du polypropylène étant de 0.91 g/cm^3.

Fabrication du béton

Le processus de fabrication du béton comprend les phases suivantes:
— malaxage du mélange gravillons-sable-fibres pendant 2 minutes. Cette opération permet de défibrilliser les fibres, ce qui permet d'avoir un meilleur accrochage dans le mortier et d'améliorer ainsi l'adhérence.
— introduction du ciment et de l'eau, suivie d'un malaxage de 2 minutes environ.

L'introduction des fibres, juste avant la fin du malaxage, ne permet pas, comme l'ont montré les essais préliminaires, d'avoir une imprégnation complète des fibres.

Par ailleurs, à partir d'une valeur du pourcentage égale à 1% on a dû abandonner le malaxage mécanique classique pour le remplacer par une méthode manuelle.

RESULTATS DES ESSAIS

Résistance à la compression

Les essais de compression ont été effectués sur éprouvettes cylindriques 110 × 220 mm, confectionnées conformément à la norme NF P 18.400.

Le tableau suivant, et le graphique de la Figure 2 (sur laquelle on a représenté la variation du rapport de la résistance en compression du béton renforcé à celle du béton témoin en fonction du pourcentage de fibres) montrent que l'incorporation de fibres de polypropylène dans le béton apporte une légère amélioration lorsque le pourcentage de fibres est faible. Par contre, pour les forts pourcentages, la résistance en compression chute. Par ailleurs, les valeurs de la résistance en compression à 28 jours sont légèrement inférieures à celles obtenues à 7 jours.

Figure 2 — Influence de la longueur et du pourcentage de fibres sur la résistance à la compression du béton renforcé.

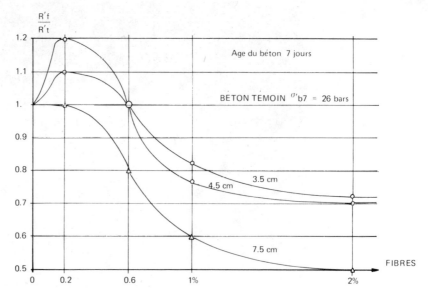

Tableau 1 — Valeurs du rapport de la résistance en compression d'un béton renforcé avec des fibres à celle du béton témoin.

Conservation	Age du béton % fibres longueur	7 jours 0	0.20	0.50	1	2	28 jours 0	0.20	0.50	1	2
sèche	35 mm	1	1.10	1	0.82	0.72	1	1.04	0.98	0.81	0.70
	45 mm		1.20	1	0.76	0.70		1.08	0.92	0.72	0.65
	75 mm		0.99	0.80	0.60	0.50		0.95	0.78	0.64	0.60
humide	35 mm	1	1.10	0.99	0.80	0.73	1	1.06	0.98	0.81	0.71
	45 mm		1.21	1	0.74	0.70		1.13	0.97	0.74	0.65
	75 mm		0.98	0.82	0.57	0.54		0.88	0.79	0.63	0.60

En ce qui concerne la morphologie de la rupture en compression celle-ci est très différente de celle obtenue avec un béton non renforcé (Figure 3). En effet, si l'on soumet, en compression, une éprouvette confectionnée en béton normal, les microfissures réparties initialement de façon aléatoire, se développent d'abord parallèlement à l'effort, alors que les fissures perpendiculaires à celui-ci se referment. Et la rupture intervient, non pas uniquement par compression, mais sous l'action de contraintes secondaires de traction et de cisaillement engendrées par les forces de compression. Dans le cas d'un béton renforcé par des fibres de polypropylène, le réseau de fissures verticales disparaît pour faire place à un écaillement qui est d'autant plus important que le pourcentage de fibres est élevé. La rupture à lieu, de ce fait, par désorganisation du matériau.

A ce sujet, il est intéressant de signaler l'influence de la vibration sur l'orientation des fibres, qui ont tendance à s'aligner suivant un plan perpendiculaire au sens de la vibration. Ainsi si l'on utilise une table vibrante on a, comme on a pu le constater, une orientation préférentielle suivant la perpendiculaire à l'axe longitudinal de l'éprouvette. De ce fait, sous l'application de l'effort de compression, les fibres bloquent le développement des fissures transversales de traction. Une certaine quantité d'énergie est absorbée par les fibres et le reste se libère sous la forme de nouvelles fissures. La généralisation de répartition d'énergie provoque la désorganisation du béton et par suite, son effondrement.

Figure 3 Morphologie de la rupture en compression d'une éprouvette de béton normal et de béton renforcé avec des fibres de polypropylène.

Résistance à la traction flexion

Les essais de traction flexion ont été réalisés sur des éprouvettes 70 × 70 × 280 mm confectionnées suivant la norme NF P 18.401.

Le tableau suivant et le graphique de la Figure 4 (sur lequel on a représenté la variation du rapport de la résistance en traction flexion du béton renforcé à celle du béton témoin en fonction du pourcentage de fibres) regroupent les valeurs obtenues.

Tableau 2 Valeurs du rapport de la résistance en traction flexion d'un béton renforcé avec des fibres à celle du béton témoin

Conservation	Age du béton % fibres longueur	7 jours					28 jours				
		0	0.20	0.50	1	2	0	0.20	0.50	1	2
sèche	35 mm	1	0.99	1.20	0.87	0.58	1	0.97	1.06	1	0.52
	45 mm		1.04	1.23	1.13	0.59		1.06	0.90	0.91	0.49
	75 mm		0.96	1.28	1.02	—		1	1.04	0.96	—
humide	35 mm	1	1.12	1.05	0.80	0.50	1	0.95	0.94	0.83	0.40
	45 mm		1.12	1.10	0.84	0.45		0.86	0.91	0.68	0.42
	75 mm		0.98	1.04	0.80	—		0.93	0.90	0.67	—

Les essais ont montré que la présence de fibres de polypropylène ne retarde que de très peu l'apparition de la première fissure. Ce qui est normal étant donné les valeurs relativement basses du module et du pourcentage de fibres.

En effet, si l'on soumet en traction un béton renforcé par des fibres supposées orientées suivant l'effort N, et si l'on désigne par:

A_b : la section transversale de béton,
E_b : le module de Young du béton, ($E_b \simeq 300{,}000$ bars),
N_b : la fraction de l'effort de traction N passant dans le béton,
A_f : la section transversale des fibres,
E_f : le module des fibres, ($E_f \simeq 90{.}000$ bars à $140{.}000$ bars),
N_f : la fraction de l'effort de traction N passant dans les fibres,

et par :

$m \bar{\omega}_o$: le rapport $m \bar{\omega}_o = \dfrac{A_f}{A_b} \cdot \dfrac{E_f}{E_b}$.

On a, en écrivant que l'allongement δ est le même pour les deux matériaux en présence, ce qui suppose une adhérence parfaite :

$$\delta = \frac{N_b}{E_b} \frac{1}{A_b} = \frac{N_f}{E_f} \frac{1}{A_f}.$$

Et les fractions de l'effort N passant respectivement dans les fibres (N_f) et dans le béton (N_b) sont données avant fissuration par les expressions suivantes:

$$N_f = \frac{\bar{\omega}_o m}{1 + \bar{\omega}_o m} N \qquad \text{avec} \quad N = N_b + N_f$$

$$N_b = \frac{1}{1 + \bar{\omega}_o m} N$$

d'où

$$N_f = \bar{\omega}_o m N_b$$

pour $\bar{\omega}_o = 2.4\%$, $m = \dfrac{100{,}000}{300{,}000} = 0.33$, $N_b = 40$ bars

$$N_f = 0.32 \text{ bars.}$$

On voit, par conséquent, que N_f est très inférieur à N_b et que les fibres ne travaillent, de ce fait, pratiquement pas avant la fissuration du béton.

Fonctionnement d'une éprouvette réalisée en béton renforcé avec des fibres de polypropylène soumise à la flexion

Le fonctionnement mécanique d'une éprouvette de béton de fibres de polypropylène peut être schématisé de la façon suivante:

— Avant la première fissuration, les fibres n'interviennent que très peu.

— Lorsque le dépassement de la contrainte de traction se manifeste dans le béton par l'apparition d'une première fissure, l'élément accuse une déformation et l'effort est repris par la fibre seule. A cet instant, deux cas peuvent se produire:

— (i) Ou bien les fibres sont en pourcentage extrêmement faible et les fibres cassent les unes après les autres,

— (ii) Ou bien les fibres sont en pourcentage suffisant et reprennent les efforts de traction. Dans ce cas, les efforts de traction qui régnaient dans le béton avant fissuration sont transmis par le réseau de fibres, mettant en jeu des efforts de liaison importants entre la fibre et le béton.

Si l'adhérence fibres-béton est suffisante, ce qui a pu être constaté au cours des essais, la contrainte dans le béton s'anule au bord de la fissure et augmente à mesure que l'on s'en éloigne, jusqu'à une certaine distance où le dépassement de la contrainte de traction du béton se traduit par une nouvelle fissure, et ainsi de suite.

Ainsi, si dans l'exemple précédent on a n fibres de longueur ℓ et de surface

Figure 4   **Influence de la longueur et du pourcentage de fibres sur la résistance à la traction flexion du béton renforcé.**

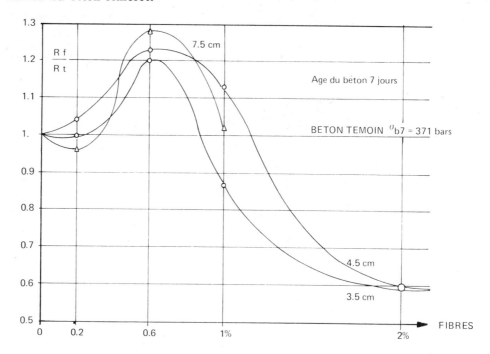

périmétrale s_p orientées dans le bon sens, la force de liaison F entre les fibres et le béton sur une longueur ℓ est donnée par l'expression suivante:

$$F = \ell \, n \, s_p \, \tau$$

en désignant par τ la contrainte d'adhérence entre les fibres et le béton, supposée constante le long de la fibre.

Cette force de liaison F est égale à la force de traction pouvant être véhiculée dans la section de béton A_b et, par conséquent, une seconde fissure intervient à un distance ℓ' pour laquelle la résistance à la traction (σ_b) du béton est atteinte,

soit $\qquad \ell' \, n \, s_p \, \tau = A_b \, \sigma_b$

d'où $\qquad \ell' = \dfrac{A_b \, \sigma_b}{n \, s_p \, \tau}$.

Par conséquent, les fissures sont d'autant mieux réparties, que:
— le produit $n \, s_p$ est grand. On a intérêt à avoir une grande surface périmétrale et, par conséquent, à obtenir un degré de défibrillisation élevé,
— l'adhérence est efficace. La taille maximale des granulats doit être réduite et la compacité particulièrement étudiée.

Cependant, comme on l'a dit précédemment, la répartition de la fissuration ne peut être réalisée qu'à la condition que l'on n'ait pas rupture simultanée des fibres et du béton, aussi l'on peut se demander à partir de quelle valeur du pourcentage la répartition est possible. On peut en avoir une idée approximative en écrivant que la force de traction $(A_b \, \sigma_b)$ amenant la rupture du béton doit être reprise par les fibres $(A_f \, \sigma_f)$,

soit $\qquad A_f \, \sigma_f = A_b \, \sigma_b$

d'où $\qquad \omega = \dfrac{A_f}{A_b} = \dfrac{\sigma_b}{\sigma_f} \simeq \dfrac{40}{700} \simeq 0.6\%.$

Figure 5   **Comportement en flexion d'une pièce prismatique renforcée avec des fibres de polypropylène.**

Par conséquent, en première approximation, le rapport de la section transversale des fibres à celle du béton doit être égal au minimum à 0.6%. Il est à remarquer, par ailleurs, que cette valeur est d'autant plus importante que la résistance du béton à la traction est élevée.

La figure 5 montre le comportement d'une petite poutre de béton renforcé avec des fibres de polypropylène soumise en flexion.

Module d'élasticité

Si l'on se reporte au tableau suivant (Tableau 3) qui donne les valeurs du module d'élasticité obtenues pour les différents bétons étudiés, on constate que les valeurs obtenues diminuent lorsque la longueur des fibres et leur pourcentage augmentent.

Tableau 3   **Valeurs du module d'élasticité en compression à 28 jours (en bars), calculées pour une contrainte appliquée de 100 bars.**

Conservation	% fibres longueur	0	0.2	0.6	1
humide	35 mm	284,000	256,000	236,000	152,000
	45 mm		249,000	240,000	139,000
	75 mm		213,000	204,000	114,000*
sèche	35 mm	260,000	244,000	185,000	116,000
	45 mm		210,000	179,000	102,000
	75 mm		213,000	182,000	94,000*

*Contrainte appliquée de 50 bars environ.

CONCLUSIONS

La diminution de la valeur des résistances en compression et en traction pour les forts pourcentages de fibres peut s'expliquer par une mauvaise homogénéité du béton et par une valeur du rapport E/C élevée. L'une comme l'autre sont inhérentes à la difficulté de mise en place du béton qui nécessite une quantité d'eau assez importante. Pour pallier cet inconvénient, il pourrait être envisagé de procéder à un essorage du béton. Par ailleurs, il semble que l'on ait intérêt à utiliser des granulats de plus faible dimension et à s'orienter vers des mortiers.

L'intérêt des bétons de fibres de polypropylène est principalement dû à leur grande déformabilité et à la possibilité qu'on a de la "doser". Aussi, de tels bétons peuvent certainement être utilisés avec profit dans l'industrie pour renforcer des éléments susceptibles de recevoir des chocs à un moment quelconque de leur existence.

5.3 Strength and deformation properties of concrete reinforced with randomly spaced steel and basalt fibres

Karol Komloš
Institute of Construction and Architecture of the Slovak Academy of Sciences, Bratislava, Czechoslovakia

Summary

The paper deals with the influence of steel and basalt fibre reinforcement on the compressive, flexural, splitting-tensile and direct-tensile strength of concrete, as well as on the stress-strain relationship of concrete in uniaxial tension. As fibre reinforcement two kinds of steel fibres, straight and wavy, having an aspect ratio of 100, as well as basalt fibres were applied. The volume percentage of the reinforcement in the case of steel fibres was 0.25; 0.50; 0.75 and 1.0, and in the case of basalt fibres 0.1; 0.5 and 1.0. The investigations have shown that the highest percentage of steel fibres causes in concretes of different composition an increase in compressive strength of 8 to 37 per cent, in flexural strength an increase of 19 to 42 per cent, and an increase in splitting and direct tensile strength varying within the range of 13 to 53 and 12 to 50 per cent, respectively. The highest content of basalt fibres caused the following increase of compressive, flexural, splitting tensile and direct tensile strength: 11 to 20 per cent, 7 to 22 per cent, 7 to 20 per cent and 17 to 31 per cent, respectively. Compared with plain concrete, concretes having the highest content of steel and basalt fibres showed an increase in the ultimate tensile strain of 44 and 15 per cent, respectively.

Résumé

Le présent article s'occupe de l'influence de l'armature en fibres d'acier et de basalte sur la résistance à la compression et à la flexion aussi bien que sur la résistance établie à l'aide des essais brésiliens. L'auteur analyse egalement la résistance à la tension directe et la relation déformation—contraintes du béton soumis à la tension uniaxiale. L'auteur a utilisé comme armature deux sortes de fibres en acier, droit et ondulé, dont le rapport longueur: diamètre était 100, aussi bien que des fibres en basalte. Le pourcentage par volume d'armatures en cas de fibre en acier s'elevait à 0.25; 0.50; 0.75 et 1.0 tandis qu'en cas de fibres en basalte à 0.1; 0.5 et 1.0. Les recherches poursuivies on prouvé que le pourcentage le plus haut des fibres d'acier produit dans les bétons de composition variée une augmentation de la résistance à la compression de 8 à 37 pour cent, une augmentation de la résistance à la flexion de 19 à 42 pour cent et une augmentation de la résistance à la tension établie à l'aide d'essais brésiliens et de la résistance à la tension directe variant entre 13 à 53 pour cent et de 12 à 50 pour cent. Le contenu de fibres de basalte a causé les hausses suivantes de résistances à la compression, à la flexion, à la tension indirecte et à la tension directe: 11 à 20 pour cent, 7 à 22 pour cent, 7 à 20 pour cent et 17 à 31 pour cent. Les bétons au contenu le plus haut de fibres d'acier et de basalte ont montré une augmentation de la limite de déformation par tension de 44 à 15 pour cent en comparaison avec les bétons ordinaires.

INTRODUCTION

In recent years a large number of papers dealing with the investigation of fibre cements and fibre concretes have been published. Also the mechanism through which fibres influence the physical and mechanical properties of these composite materials has been studied (1,2,3,4,5,6,7). These investigations made possible the development of fibre composites based on cement and their introduction into practice. However, there are still some theoretical and technological problems to solve in order to widen the present uses and, further, to make these composite materials attractive for other fields as well. The experimental work carried out was to establish the influence of the type of fibres and their geometry on strength and strain properties of fibre concretes.

MATERIALS AND MIXES

A grade 400 Portland cement—according to Czechoslovak Standard ČSN 722117—was used. Three cement/aggregate ratios by weight equal to 1:7.5, 1:5 and 1:4 were used. Natural Danube river sand and gravel were used with the following single particle size distribution: 0-1 mm = 19%, 1-2 mm = 12%, 2-4 mm = 16%, 4-7 mm = 15%, 7-15 mm = 38%. Details of the mixes are given in Table 1.

Table 1   **Concrete matrix compositions**

Mix No	Cement/aggregate ratio	w/c	Vebe time (seconds)
1	1:7.5	0.58	20
2	1:7.5	0.63	8
3	1:7.5	0.70	5
4	1:5	0.36	58
5	1:5	0.44	25
6	1:5	0.52	10
7	1:4	0.32	28
8	1:4	0.40	10
9	1:4	0.48	2

Three types of fibres were employed, two of which were of steel with a circular cross section and one a basalt fibre. One of the steel fibres consisted of straight wires with $d = 0.30$ mm, $\ell_f = 30$ mm, and $f_{st} = 1860$ N/mm²; the other was wavy wires with $d = 0.20$ mm, $\ell_f = 20$ mm, and $f_{st} = 2450$ N/mm². The geometry of the applied basalt fibres was: $d = 17 - 25 \mu$m, $\ell_f \sim 10$ mm. The volumetric percentage of the fibre-reinforcement in the case of steel fibres was 0.25; 0.50; 0.75 and 1.0, and in the case of basalt fibres 0.1; 0.5 and 1.0.

All the concretes were mixed in a pan-mixer. In addition to the fibre concrete mixes, a control mix without fibres was also prepared. A total of 126 mixes were tested.

PREPARATION OF TEST SPECIMENS

Cubes were cast in 200 mm steel moulds. Wooden moulds, the inner surfaces of which were coated with a film of hard plastic material, were used for casting specimens for direct tensile testing. This specimen is 560 mm in length and has a central portion 200 mm long and 100 mm square cross-section. Steel cylinders, 150 mm in diameter and 300 mm high, were used for the specimens for the indirect-cylinder-splitting-tensile strength tests, and prisms cast in 100 x 100 x 400 mm steel moulds were used for the flexural strength test. The concrete was compacted on a vibrating table at 50 Hz and

Figure 1 Strength ratio versus V_f relationship (steel fibres $\ell_f = 30$ mm, d = 0.30 mm, $\ell_f/d = 100$).

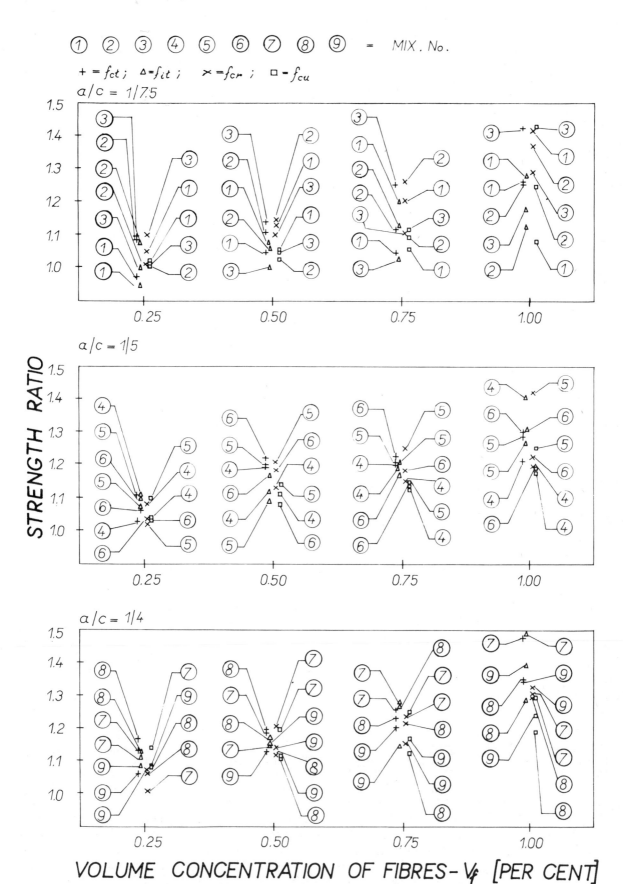

Figure 2   Strength ratio versus V_f relationship (steel fibres $\ell_f = 20$ mm, $d = 0.20$ mm, $\ell_f/d = 100$).

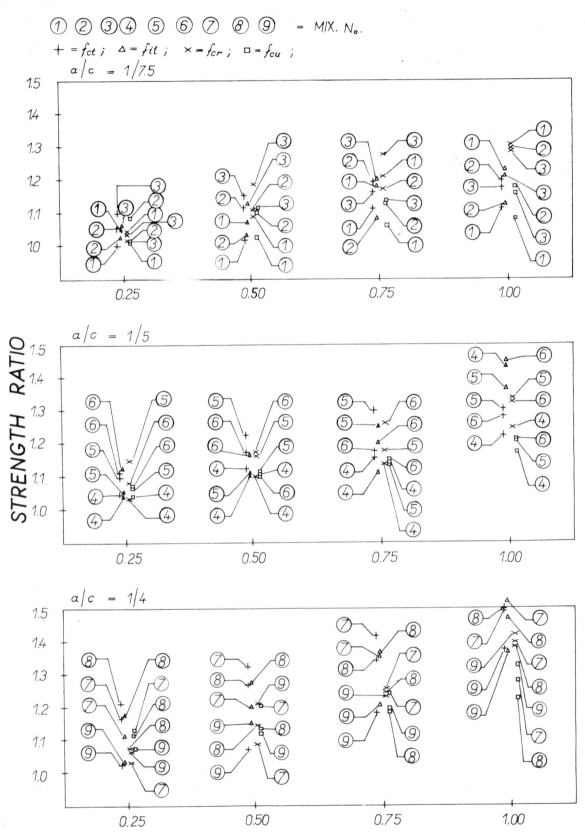

Figure 3 Strength ratio versus V_f relationship (basalt fibres $\ell_f = 10$ mm, $d = 17\text{-}25\,\mu$m).

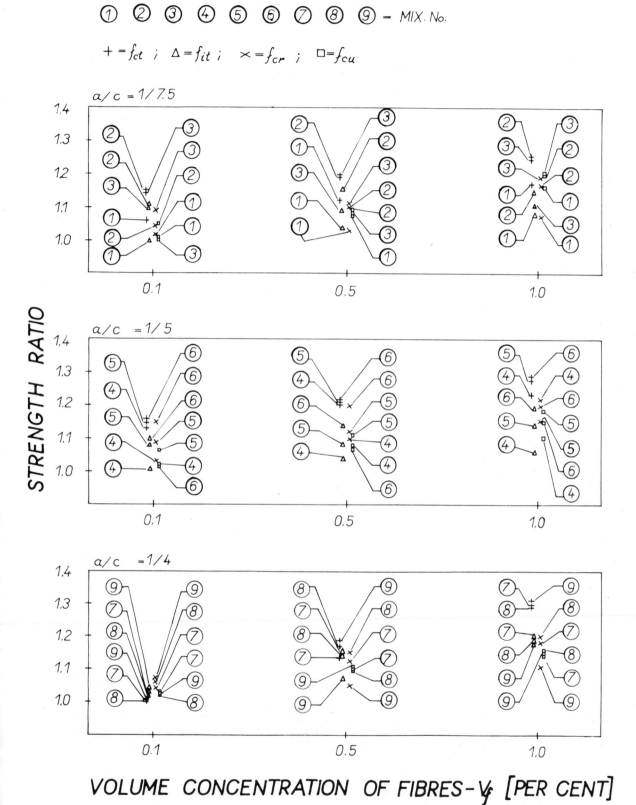

0.35 mm amplitude the vibration time being dependent upon the composition and the consistence of the mix. Three specimens of each type were cast and were stored in moulds for 48 hours in a curing chamber at 20 °C and 100 per cent relative humidity. The moulds were then removed and the specimens were stored at the same curing conditions until testing at 28 days.

TESTING PROCEDURES

After the appropriate curing period, the cubes were tested in a 300 tonne machine. The cylinders were tested for splitting tensile strength, 3 x 12 mm plywood strips being used as packing material. The beams were tested for flexural strength, thirdpoint loading being used. Self-centering clamping jaws were applied for testing specimens in direct tension.

The investigation of the strain properties of concrete in direct tensions was carried out with the aid of strain-gauges. Each specimen tested in this way was provided with eight strain-gauges which had been stuck onto the specimens in the direction of their longitudinal axes. In the same way eight strain-gauges had been stuck onto the comparative specimen. The strain-gauges were coupled in such a way that a pair of strain-gauges stuck on the tested specimen and a pair of strain-gauges stuck on the comparative specimen were coupled together. In this way the accuracy of the measured strains was doubled. The strains were measured with the aid of a tensometric apparatus. The records were graphically recorded by a moving-coil oscillograph.

TEST RESULTS

The test results obtained are summarized in Table 2 and Figs 1, 2, 3, and 4. Figs 1, 2 and 3 show the relationship between strength ratio and the volume concentration of fibres for steel fibres d = 0.30 mm (Fig 1), steel fibres d = 0.20 mm (Fig 2), and basalt fibres (Fig 3), as well as for all nine concrete compositions and strength characteristics ie direct tensile, splitting tensile, flexural and cube strength. The strength ratio is the ratio of strength of the fibre concrete to strength of the concrete matrix. Each value plotted in the figures represents the mean value of three tests. The investigations have shown that the highest percentage of steel fibres, 1.0 per cent, causes in concretes of different composition an increase in cube strength of 8 to 37 per cent, in flexural strength an increase of 19 to 42 per cent, and the increase in splitting and direct tensile strength varied within the range of 13 to 53 and 12 to 50 per cent, respectively. The highest content of basalt fibres, 1.0 per cent, caused the following increase of cube, flexural, splitting tensile and direct tensile strength: 11 to 20 per cent, 7 to 22 per cent, 7 to 20 per cent and 17 to 31 per cent, respectively. The results of the investigation of the influence of fibre reinforcement on the stress-strain relationship of concrete in direct tension are given in Table 2 and Fig 4. Concretes having the highest content of steel and basalt fibres, 1.0 per cent, showed, compared with the concrete matrix, an increase in the limited tension strain of 44 and 15 per cent, respectively.

CONCLUSIONS

The investigations have shown that the rate of influence of both kinds of steel fibres on concrete strength and strain characteristics was very much the same though in one case straight and in the other case wavy fibres were used. These tests have confirmed the results obtained by several other authors that one of the main parameters influencing the mechanical characteristics of fibre concretes is the aspect ratio of the fibre (ℓ_f/d), which was the same (100) in both cases.

The investigations have further shown that the efficiency of randomly spaced wire reinforcement increases with the increasing cement content of the concrete.

The results indicate clearly that the tension test characteristics are more affected by the presence of the randomly spaced reinforcement than the cube strength of concrete. This phenomenon could be strongly influenced by the kind of concrete compaction which affects the fibre orientation as described by Edgington and Hannant (8).

Table 2   Stress-strain relationship of fibre concrete in uniaxial tension

Kind of fibres	V_f Per cent		σ_{ct} (N/mm²) – ϵ_{ct} (10^{-3}) relation					
	0	σ_{ct}	1	2	3			
		f_{ct}				3.1		
		ϵ_{ct}	0.031	0.0615	0.0985	0.104		
Steel	0.25	σ_{ct}	1	2	3			
		f_{ct}				3.2		
		ϵ_{ct}	0.03	0.06	0.099	0.112		
ℓ_f = 30mm	0.50	σ_{ct}	1	2	3		3.52	
		f_{ct}				3.6		
d = 0.30mm		ϵ_{ct}	0.0296	0.059	0.0897	0.127	0.1314	
	0.75	σ_{ct}	1	2	3	4		4.01
		f_{ct}					4.01	
		ϵ_{ct}	0.029	0.057	0.085	0.13	0.135	0.138
	1.00	σ_{ct}	1	2	3	4		4.24
		f_{ct}					4.3	
		ϵ_{ct}	0.0275	0.0545	0.0815	0.1135	0.137	0.148
	0	σ_{ct}	1	2				
		f_{ct}			3			
		ϵ_{ct}	0.0315	0.0645	0.207			
Steel	0.25	σ_{ct}	1	2	3			
		f_{ct}				3.1		
		ϵ_{ct}	0.03	0.063	0.1015	0.109		
ℓ_f = 20mm	0.50	σ_{ct}	1	2	3		3.72	
		f_{ct}				3.74		
d = 0.20mm		ϵ_{ct}	0.03	0.0605	0.0915	0.127	1.32	
	0.75	σ_{ct}	1	2	3	4		4.15
		f_{ct}					4.17	
		ϵ_{ct}	0.0285	0.056	0.0845	0.12	0.137	0.144
	1.00	σ_{ct}	1	2	3	4		4.38
		f_{ct}					4.42	
		ϵ_{ct}	0.0265	0.0524	0.0776	0.1094	0.142	0.147
	0	σ_{ct}	1	2	3			
		f_{ct}				3.04		
		ϵ_{ct}	0.0315	0.062	0.107	0.113		
Basalt	0.10	σ_{ct}	1	2	3			
		f_{ct}				3.11		
		ϵ_{ct}	0.0315	0.062	0.104	0.113		
ℓ_f = ~10mm	0.50	σ_{ct}	1	2	3			
		f_{ct}				3.35		
d = 17-25 μmm		ϵ_{ct}	0.0315	0.062	0.099	0.117		
	1.00	σ_{ct}	1	2	3		3.43	
		f_{ct}				3.43		
		ϵ_{ct}	0.0315	0.062	0.099	0.127	0.129	

Figure 4   **Relationship between relative stress and strain increments of fibre concretes in uniaxial tension.**

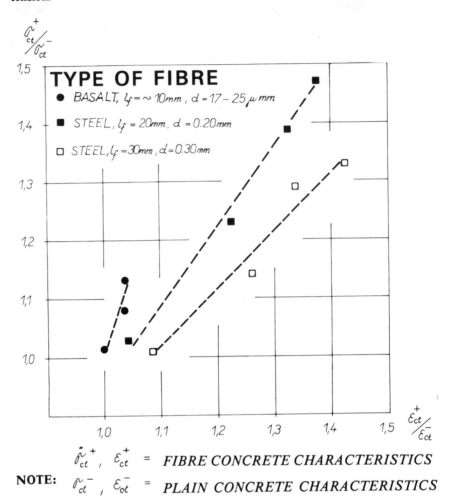

NOTE: $\tilde{\sigma}_{ct}^{+}, \varepsilon_{ct}^{+}$ = FIBRE CONCRETE CHARACTERISTICS
$\sigma_{ct}^{-}, \varepsilon_{ct}^{-}$ = PLAIN CONCRETE CHARACTERISTICS

The studies made have shown that the concrete reinforced with randomly dispersed basalt fibres show a smaller increment of the direct tension limit strain in comparison with plain concrete than concretes reinforced with randomly dispersed steel fibres. It has further been found that there was no notable difference in the stress-strain curve of concrete reinforced with both types of steel reinforcement. The results indicate further that the steel fibre reinforcement causes an increase of the elastic modulus of the composite. On the other hand, the basalt fibre reinforcement does not seem to influence the elastic modulus of the composite.

It has further been found that there exists a relationship between relative stress and strain increments of concrete reinforced with randomly dispersed fibres. The above mentioned relationship is represented in Fig 4. The presented plot (Fig 4) in which σ_{ct}^{-} and ε_{ct}^{-} are the stress and strain values, respectively, of the concrete matrix and σ_{ct}^{+} and ε_{ct}^{+} the stress and strain values of fibre concretes, shows that the relationship between the relative stress and strain increment of concrete reinforced with individual fibre types is practically linear.

REFERENCES

1 Romualdi, J P and Batson, G B, Mechanics of crack arrest in concrete. Journal of the Engineering Mechanics Division, ASCE, Vol 89, No EM 3, June 1963, pp 147-168.

2 Romualdi, J P, The static cracking stress and fatigue strength of concrete reinforced with short pieces of thin steel wire. International Conference on the Structure of Concrete, London, 1965, paper D2.

3　Shah, S P and Rangan, B V, Fibre reinforced concrete properties. ACI Journal, Proceedings Vol 68, No 2, February 1971, pp 126-135.

4　Aveston, J and Kelly, A, Theory of multiple fracture of fibrous composites. Journal of Materials Science, Vol 8, 1973, pp 352-362.

5　Majumdar, A J, The role of the interface in glass fibre reinforced cement. Cement and Concrete Research, Vol 4, No 2, March 1974, pp 247-266.

6　Swamy, R N and Mangat, P S, A theory for the flexural strength of steel fibre reinforced concrete. Cement and Concrete Research, Vol 4, No 2, March 1974, pp 313-325.

7　Swamy, R N and Mangat, P S, Influence of fibre geometry on the properties of steel fibre reinforced concrete. Cement and Concrete Research, Vol 4, No 3, May 1974, pp 451-465.

8　Edgington, J E and Hannant, D J, Steel fibre reinforced concrete. The effect on fibre orientation of compaction by vibration. RILEM, Matériaux et Constructions, Vol 5, No 25, 1972, pp 41-44.

9　Komloš, K, Sklenné a čadičové vlákná ako výstuž betónov. Stavivo, Vol 51, No 2, February 1973, pp 51-56.

10　Komloš, K, Náhodne rozptýlené ocel'ové vlákno ako výstuz betónov. Stavivo, Vol 51, No 3, March 1973, pp 59-65.

5.4 The effects of fibre reinforcements on lightweight aggregate concrete

A G B Ritchie and O A Al-Kayyali
Civil Engineering Department, University of Strathclyde, Glasgow, United Kingdom

Summary

This paper details an investigation into the influence of two types of fibre, namely polypropylene and wire on the rheological and structural properties of lightweight concrete made from three contrasting forms of lightweight aggregate, foamed slag, Lytag and Leca. These materials were also compared with a natural gravel aggregate. This provided a comprehensive cover of surface texture, porosity and density.

The proportions used for the individual mixes were based generally on the manufacturers' recommendations confirmed by laboratory testing with the initial water content being determined on the basis of observed medium workability.

This gave a nucleus of acceptable mixes related to the characteristics of the aggregate form and grading. The variation in the water content and aggregate particle strength gave differing characteristic strengths but this did not present any real problem since the basic mix strength was used as a datum from which to compare the effects of fibre reinforcement on each lightweight concrete in turn. The percentage of each fibre added was increased until a stage was reached where compaction became very difficult or balling occurred.

The effect of the fibre inclusions on the flow characteristics within each aggregate type and the comparison between the relative behaviour of all the different aggregate forms was investigated initially using the standard workability tests of slump, Vebe, and compacting factor. The limitations of these tests when applied to the combination of lightweight concrete stiffened by the addition of fibres was clearly indicated. The application of the vane test to these conditions was illustrated and the relative stiffening effect of the fibre type and content as influenced by the characteristics of the aggregate particles was investigated.

The corresponding range of hardened concrete embodying the above variables was then studied and compared structurally using as a basis, tests for crushing strength, indirect tensile strength, modulus of rupture, modulus of elasticity and impact resistance.

The investigation illustrated that there appears to be no real constructional problems associated with fibre reinforced lightweight aggregate concrete and that significant improvements can be made thereby to specific structural properties particularly if a judicial choice is made of type of fibre and form of lightweight concrete.

Résumé

    Cette communication décrit une étude sur l'influence de deux types de fibres, notamment le polypropylène et le fil d'acier, sur les caractéristiques rhéologiques et structurelles d'un béton léger fait avec trois granulats légers contrastants: le laitier mousse, le Lytag et le Leca. Ces matériaux ont été comparés également avec un gravillon naturel. Cette combinaison a donné une gamme complète de textures de surface, de porosités et de densités.

    Les proportions utilisées pour les différents mélanges étaient généralement basées sur les recommandations des fabricants confirmées par des essais de laboratoire, la teneur en eau initiale étant déterminée sur la base de l'ouvrabilité moyenne constatée.

    Cela a donné un nombre de mélanges acceptables et leur rapport aux caractéristiques de granulométrie et de forme des granulats. La variation de la teneur en eau et de la résistance des grains a donné des résistances caractéristiques différentes, mais cela ne présentait aucun problème réel puisque celle du mélange de base était utilisée comme donnée de base pour comparer l'influence de l'armature de fibre pour chacun des bétons légers.

    On a augmenté le pourcentage de chaque addition de fibre jusqu'à ce qu'on ait atteint une phase où le damage devenait très difficile où des grumeaux se formaient.

    L'effet des inclusions de fibres sur les caractéristiques d'écoulement pour chaque type de granulat et la comparaison entre le comportement ont été étudiés d'abord au moyen des essais traditionnels d'étalement, Vebe et facteur de damage. On a constaté nettement les limites de ces essais quand on les applique à un béton léger raidi par une ajoute de fibres. On a démontré l'essai à palettes appliqué dans ces conditions et on a étudié l'effet du type de fibre et de la teneur en fibres par rapport à l'influence des caractéristiques des granulats.

    On a alors étudié la gamme correspondante de béton durcis représentant toutes ces variables en comparant la résistance à l'écrasement, la résistance à la traction indirecte, le module d'élasticité et de rupture et la résistance au choc. Cette étude a montré qu'il ne semble pas y avoir de vrais problèmes structurels associés à l'emploi de béton de granulats légers armé de fibres et qu'on peut obtenir une amélioration significative de certaines caractéristiques spécifiques de la structure particulièrement en faisant un choix judicieux du genre de fibre et de béton léger.

INTRODUCTION

A considerable amount of research has now been completed on fibre reinforced concrete using natural aggregates but very little work has been recorded on the effect of fibre reinforcement on lightweight aggregate concrete.

The theory and construction techniques of fibre reinforcement are well documented as are the characteristics of lightweight aggregate with respect to its constructional and structural properties.

The authors considered it would be interesting and perhaps valuable to try to combine these two fields of activity and investigate the rheological and structural characteristics of fibre reinforced lightweight aggregate concrete mixes. The concept of reinforcing lightweight concrete with fibres has not yet received much research attention. The subject has been touched on by Hannant (1) with reference to steel fibres and steel reinforced lightweight aggregate beams and also by Gunasekaran and Ichikawa (2) with reference to steel fibres and lightweight concrete made with regulated set cement and sintered fly ash aggregates.

In this present investigation a range of structural properties was examined with respect to concrete test specimens made using three types of contrasting lightweight aggregates in conjunction with two types of fibre.

MATERIALS

The basic materials used were blended ordinary Portland cement and contrasting types of lightweight aggregate, namely foamed slag, sintered pulverised fuel ash and expanded clay. A natural aggregate was also used as a general basis for comparison. The reinforcing fibres were polypropylene and steel.

Aggregates

(a) Foamed Slag (Foamag). Was a crushed angular and porous material produced by treating molten blastfurnace slag with water. The slag was used in two sizes; medium (12 mm down) and fine (5 mm down) with loose bulk densities of 725 and 960 kg/m^3 respectively.

(b) Sintered Pulverised Fuel Ash (Lytag). Was a very rounded aggregate manufactured by pelletising and sintering pulverised fuel ash. The material was used in three sizes; 12 mm max, 6 mm max and fine having loose bulk densities of 872, 874 and 976 kg/m^3 respectively.

(c) Expanded Clay (Leca). Was an irregularly rounded material produced in a rotary kiln. It has a dense skin and a honeycombed interior. The aggregates were supplied in three grades; coarse, medium and fine with loose bulk densities of 450, 450 and 570 kg/m^3 respectively.

(d) Natural Aggregate (Mid Ross Gravel). This was an irregular material with a water worn surface texture and was combined to give a Type No 2 overall grading (3).

Fibres

(i) Polypropylene. These fibres were of 1/700 type (one pound per 700 yards, or 1 kg per 1420 m), split film single strand form. They had a cross sectional area of 0.9 mm^2, a length of 35 mm and were chopped manually in the laboratory.

(ii) Wire. These were selected from those commercially available and were 0.38 mm dia x 25 mm long (Duoform Type).

MIXES

Foamag

A series of preliminary tests was carried out to determine the best proportions of slag to cement and the percentage of fines to suit the twin demands of workability and strength. The final selection made, bearing in mind the intended incorporation of fibres, was an aggregate/cement ratio of 2.0 by weight with a fines content of 50% by volume of the total aggregate and a water/cement ratio of 0.70.

Lytag In a similar manner an equivalent basic mix using Lytag aggregate was finally adopted as an A/C of 2.87 with 38% fines content and a W/C of 0.725.

Leca A comparable basic mix using Leca was adopted as an A/C of 1.42 with 50% fines content and a W/C of 0.55.

Natural aggregate A final basis for comparison of aggregate characteristics was provided by using a standard aggregate with an A/C of 4.8 and a W/C of 0.50.

Fibre inclusions: To each of the basic mixes fibre reinforcements were added without any increase in water content until an amount was reached that made compaction very difficult or started balling. With the Foamed Slag and Lytag mixes the range of the investigation was extended by establishing additional sets of mixes with increased W/C ratios.

All the materials were batched by weight and mixed in a pan mixer. The polypropylene fibres were dispersed into the wet mix in the final stages of mixing to prevent unravelling due to friction with the aggregates. The wire fibres were dispersed during the mixing of the dry materials. The samples were compacted on a standard vibrating table and finished off with tamping rod and float.

RHEOLOGICAL CHARACTERISTICS

Standard workability tests The workability of all the basic mixes was established in the category of medium as determined by visual observation and feel rather than by any of the standard tests of BS 1881 (4). This was necessary and desirable because the lightweight mixes did not produce a satisfactory slump and the end point of the Vebe test was not clearly defined. The compacting factor test did however yield a measure of the reduction in compactability as the fibre content was increased relative to the observed workability condition of the mix. The comparison between the results of this test and the observed workability is illustrated in Fig 1. This measure was only comparable within one material since the basic compacting factor for the unreinforced mix changed with the type of aggregate used as shown in Fig 2, all basic mixes having observed medium workability.

The progressive stiffening of a mix cannot be readily monitored using any standard workability test since the mix soon becomes too stiff for the designed range of the instrument and the results cease to be meaningful. In order to try to measure and compare stiffening more precisely it was decided to monitor the development of the internal resistance of the fibre reinforced concrete by using the vane test.

Figure 1 **Comparison between the standard workability tests and the observed condition of the mix with increasing percentages of fibre.**

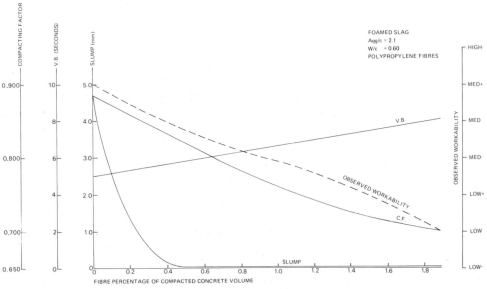

Figure 2   **Comparison of the results of the compacting factor test for different types of aggregate and with increasing content of polypropylene fibres.**

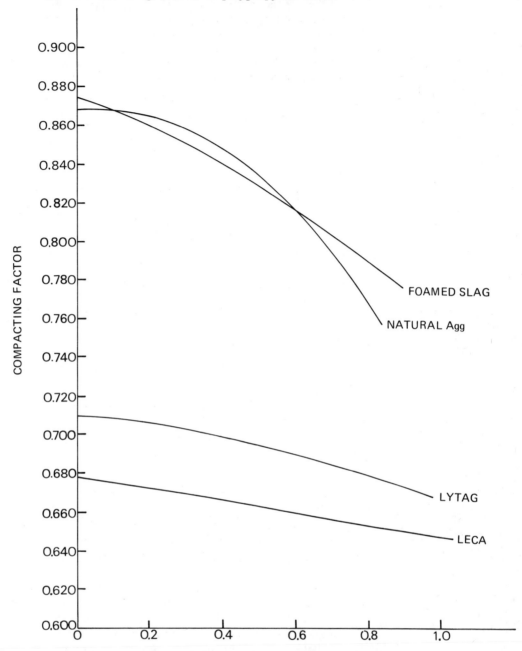

The vane test

The apparatus and test procedure have been fully detailed previously (5) and will not be repeated here. A typical set of results is shown in Fig 3 which includes the same range of polypropylene mixes as used in Fig 2. This illustrates how the vane test can be used to compare the relative workability of all the basic mixes taken together and also provides an indication of the relative stiffening of each as the fibre content is increased. This test was also used to show the relative rate of increase of internal resistance with time.

STRUCTURAL STRENGTH

This was investigated in terms of the following properties:

(a) Crushing Strength

(b) Indirect Tensile Strength

(c) Flexural Strength

(d) Modulus of Elasticity

(e) Impact Resistance

Figure 3   **Comparison of internal resistance as measured by the vane test for different types of aggregate and with increasing fibre content.**

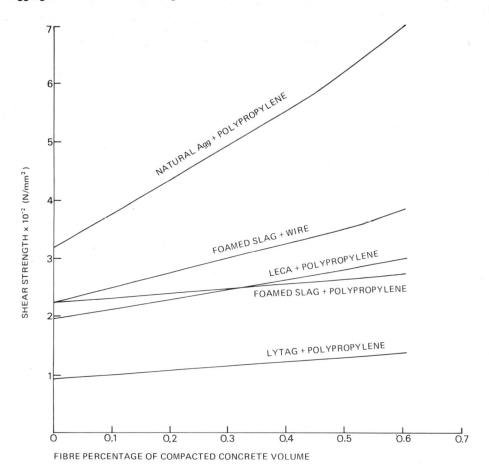

These tests were systematically carried out on the unreinforced units and then on samples with increasing percentages of fibres. Since the basic set of mixes using different types of aggregate had as its common factor the observed workability it followed that each would have a different water/cement ratio and particle strength and hence a different unreinforced characteristic strength. Because of this it was decided to present the various physical characteristics and the effect thereon of fibre reinforcement together in the one figure for each type of aggregate. This would then enable the pattern of the various strengths to be observed and compared with the trends exhibited by a contrasting mix. The results obtained for Foamed Slag, Lytag, Leca and Mid Ross aggregates combined with various percentages of polypropylene and wire fibres are shown in Figs 4-8 for tests made at 28 days. The related 7 day strength results in each case confirmed the pattern of the final results. Observations on the individual tests are as follows:-

Crushing strength

All the unreinforced samples exhibited spalling at failure while the fibre reinforced units held together well and complete crushing did not occur until after the load platen was made to follow through. It can be observed from the figures that the effect of fibre inclusion on crushing strength is different for different types of aggregates and fibres. Generally speaking the addition of polypropylene fibres had no beneficial effect on

Figure 4 Summary of the results of various strength tests at 28 days for foamed slag–polypropylene mixes.

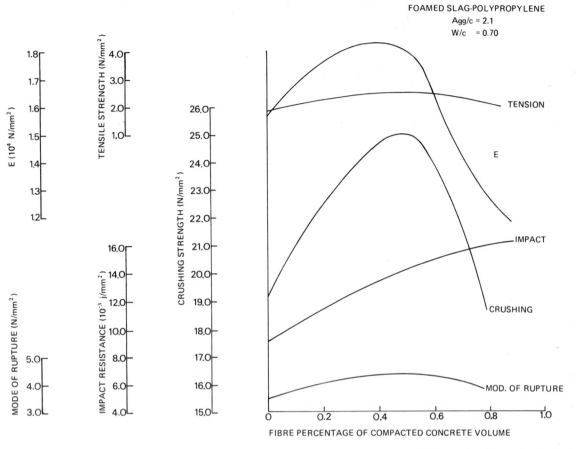

Figure 5 Summary of the results of various strength tests at 28 days for foamed slag–wire mixes.

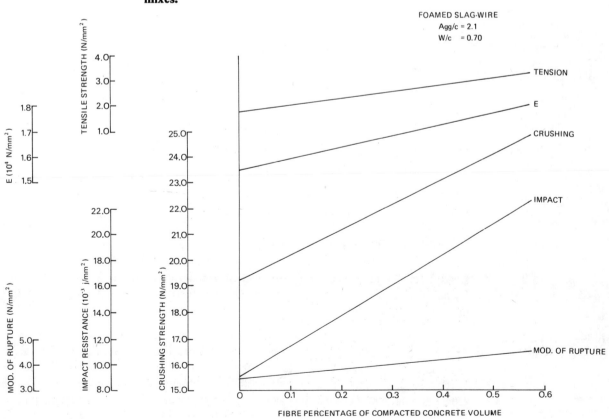

Figure 6   Summary of the results of various strength tests at 28 days for Lytag–polypropylene mixes.

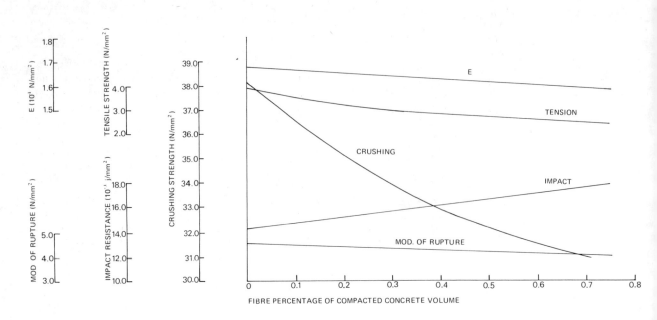

Figure 7   Summary of the results of various strength tests at 28 days for Lytag–wire mixes.

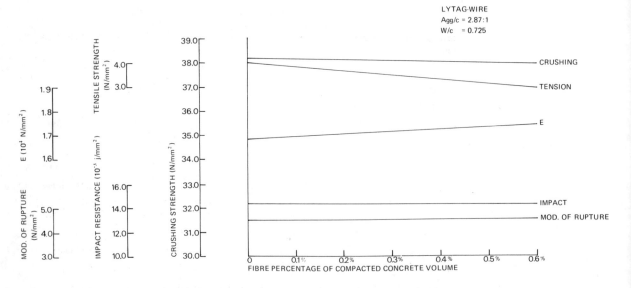

crushing strength. With the Foamed Slag mixes a small percentage of fibre increased the strength but this dropped off again with further additions. Wire fibres produced a steady increase in strength with Foamed Slag concrete but little or no increase with Lytag. The natural aggregate mixes exhibited a decrease with polypropylene and an increase with wire fibres.

Indirect tensile test (cylinder splitting)

The unreinforced cylinders failed suddenly and were split into separate halves. The fibre reinforced specimens merely cracked at failure without any sign of collapse or separation. The pattern of results over the various aggregate types followed a similar pattern to that for the crushing strength. Wire fibres had a significant effect on the tensile strength of Foamed Slag concrete but had little influence on Lytag or Leca mixes.

Flexural strength (modulus of rupture)

Again the same relative pattern of results emerged illustrating that wire fibres in Foamed Slag mixes gave a significant improvement in physical properties.

Modulus of elasticity

This was determined using 150 mm diameter cylinders and Demec gauges. The results indicated that of the three types of lightweight aggregate concrete studied Foamed Slag was most clearly influenced by the effect of fibre reinforcement. The modulus of elasticity was slightly improved with small percentages of polypropylene fibres then dropped sharply with further additions. Wire fibres however steadily improved this property for this form of lightweight concrete.

Impact resistance

The impact resistance of fibre reinforced concrete has been studied by several investigators (6) (7). Hoff (8) has investigated the impact resistance of plain lightweight concrete in relation to its use in shock dissipating construction schemes.

The methods of testing to determine impact resistance of concrete differ from one research worker to another. However this does not affect the relative pattern of results. The method of test used in this project was developed by Mackintosh (9) using a modified Izod test machine. The specimens used were standard beams 100 x 100 x 500 mm. Prior to testing, a notch of 20 mm depth and 3 mm thickness was cut across the face of the specimen at a distance of 125 mm from one end in order to provide a plane of weakness, and locate the test section. The specimen was clamped in position so that the top 60 mm would be struck by the swinging hammer. As can be seen from the results obtained both polypropylene and wire fibres improved the impact resistance of all the lightweight concrete mixes investigated. The improvement was most significant in the case of Foamed Slag where the addition of polypropylene fibres could double the impact resistance and wire fibres tripled it. Therefore despite the doubtful improvement in certain cases with some of the previously examined properties there was a clear indication that the impact resistance always benefited from the use of fibre reinforcement.

CONCLUSION

It would appear from the investigation that fibres can be introduced and successfully incorporated in lightweight aggregate concrete mixes. The recording of workability presents a problem when trying to achieve a test sensitivity that matches the appearance and handability of the mix and can apply to all types of lightweight concrete. The compacting factor has a limited amount of success but can only apply within an individual material.

The vane test does give a useful overall comparison of the relative stiffness of the various mixes and gives the feel of the concrete as it stiffens up with the change of fibre, aggregate or mix characteristics and with increase in time.

As far as structural strength is concerned this is very dependent on the type of aggregate and fibres involved. Generally speaking the pattern of strength development with increasing percentage of fibre for one form of test, say crushing, is closely followed by the other forms of test used, the only exception to this being impact resistance which apart from Lytag and wire fibres always seems to be improved to some extent by the addition of fibres at least over the scope of this investigation.

This was also borne out by the natural aggregate mixes. One important factor which appears to emerge is that not all forms of lightweight aggregate respond to fibre reinforcement and therefore discretion has to be exercised when selecting a suitable mix for a structural unit. This is made clear particularly with the Lytag mixes which appeared to derive no real increase in any of the structural properties including impact even with increasing percentages of wire fibres. On the other hand significant improvements can be achieved with Foamed Slag mixes, extensively with wire fibres and to a limited extent with small percentages of polypropylene.

REFERENCES

1 Hannant, D J, "Steel Fibres and Lightweight Beams", Concrete, Vol 6, No 8, August 1972, pp 39-40.

2. Gunasekaran, M and Ichikawa, Y, "The Strength and Behaviour of Steel Fibre Reinforced Lightweight Concrete made with Regulated Set Cement and Sintered Fly Ash Aggregates", Proceedings, International Symposium on Fibre Reinforced Concrete, Ottawa, October 1973, ACI Publication SP44, 1974, Part 1, Properties of Fibre Reinforced Concrete, pp 113-126.

3. Road Research Laboratory, "Design of Concrete Mixes", Road Note No 4, HMSO.

4. British Standards Institution, "Methods of Testing Concrete", BS 1881, Part 2: 1970.

5. Ritchie, A G B and Rahman, T A, "Effect of Fibre Reinforcement on the Rheological Properties of Concrete Mixes", Proceedings, International Symposium on Fibre Reinforced Concrete, Ottawa, October 1973, ACI Publication SP44, 1974, Part 1, Properties of Fibre Reinforced Concrete, pp 29-44.

6. Green, H, "Impact Strength of Concrete", Proceedings, Institution of Civil Engineers, Vol 28, July 1964, p 383.

7. Hughes, B P and Gregory, R, "The Impact Strength of Concrete using Green's Ballistic Pendulum", Proceedings, Institution of Civil Engineers, Vol 41, December 1968, pp 731-750.

8. Hoff, G C, "New Applications for Low-Density Concretes", ACI Publication SP 29-11, 1971, Lightweight Concrete, pp 181-220.

9. Macintosh, D M, "Certain Aspects of Polypropylene Fibre Reinforcement of Concrete", MSc Thesis, Civil Engineering Department, University of Strathclyde, 1971.

Section 6

Properties and Testing of Asbestos Fibre Cement

6.1 Opening Paper: Properties and testing of asbestos fibre cement

Harald G Klos
Dipl Ing Dr Tech,
Consultant on asbestos cement

Summary

Asbestos cement has been known since 1879. The break-through to large scale technical production and application took place by the invention of the wet process-sieve cylinder machine by Ludwig Hatschek in 1900.

Basically, asbestos cement is made of asbestos fibres, cement and water. Besides this, many variations by replacing parts of the fibres by other fibres and by substituting parts of cement by other binding agents or fillers are known.

Asbestos cement is a product of high corrosion and abrasion resistance. By using cement based on blast-furnace slag its corrosion resistance can be increased considerably.

By using special fillers or agents and/or by applying special curing processes the properties of asbestos cement can be varied in a wide range.

The main products are corrugated sheets, flat boards and slates for roofings and wall coverings, as well as pressure and sewer pipes. Moulded pieces are fabricated as accessories for roofings and pipings and besides that in ample varieties for various applications.

Unlike other artificial stones, and of course in contrast to all plastic materials, asbestos cement consists only of basic natural components. Its aging does not mean decomposition, but on the contrary, an increase of inherent strength. Asbestos cement does not pollute the environment.

Résumé

L'amiante ciment est connu depuis 1879. La production et l'application technique sur grande échelle ont débuté par l'invention du procédé humide avec la machine à cylindres tamis par Ludwig Hatschek en 1900.

En principe l'amiante ciment est fabriqué à partir de fibres d'amiante, de ciment et d'eau. Il y a beaucoup de variations en remplaçant une partie des fibres par d'autres fibres ou en substituant une partie du ciment par d'autres lients ou matières de remplissage.

L'amiante ciment est un produit de haute résistance à la corrosion et à l'abrasion. En utilisant du ciment de haut fourneau on peut améliorer la résistance à la corrosion considérablement.

On peut modifier les caractéristiques de l'amiante ciment à un haut degré en utilisant des matières spéciales de remplissage ou agents et/ou en employant des procédés spéciaux pour faire la prise.

Les produits les plus importants sont les plaques ondulées, les plaques plates, ardoises et tuiles pour le revêtement des parois, et aussi les tuyaux à pression et tuyaux d'assainissement.

Les pieces façonnées sont fabriquées comme accessoires pour toits et pour tuyaux et dans une ample gamme pour des applications diverses.

Contrairement aux autres pierres artificielles et bien entendu a toutes les matières plastiques, l'amiante ciment se ne compose que de matériaux à base naturelle.

Son vieillissement ne signifie pas décomposition, mais au contrairs une augmentation de sa résistance inhérente. L'amiante ciment ne pollue pas l'environnement.

Asbestos cement (1) is an artificial stone consisting of a thorough and homogeneous mixture of asbestos fibres and cement. Whilst Portland cement PZ 275 (German Standard DIN) which is or has been used predominantly as a basic material, has a minimum resistance to compression of 27.5 N/mm² and a minimum bending strength of 5 N/mm², for asbestos cement sheets a bending strength up to 38 N/mm² and for pressure pipes a bending strength and an annular tensile strength of 25 N/mm² are stipulated. These values exceed by far the strength of the basic cement.

Asbestos *Chrysotile* which is used mainly for asbestos cement has a substance strength of 5.6×10^6 to 7.5×10^6 N/mm². For *Blue Asbestos* values up to 22.5×10^6 N/mm² are known. Thus the strength of asbestos, which for the rest varies widely according to its physical and chemical constitution, has the same magnitude as special steel (2).

The chemical resistance of asbestos cement is very high, too, and has been increased considerably since using special blast-furnace cement with an increased sulphate resistance (3). We will return to this question later on.

Pipes of asbestos cement have a sufficiently high corrosion resistance, which is supported by the comparatively big wall thickness and by the fact that no inner protective layer, no gel coat or enamel determines the life time of this material.

The main fields of application of asbestos cement are:
—corrugated and flat sheets and shingles for roof covering and wall lining
—pressure pipes, inner diameter 50 to 1000 mm (exceptionally up to 2000 mm), maximum length of pipes 3 to 6 m, working pressure up to 16 bar (special manufacture up to 40 bar)
—sewer pipes
—various accessories for roofings and for pipings
—hand or machine moulded pieces like water reservoirs, flower boxes, basins, parts for mini-golf and many others.

In a wider sense, one understands by the asbestos cement industry any industry analogous or related in the final product and similar in the production process, where for instance portions of asbestos are substituted by other vegetable, mineral or synthetic fibres, or portions of cement are replaced by other binding agents or fillers. By adding cellulose, flexible boards are obtained. Curing processes comprise the normal water hydration of cement with several variations as well as steam-curing according to the Morbelli-process where approximately 40% of cement is substituted by pure silica.

Besides, there are fabricated several composite materials. One material of special practicability is *Ferro-Eternit,* which takes an inserted netting of steel wire, and has a wide range of applications for all channels, containers and other forms which can be rectified geometrically. Another application is sandwiched panels of different materials.

The most important property of this artificial stone is stipulated in the definition of fibre cement as it was drawn up in the patent application by Ludwig Hatschek (4) from March 30, 1900. Ludwig Hatschek is the inventor of the most economic process of production of asbestos cement sheets on a sieve cylinder machine, which is also the antecedent of the pipe making machine system Mazza: "Fibre cement is an artificial stone made from fibres, binding materials and possibly bonding agents of the fineness of cement, containing at least a quantity of fibres which guarantees, due to their carrying capacity for materials of the fineness of cement, a homogeneous distribution in the artificial stone".

In the following extract of German Standards the main requirements of asbestos cement are quoted:

DIN 274 Roofing sheets and plates:

Specific weight of air dried sheets and boards:
Asbestos cement roofing sheets   1.8–2.2 g/cm³
Compressed asbestos cement boards   1.7–2.1 g/cm³

Water absorptive capacity:
as per cent of dried weight
asbestos cement roofing sheets   maximum 18%
compressed asbestos cement boards   maximum 20%
uncompressed asbestos cement boards   maximum 27%

Resistance to frost and heat:
Under the effect of frost and heat, asbestos cement roofing sheets and boards must show neither splintering nor cracking.

The requested bending strength of asbestos cement sheets is indicated in Table 1.

Table 1   **Bending strength of air dried samples taken from sheets or boards**

	Minimum bending strength normal to fibre N/mm^2	Maximum bending strength parallel to fibre N/mm^2
asbestos cement roofing slates	38	29
compressed asbestos cement boards up to 8 mm thickness	30	22
more than 8 mm thickness	28	20
uncompressed asbestos cement boards up to 12 mm thickness	20	15
more than 12 mm thickness	17	15
asbestos cement corrugated sheets	17	—

DIN 19800 asbestos cement pressure pipes:

Terms of delivery refer to asbestos cement pressure pipes fabricated to machine and seamless under pressure which consist of an intimate mixture of asbestos fibres and cement. Working pressures: 2.5, 6, 10, 12.5 bar.

Dimensions and tolerances:
    ... wall thicknesses of asbestos cement pressure pipes are rated to a bursting pressure multiple of the working pressure as follows:

Nominal diameter up to 100 mm   ...   4 times
                         125–200 mm   ...   3.5 times
                         200–400 mm   ...   3 times
              more than 400 mm   ...   subject to agreement

Strengths:
Annular tensile strength   20 N/mm²
Crushing strength   45 N/mm²
Bending strength   25 N/mm²

Impermeability to water:

All pipes have to pass before applying an eventual coating an internal pressure test with 2 times the working pressure, or maintaining the testing pressure 30 seconds, no leaking or water stains must show up.

DIN 19830 sewer pipes:

Strengths:
Annular tensile strength	13 N/mm^2
Crushing strength	35 N/mm^2
Bending strength	15 N/mm^2

Testing of water impermeability:

The samples have to be subjected to an internal pressure of 2 bar. Pipes and specials must not show formation of drops.

ASBESTOS CEMENT

Perhaps more than with any other material, the properties of asbestos cement represent an addition of the properties of both basic materials, asbestos and cement. To understand what is asbestos cement, a profound knowledge of the nature and the peculiarities of these basic materials is necessary.

The following paragraphs will point out some of these peculiarities.

Asbestos

Basically each mineral which can be processed to fibres can be called asbestos. For practical reasons the numerous different varieties are classified into two main groups, the group of amphibole asbestos and the group of chrysotile asbestos. In any case asbestos is a magnesium silicate containing lime and alkalis where magnesium can be exchanged for iron. The differences in the most important technical properties of the diverse varieties, like strength, flexibility, hardness, chemical and heat resistance, filtrability and spinnability, derive from the mineralogical structure.

Chrysotile asbestos has the composition $Mg_6(OH)_6.(Si_4O_{11}).H_2O$, amphibole asbestos $Ca_2Mg_5(OH)_2.(Si_4O_{11})_2$. Even the finest chrysotile fibres (diameter 20×10^{-6} mm) are hollow, whilst amphibole fibres (diameter of finest fibres 10^{-4} mm) are solid. This explains why fibres of amosite and blue asbestos are hard and springy whilst the fibres of chrysotile are soft, flexible and absorptive.

Crocidolite or blue asbestos and amosite asbestos are the two amphibole varieties important for the asbestos cement industry (5).

Chrysotile asbestos is not resistant against mineral acids, resists better organic acids and is highly alkali-proof. On heating up to 400 °C its strength increases slightly, over that temperature it diminishes rapidly. Melting point is 1500 °C. Above 700 °C no asbestos can be used successfully. Chrysotile is mostly soft, sometimes silky. It resists mechanical defibration by card or beater mill better than amphibole asbestos. Fibre length is mostly below 40 mm. Its colour may be white, light-grey, yellowish white, reddish white to grey-green (2).

Blue asbestos is the most acid-proof asbestos; its alkali resistance is inferior to chrysotile asbestos. It is the strongest, but also hardest, asbestos and wears the tools strongly. Its brittleness is lower than that of amosite asbestos. Having at room temperature a considerably higher strength than chrysotile, its strength decreases above 200 °C rapidly and surpasses the strength of chrysotile at 400 °C. Melting point is 1150 °C (2).

The chemical resistance of amosite asbestos is almost as high as that of blue asbestos. Its strength is lower than that of chrysotile. Its fibre length may range to 300–400 mm. Its colour varies from greyish white to dark brown. Amosite asbestos is brittle. It can be used profitably for filtration purposes and for wet processes.

Table 2 shows some technical data for asbestos and for better comparison the corresponding data for other materials (2).

Table 2 Technical data for asbestos

	Fibre diameter mm	Fibre surface mm^2/g	Specific weight g/cm^3	Tensile strength N/mm^2	Modulus of elasticity N/mm^2
Chrysotile	0.000018-0.000029	130-2200	2.50	5.6-7.5	352
Blue asbestos	–	–	3.40	7.5-22.5	1075
Amosite	–	–	3.30	1.1-6.3	–
Polyamid fibre:					
nylon	0.0075	31	1.15-1.21	4.8-8.0	–
Cotton	0.01	72	1.47-1.55	3.6-5.7	–
Human Hair	0.0395	–	–	–	–
Aluminium wire hard drawn	–	–	2.70	1.7-2.3	–
Steel wire hard drawn	–	–	7.70	11	2100

TESTS AND INSPECTION CONTROL (7)

The quality and workability of the basic materials influence the quality of the asbestos cement products. To guarantee a satisfactory quality of the products, as well as to investigate previously the suitability of the basic materials for a certain process, continual laboratory tests are advisable.

In the case of asbestos, one investigates the length of fibres and/or the proportion of the different lengths, density dry and wet, dust content, specific surface and filtrability. A series of other tests of scientific value has no practical importance for the asbestos cement industry.

In the case of cement, one investigates the fineness of grain, setting time, increase of strength or progress of curing respectively and specific surface. Though the chemical analysis and the determination of the composition of phases of cement (8) have dominant importance for the production process these tests are executed only by very large companies by themselves due to the complicated nature of the method and the relatively high costs of the necessary laboratory equipment.

The most direct information on the performance during production may be obtained from test specimens made from the asbestos and cement to be checked, where of course one has to consider the different fabrication methods in laboratory and in technical manufacture. However, from these tests on asbestos cement specimens one can deduce valuable conclusions about the influence of the several composites by changing only one composite at one time or modifying the conditions of the process (for instance vacuum pressure or time of action or compression by forming roller) while maintaining an equal mixture. By a systematic series of these tests one can optimize the mixture and the technology.

I will quote in the following briefly the most usual and most important laboratory and inspection tests without going into the details of a manual:

Quebec standard test or Canadian sieve test (9)

Though this test is the most inaccurate method to determine fibre length and distribution it is the most frequently used to classify the grades of the delivery. The Quebec testing machine consists mainly of 4 boxes mounted one above the other and separated by different sieves of specified size. A certain amount of fibres (1 pound or

100 grams according to the bulkiness of the fibres) is fed to the upper box. Then the machine vibrates the boxes at a specified frequency with a specified amplitude for a determined time or number of strokes.

When the test has been completed one weighs the portions in the different boxes. A table of the standardized proportions allows then the classification of the fibre.

Bulky or matted fibres may simulate in this test a bigger fibre length.

Bauer McNett test (10) This is a wet test to determine fibre length distribution which needs a very small quantity of asbestos (10 or 20 grams). The fibres are fed into a container and are passed at an exactly specified flow rate through 5 boxes which are equipped with different sieves. Each sieve retains a certain amount of fibre, the dust being collected on a filter paper or cloth in an additional bottom box. As this test disintegrates the fibres to a very high degree the result is very accurate.

Ro-tap-sieve-analysis This is a dry test with a feed of 50 or 100 grams according to the presumed classification of the fibres. 6 sieve boxes and a bottom pan separated by sieves of different mesh size are mounted vertically above each other and are mechanically vibrated at a certain frequency. The test is comparable to the Canadian Sieve Test with a slightly higher accuracy due to the greater number of boxes.

Dust content (7) An amount of 10 or 20 grams of fibres according to their classification is compressed between two sieves. Then a flow of compressed air is directed to pass the sieves. After 2 minutes the loss of weight is determined.

Buoyancy test or wet volume test A graduated cylinder is filled with 500 ml of water and 5 grams of asbestos.

The tapped cylinder is rotated 10 or 40 times. After putting down the cylinder without further movement the sedimentation heights of the fibres are measured in relation to the setting time. This test indicates the wet bulk or the grade of defibration by a former treatment of the fibre.

Dry volume test (7) In a graduated cylinder one feeds 100 grams of fibres and measures the volume without any compression.

Resistance of asbestos to defibration Asbestos fibres of different groups and origin react distinctly to different defibration processes. To judge the progress of defibration in a certain machine comparative tests before and after treatment are performed. One of these tests is the buoyancy test as mentioned above. Samples were taken for instance from disintegrator, perplex mill edge runner, pulp engine etc. after different times of treatment or with different adjustment. Unfortunately, the buoyancy test does not reveal the shortening of fibres during defibration and the corresponding loss of strength.

Air permeability test according to Dyckerhoff (7) The testing apparatus designed by Blaine, system Dyckerhoff (7) indicates the time needed for a certain quantity of air at specified pressure to pass a compressed sample of asbestos. This time is in mathematical relation to the inner surface of the sample which can be calculated or read from corresponding tables.

The surface of the fibre is an important feature and indicates the degree of defibration. Of course, similarly to the wet volume test, it gives no data on the length or the strength of the fibre.

Filtration test For all wet processes a good filtrability of the slurry is necessary. One of the deciding factors is the filtrability of the asbestos fibres. With the testing apparatus of Asarco Central Research Laboratories, one uses a sample of 10 grams of asbestos immerged and mixed with 500 ml water and measures the time of draining 100 ml, which gives a comparable figure for the filtrability. More realistic results are obtained by testing a slurry of asbestos and cement.

Humidity test (7) A sample of asbestos is dried in a drying oven and the loss of weight is measured and calculated in percent.

Marchioli's resistance test To investigate how much of the asbestos delivered from the mines is already disintegrated and how it will behave on further treatment Marchioli has developed a volumetric cyclon. The asbestos is passed through a disintegrator; at the beginning, the fibre will become softer and more fluffy after each passage and after a certain number of passages will reach its largest volume. Continuing the passages, the volume will decrease, as a result of the destruction of the asbestos parallel to further opening. The number of passages up to the maximum specific volume indicates the elasticity, the mechanical resistance and the breaking resistance of the fibre. Marchioli has established particular index-numbers relating to the Canadian Sieve Test.

Testing of cement

The suitability of cement for the production of asbestos cement depends especially on its chemical analysis, the composition of phases and the fineness of grinding.

The calculation of composition of phases of clinker of Portland cement is treated in the periodical "Zement-Kalk-Gips", October 1965 by A. Glausner (8). Other usual tests can be performed easily.

Testing of fineness of grain (7)

The fineness of grinding or of grain respectively is tested by screening by hand or by machine. According to DIN 1164 the residues on a testing sieve 0.09 must not exceed 20%. Actually mostly finer cement with residues below 10% are used. Finer cement improves the setting behaviour and the development of strength, but reduces the filtrability.

Specific surface-test Blaine-Dyckerhoff (7)

The air permeability as an index of the specific surface of cement is measured basically in the same way as mentioned for asbestos.

Testing of setting (7)

The initial and final set times determine the suitability of asbestos cement for the formation process and the necessary number of moulds or mandrels.

Evolution of strength (7)

The evolution of strength (hydration) is measured through the bending strength of test pieces after specified periods. This test reveals possible changes in the quality of the delivered cement. For the bending test exist a number of simple or sophisticated testing machines. Quick testing after 5 to 6 hours is possible by manufacturing samples with addition of silica and curing in an autoclave. Another quick test with a small cylinder method according to F. Keil and H. Mathieu is described in "Zement-Kalk-Gips" July 1964 (11). This method delivers after 6 hours a compression strength corresponding to the 3-day standard strength. A. Meyer describes in the same periodical in November 1965 an improved small cylinder method where after 5 hours the 28-day strength is obtained, with a medium error of 4% (12).

Testing of asbestos cement

Independently of the inspection controls as described below, laboratory tests are performed to check an equal quality of the product and to investigate the influence of various factors. Test specimens are manufactured in the laboratory; the tests are in the main similar to those on cement. The strength of these test specimens can be measured after 3, 7 and 28 days. By adding silica and curing in autoclave corresponding values can be obtained within 24 hours.

INSPECTION CONTROL

The acceptance tests are specified in the corresponding national or international standards.

For the control of sheets are prescribed controls of:
dimensions and shape
specific weight
water absorptive capacity
resistance to frost
resistance to heat
bending strength
water impermeability
punching strength.

For pressure and sewer pipes the following controls are required:
dimensions
watertightness

annular tensile strength
crushing strength
bending strength
resistance to frost (tested by cyclic freezing and thawing).

REFERENCES

1. Klos, H G, Asbestzement—Technologie und Projektierning 1967 pp 2–6, 12, 13, 46–53, 61, 63, 65, 222–233, 238, 241–247.
2. Harl, Frank, Asbest 1952.
3. Duritwerke Kern & Co, Asbestzement Druck–und Kanabrohre.
4. Hatschek, Ludwig, Osterr Patent 5970, 1901.
5. Cape Asbestos Co Ltd, Asbestos the raw material.
6. Kühl, Hans, Zementchemie I, 1956.
7. Asbestos Products Association, Manual of Testing Procedures for Chrysotile Asbestos Fibre, 1962.
8. Bogue, R H, The Chemistry of Portland Cement, New York 1963.
9. Asbestos Corp Ltd, Quebec-Asbestos Testing Machine.
10. The Bauer Bros Co, USA, Bauer McNett test, Laboratory Report 1964.
11. Meyer, A, Weiterent-wicklung des Kleinzylinderverfahrens, Zement-Kalk-Gips 11/1965.
12. Keil, F and Mathieu, H, Schnellprüfung von Zement nach dem Kleinzylinderverfahren, Zement-Kalk-Gips 7/1964.

6.2 Investigation of the "corrosion" of asbestos fibres in asbestos cement sheets weathered for long times

Ludmilla Opoczky and László Péntek jnr
*Central Research and Design Institute for the Silicate Industry,
Budapest, Hungary*

Summary

The condition of asbestos fibres in asbestos cement sheets of different ages (the oldest one being 58 years) exposed to atmospheric effects was investigated by modern methods. In all of the asbestos cement sheets investigated, cement hydration products had crystallized on the surface of the asbestos fibres and along cleavage planes. This latter phenomenon is connected with the high affinity of the asbestos fibres to $Ca(OH)_2$ and the cement hydration products. The existence of chemical interactions taking place on the boundary surface of the asbestos fibre and the adsorbed materials in asbestos cement sheets weathered under atmospheric conditions is, however, problematic.

The existence of these surface reactions, ie the "corrosion" of asbestos fibres, is proved by the series of investigations carried out on asbestos cement sheets hardened for 16 years and over. Two types of process can be observed:

A partial carbonation on the asbestos fibres observable on the surface but not inside the fibres. In this process, the asbestos fibre acted only as the adsorbent of $Ca(OH)_2$;

The other process can be seen exclusively in older sheets. It mainly takes place along the cleavage planes between the fibres, and a colourless, optically oriented material having a lower refractive index and birefringence than the asbestos, is formed. (It is probably brucite—$Mg(OH)_2$.)

In the "corrosion product" magnesite ($MgCO_3$) and calcium silicate hydrates of very low lime content could also be identified. The formation of secondary magnesium silicate hydrates is also probable.

The "corrosion products" of asbestos fibres are crystallized on the surface of asbestos fibres and along them, respectively, as well as in the spaces between the fibres, thus ensuring the bond between the asbestos and the cement matrix and between the individual fibres of the asbestos bundle, compensating the decrease of strength that would otherwise be expected from the "corrosion". So, even the sheet aged 58 years, in which 50-55% of the chrysotile asbestos is in a corroded state, has excellent mechanical properties.

Résumé

Nous avons étudié par des méthodes modèrnes l'état des fibres d'amiante dans des plaques d'amiante de différents âges (le plus agé 58 ans) exposées—conformément à la pratique—à l'effet atmosphérique.

Dans toutes les plaques d'amiante étudiées, des produits hydratés formés au cours de l'hydration du ciment, sont cristallisés sur les surfaces des fibres d'amiante, ainsi que tous le long des clivages des plaques.

Ce dernier phénomène est en corrélation avec l'aptitude d'adsorption— notoirement grande—des fibres d'amiante à l'entroit de $Ca(OH)_2$ ainsi que des produits hydratés. Cependant ont remet en question si, dans les plaques d'amiante durcies en condition normale, il y aura des interactions chimiques aux surfaces limites des amiantes et des matériaux adsorbés.

Nos études démontrent l'existence de pareilles réactions des surfaces, c'est à dire, de l'existence de la corrosion des plaques d'amiante durcies jusqu'à 16 ans ou à plus longues échéances.

On peut remarquer deux processus:
—la carbonation sur les fibres d'amiante, visible au bord des fibres et pas observable à l'intérieur;

—l'autre processus—remarquable surtout pour les plaques plus agées—se déroule entre les plans de clivage entre les fibres, il produit une matière, incolore avec réfraction et biréfringence, qui est, selon toute probabilité, de la brucite— $Mg(OH)_2$. Entre outre, dans le produit de corrosion on a réussi à identifier de la magnésite ($MgCO_3$) et des hydrosilicates de calcium pauvres en chaux. Aussi la formation des hydrosilicates secondaires de magnésium est probable.

Les produits de la corrosion des fibres d'amiante se cristallisent sur les surfaces de celle-ci et aux interfaces des fibres d'amiante, assurant de cette manière la liaison entre l'amiante et le matrice, de même qu'entre des faisceaux de fibres d'amiante effilochés; c'est ainsi qu'ils compensent l'abbaissement des résistances présumées, provoquées par la corrosion. Les plaques d'amiante de 58 ans, dans lesquelles le 50-55% de crysotilasbest est en état corrodé, ont encore des propriétés mécaniques excellentes.

INTRODUCTION

The problem of the chemical interaction between the asbestos and cement as well as the "corrosion" of the asbestos fibres has been a subject of research for a long time. According to some authors (1,2) a surface reaction takes place in autoclaved asbestos cement products, between the chrysotile asbestos and the $Ca(OH)_2$ formed during cement hydration. Due to this reaction the strength of the asbestos fibres is reduced (3). Other authors (4) have stated, that although the reactions between the chrysotile and $Ca(OH)_2$ are of negligible importance at the temperatures used in practice, they may have a significant effect on the bond between the asbestos fibre and cement.

The interaction between asbestos and cement has generally been investigated under hydrothermal conditions, usually at elevated temperatures and pressures, ie under conditions different from those used in traditional industrial practice. The authors of the present paper have investigated this problem for asbestos cement sheets exposed for long times to atmospheres experienced in practice.

MATERIALS AND METHODS

Technological asbestos cement sheets of different classes stored under normal conditions were used as experimental materials; their important technical characteristics are shown in Table 1.

Table 1   **Technical characteristics of asbestos cement sheets**

Product age (years)	Cement quality	Cement quantity (% by weight)	Quantity of chrysotile asbestos (% by weight)	Bulk density (kg/m^3)	Porosity (%)	Bond strength (N/mm^2) I	Bond strength (N/mm^2) II
2	500*	86	14	1790	13.30	38.8	26.4
8	500*	86	14	1900	9.79	36.0	27.0
16	500*	86	14	1750	14.06	30.0	21.5
40	LSF 0.8**	86	14	1800	12.52	42.0	30.0
58	LSF 0.8**	86	14	1890	11.22	54.0	38.8

*   Hungarian Standard (MSZ) No. 523.
** LSF = Lime Saturation Factor.

Corrosion of asbestos fibres in these tests was observed mainly by an optical microscope. The investigation of the thin sections was carried out by an "Opton" polarizing microscope, with parallel and convergent polarized light, between parallel and crossed nicols. The relative refractive index was generally measured by the Becke-method and in some cases by using immersion liquids. In addition, for the identification of the products formed, X-ray diffraction, scanning electron microscopy and thermal derivatographic methods were used.

RESULTS

Optical investigations

In all the thin sections of asbestos cement sheets investigated there are chrysotile asbestos fibres and bundles, embedded in a matrix, which have the following characteristics: colourless; fibrous or platelike shape depending on the orientation; low refraction; low birefringence ($\sim$0.01-0.008); straight extinction; negative principal zone character.

The texture of the matrix (ie of the sheet) as well as the state of the asbestos fibres showed the following changes with age:

Figure 1   Asbestos cement sheet, 2 years old. A corroded asbestos fibre. Exposure taken between crossed nicols.

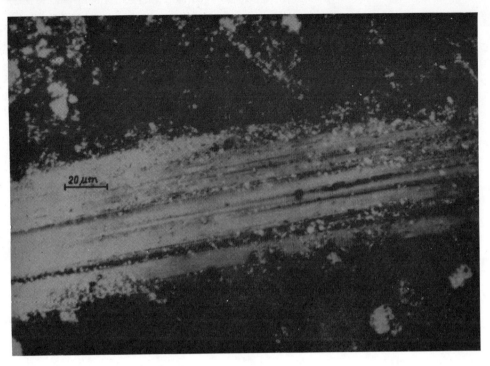

Asbestos cement sheet 2 years old (see Fig 1)

In a decidedly amorphous matrix, asbestos fibres, present as porphyritic inclusions, are coated with the hydration products of cement. The product of the "corrosion" can be observed particularly on the edges of the fibres, which often show no definite boundary towards the matrix and consists of colourless, highly-refractive and very strongly birefringent (0.17) small calcite and very small $Ca(OH)_2$ crystals, having spherolitic shapes (Ng-Np $\sim$ 0.029-0.030).

The matrix, for the most part, consists of brownish-grey, weakly refractive, isotropic spots of blurred outline, sometimes having very low birefringence. They are colourless aggregations of grains consisting of slightly crystallized isotropic calcium silicate hydrates of low refractive index and calcite crystals.

Asbestos cement sheet, 16 years old

The texture of the sheet is crystalline and porphyritic. The matrix for the most part consists of calcium silicate hydrates in which angular, highly-refractive, very weakly birefringent grains coexist with isotropic, roundish grain-aggregations of low refractive index.

Corroded asbestos fibres can be seen as porphyritic inclusions. The "corrosion" can be observed particularly on the edges of the fibres (mainly the small calcite crystals), but it can already be seen along the cleavage planes between the fibres where calcium silicate hydrate flakes of 0.1-0.001 mm diameters can be identified (Ng-Np $\sim$ 0.036-0.040).

Asbestos cement sheet, 58 years old

The texture of the sheet is fully crystalline and porphyritic. The matrix consists predominantly of calcium silicate hydrates and of colourless, angular, in some cases easily observable, calcite crystals showing rhombohedral cleavage planes.

Two types of crystalline calcium silicate hydrates could be distinguished in the matrix:

1. Calcium silicate hydrate crystals having the following characteristics: formed from clinker grains probably by a topochemical process; colourless or slightly yellowish-brown colour; high or medium refraction (1.640-1.649); low birefringence ($\sim$ 0.005). These crystals, often present in distinct orientations, show wavy extinction, have one (or two, but low-angled) optical axis, are optically positive and their concentration in the thin section is about 20% (5).

Figure 2   Asbestos cement sheet, 58 years old. A corroded asbestos fibre. Exposure taken between crossed nicols.

2. Calcium silicate hydrate crystals, crystallized from the gel which are colourless, isotropic or of very low birefringence, forming small bunchy or laminated aggregations and often having a particle size in the range 20-30μm. They generally are surrounded by small carbonate grains (5).

50-53% of the chrysotile asbestos fibres found as porphyritic inclusions are in a corroded state (see Fig 2).

The "corrosion" process took place along the cleavage planes between the fibres. In this process a colourless material consisting of thick crystals having lower refractive

Figure 3   Asbestos cement sheet, 58 years old. SEM-photograph.

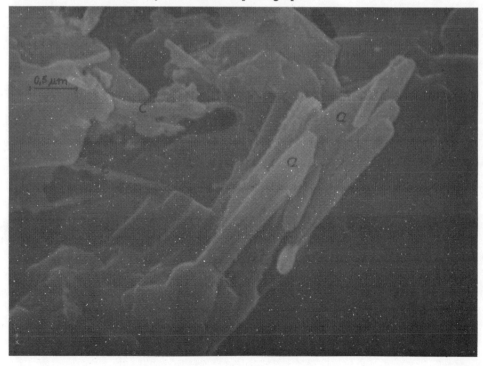

273

indices and birefringence (~0.004) than asbestos, was formed. These grains, displacing the asbestos, show optically oriented coalescence, have a positive principal zone character, ie the direction of the axis "c" of the individual grains is parallel to the direction of cleavage of asbestos. This material is supposedly brucite or a mixture of secondary magnesium silicate hydrates. The decomposition products contain submicroscopic calcium silicate hydrate crystals, often of 0.1 mm diameter (Ng-Np~ 0.036-0.040).

Investigations by SEM   Only one print is presented here showing the sheet aged 58 years (see Fig 3).

In the figure bundles of asbestos (a) are seen corroded along the cleavage planes. They are coated by the hydration products of cement, as well as by quasicoalesced calcite crystals in close proximity (b). On the left side of the picture there are thinner fibres of wave-like tracing due to the above mentioned change.

Other SEM photographs prove the existence of the surface-reactions and, also, that asbestos fibres and bundles act as crystallization nuclei during cement hydration, as

Figure 4   Asbestos cement sheet, 58 years old, TGc, DTG and DTA patterns.
a,b = original sheet
c,d = "fibre-enriched material".

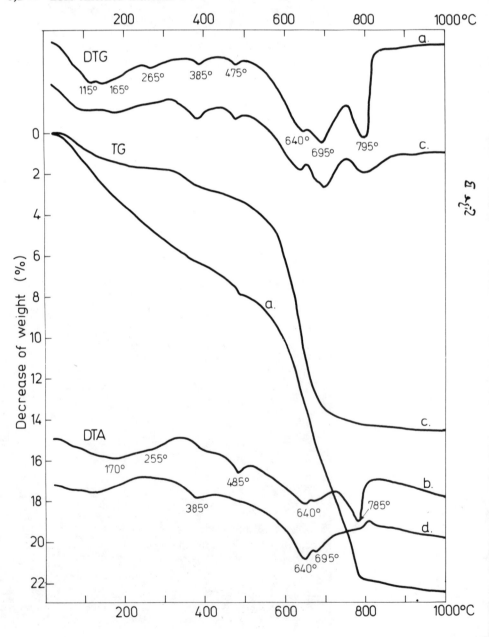

the hydration products are mainly crystallized on the surface of the crystals as well as along the cleavage planes. The products on the fibres are formed at uniform spacings, preserving the original shape and continuity of the asbestos fibres.

THERMAL AND X-RAY DIFFRACTION TESTS

For the identification of the "corrosion products" formed in asbestos cement sheets of 40-58 years of age, thermal (thermogravimetric and a differential thermal) and X-ray diffraction investigations were carried out on the original sheet as well as on the fibre-containing parts separated from the matrix. The separation of fibres from the matrix is a very difficult process and usually only "fibre-enriched material" can be obtained. Fig 4 shows the thermoanalytic patterns for the samples.

On the DTA curves made of the original asbestos cement sheet, in addition to the (115-265 °C) peaks (caused by the decomposition of the cement hydration products) the following peaks, important for the identification of asbestos "corrosion" products are seen:

At approx 385 °C an endothermic peak (a,b) due to the decomposition of brucite is visible. The amount of $Mg(OH)_2$ calculated from the decrease of weight that corresponds to this peak, is 2.78%. The same peak is seen in the pattern (c,d) made of the "fibre-enriched material".

The endothermic peak at 640 °C is due to the decomposition of magnesite and corresponds to 3.82% $MgCO_3$; the same peak is seen in the pattern of the "fibre-enriched material". The amount of the MgO combined in the $Mg(OH)_2$ and $MgCO_3$ is according to our calculations 3.74%. At the same time chemical analysis of the asbestos cement sheet gives a total MgO-amount of 5.73%. Consequently, the higher amount of MgO combined in the $Mg(OH)_2$ and $MgCO_3$ originates from the asbestos. The cement also may contain some MgO.

The peak at 695° shows the decomposition of chrysotile asbestos. In the original pattern the evaluation of (a,b) is made more difficult by the possible decomposition of calcium silicate hydrates. In the curves (c,d) of the "fibre-enriched material" the appearance of similar double peaks suggests the decomposition of two kinds of magnesium silicate hydrates: the first is a primary one from the chrysotile asbestos, the other is from a secondary cryptocrystalline one, containing more water than the first.

In addition to the previously mentioned peaks in curves (a,b) of the original sheet, the endothermic peak at 475 °C shows the decomposition of portlandite, corresponding to about 2.4% $Ca(OH)_2$. The endothermic peak at 795 °C is due to the decomposition of calcite ($CaCO_3$), and the decomposition of certain calcium silicate hydrates formed during cement hydration.

According to the X-ray diffraction investigations, the sheet aged 58 years contains clinker minerals and a small amount of portlandite (lattice spacings: 4.90 Å, 2.62 Å, 1.92 Å) and calcite (3.03 Å, 2.49 Å, 2.28 Å). The "tobermorite-phase" (9.85 Å, 3.07 Å, 2.80 Å) and chrysotile asbestos (3.65 Å, 7.24 Å) can also be identified.

Peaks of brucite—$Mg(OH)_2$ (4.77 Å, 2.37 Å) and magnesite—$MgCO_3$ (2.74 Å, 2.10 Å and 1.70 Å) appeared in the X-ray diffraction pictures of "fibre-enriched materials".

DISCUSSION

In all the asbestos cement sheets investigated, the cement hydration products crystallized on the surface of the asbestos fibres and along the cleavage planes. This latter phenomenon is connected with the high affinity of the asbestos fibres for $Ca(OH)_2$ and cement hydration products. So, according to our investigations, 1 gramme of the P-3-60 Ural asbestos is able to adsorb about 60 mg of calcium hydroxide during a 1-month period. Due to pulping, the adsorptivity of the asbestos is further increased, since the BET surface area and inner pore volume of the asbestos increase. Surface chemical reactions take place between the asbestos and cement in the sheet during which the cement behaves predominantly as a source of $Ca(OH)_2$.

Two types of processes can be observed:

A partial carbonation of the asbestos fibres on their surface and edges but not within the fibre. In this process the asbestos fibre behaves predominantly as the adsorbent of $Ca(OH)_2$, which, with the passing of time, is converted into calcite by airborne CO_2.

The other process, that is observed exclusively in older (over 16 years) sheets, takes place mainly along the cleavage planes between the fibres, when a colourless product—probably brucite ($Mg(OH)_2$)—having a lower refractivity and birefringence than asbestos, is formed. In addition, $MgCO_3$ and calcium silicate hydrates of very low lime content were identified. The formation of secondary magnesium silicate hydrates is also probable, although this has not yet been fully proved. The reaction takes place in the following way:

The surface of chrysolite is affected by the lime-rich solution diffused into the pores and spaces between the fibres, especially at the "active" places introduced during mechanical pulping and activation of the asbestos. The superficial regions of chrysotile asbestos are then converted into oriented brucite. Surplus magnesium and silicon are released, the latter combining with calcium to form calcium silicate hydrates. It can be assumed that the surplus silicon remaining after the reaction with calcium combines with the magnesium to give secondary magnesium silicate hydrates.

The corrosion of asbestos fibre is promoted by the presence of airborne CO_2.

REFERENCES

1 Berkovich, T M, Meiker, D M, Gracheva, O I, Kupreeva, N I, DAN SSSR, 120, 372 (Chem. Abstr., 53, pp 2950,) 1958.
2 Smirnov, N N, Petrographia Asbestocementa, Gostroizdat Moscow, 1962.
3 Fateeva, I I and Trubicina, L B, , O himicheskom vsaimodeistvee asbesta i cementa pri tverdenee asbestocementa. Izv. VUZ Straitelstvo i Architectura, Vol 17, No 2, p 77, 1974.
4 Ball, M C and Taylor, H F W, An X-ray study of some reactions of chrysotile, J. Appl. Chem., 13, p 145, April 1963.
5 Opoczky, L and Böcs, A, Azbesztszálák korróziójának vizsgálata, Épitöanyag, Vol 24, No 1, p 1.
6 Gaze, R and Robertson, R H S, Unbroken tobermorite crystals, Mag. Concr. Res., Vol 9, p 25, March 1957.

Section 7

Properties and Testing of Cement containing Fibres other than Asbestos

7.1 Opening Paper: Properties of fibre cement composites

A J Majumdar
Building Research Establishment, Garston, United Kingdom

Summary *Naturally occurring asbestos fibres have been commercially used for fifty years or more to strengthen cement. In recent years the potentialities of several other fibres as reinforcement have been examined in considerable detail. Foremost among them are metal fibres such as steel wire, polymer fibres such as nylon or polypropylene and graphite (carbon) fibres and special glass fibres have been developed to combat the chemical attack by the highly alkaline cement paste.*

A summary of the properties of these new fibre cement composites is given in this review. The cement paste, with or without fine aggregates, is considered here as the matrix material and the discussion covers most of the important fibres but excludes asbestos.

Résumé *Les fibres de l'amiante naturelle ont été utilisées commercialement pour renforcer le ciment depuis plus d'une cinquantaine d'années. Plus récemment, on a examiné de façon approfondie les possibilités d'emploi de plusieurs autres genres de fibres à cet effet. Parmi celles-ci, les fibres principales sont les fils d'acier, les fibres polymères telles que de graphite (carbone), de nylon ou de polypropylène, et les fibres de verre spéciales qui ont été mises au point pour résister à l'attaque chimique par la pâte de ciment très alcaline.*

On donne un résumé des caractéristiques de ces nouveaux composés de ciment de fibres. On considére le ciment, avec ou sans granulats fins, comme la matière dont se compose la matrice, et l'expose traîte de la plupart des fibres les plus importantes, à l'exception de l'amiante.

Copyright—Building Research Establishment, Department of the Environment.

NOTATION

E	=	Young's modulus
V	=	Volume fraction
α	=	$E_m V_m / E_f V_f$
ϵ	=	Strain
σ	=	Stress
τ	=	Average interfacial shear strength
d	=	Fibre diameter
r	=	Fibre radius
ℓ_f	=	Fibre length
ℓ_c	=	Critical fibre length

Subscripts: c Composite

f Fibre

m Matrix

u Ultimate stress or strain

INTRODUCTION

Cement pastes are not used as a material for construction as they crack easily due to dimensional instability which may be caused by changes in environmental conditions. The inclusion of aggregates reduces this effect and makes the material economically viable, but the resulting products, mortar and concrete, are still very weak in tension and against impact, and they fail in a brittle fashion. The current interest in fibre additions to such a matrix stems from a need to overcome these deficiencies in cementitious materials which otherwise possess the important properties of high compressive strength and stiffness and are very cheap and durable. They are also attractive from the point of view of fabricating components of complicated shapes if advantage is taken of the initial plasticity of the cement paste.

Asbestos cement products provide examples in which cement pastes are strengthened by the naturally occurring asbestos fibres and these have been used widely in the building industry for the last fifty years or more. There is a similar example in concrete reinforced by steel bar or wire, which is perhaps the most important material of construction in the world today. Serious explorative studies with other fibres commenced roughly two decades ago, and these were probably inspired by the success of the glass fibre plastics (grp) industry developed during the second world war. In comparison with asbestos, these synthetic fibres (eg glass, steel or polymer fibres) are more expensive at their present price level, some (eg graphite) prohibitively so. This is the main limiting factor in the large-scale application of fibre-cement composites. Nevertheless, composites using steel wire and glass and polymer fibres have already reached the market place and the properties of these materials are being investigated both extensively and intensively in many countries.

THEORETICAL CONSIDERATIONS

Much work has been done in recent years in elucidating the scientific principles which govern composite action in brittle matrix composites such as fibre cement and concrete. Various theories which have been proposed are discussed in other papers given at the symposium and will not be repeated here. The main objective of the theoretical work has been the prediction of the tensile stress-strain behaviour of the composite. At least three regimes have been recognised in the tensile stress-strain curves of fibre cement, illustrated in Fig 1 with aligned composites containing continuous fibres:

Figure 1   Idealised tensile stress-strain curve of fibre cement composites (1).

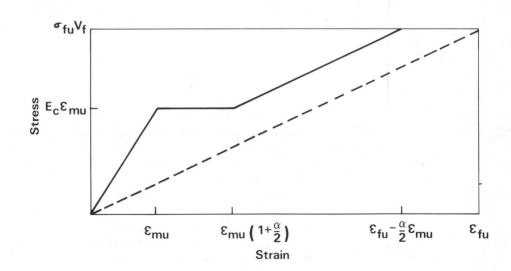

(a) Under stress, the composite deforms initially in a linear elastic manner and the slope of the curve is practically the same as that predicted by the rule of mixtures. This regime terminates at the limit of proportionality (LOP) where the stress-strain diagram deviates from linearity. It is not entirely clear whether the cracking of the composite, signifying that the matrix has already reached its failure strain, always begins at this point.

(b) Provided that there are sufficient fibres to support the stress after the matrix has failed, a regime of multiple cracking of the matrix then ensues. The critical fibre volume fraction which has to be exceeded for this purpose is defined by the inequality $\sigma_{fu} V_f > E_c \epsilon_{mu}$. At V_f lower than the critical the composite will fail in a single fracture mode.

(c) At the completion of multiple cracking, the cement matrix is divided into blocks and the load is supported by fibres which stretch and slide until the composite fails. The slope of the curve in this regime is given by $E_f V_f$ and the composite failure stress by $\sigma_{fu} V_f$. The failure strain of the composite is less than ϵ_{fu} by an amount which depends on the extension suffered during multiple cracking.

The phenomenological description of the stress-strain behaviour of the composite in tension as outlined above was first put forward by Aveston, Cooper and Kelly (1) who have based their theory of fracture in brittle matrix composites on force and energy balance criteria. At first the 'debonded' case, where the fibre/matrix interface is assumed to be completely frictional in character (an assumption which is perhaps true for most of the fibre cement composites), was analysed and subsequently the fully 'bonded' or the elastic case has been treated (2). An extension of the latter work shows (3) that there is no contradiction in principle between the prediction of the composite cracking stress from the new theory and that put forward by Romualdi and Batson (4) a decade earlier. Both are based on fracture mechanics concepts and both predict that the cracking stress of fibre cement and concrete are inversely proportional to the square root of the fibre spacing. Both theories predict an increase in the cracking strain of the composite at the limit of proportionality compared to that of unreinforced cement. However, the predicted increases are rather large and they are not realised in practice.

Fibre arrays of particular importance in reinforced cement and concrete are those with discontinuous fibres arranged randomly in two or three dimensions. To deal with such cases, the concept of efficiency factors for orientation and length was introduced by Krenchel (5). These are basically estimates of how efficient discontinuous random fibres are as reinforcements in comparison with continuous aligned ones. When the fibres are sufficiently long relative to ℓ_c, the shortest length of fibre which can be loaded to failure in the composite, the failure stress of the composite is given by

$$(1 - \ell_c/2\ell_f) \sigma_f V_f,$$

$(1 - \ell_c/2\ell_f)$ being the length efficiency factor. Other values have been proposed for this correction term and there is disagreement also as to the numerical value of the orientation efficiency factor to be used when the reinforcement is randomly distributed in the matrix.

The concept of multiple fracture is of great significance in fibre cement composites as it enables the material to behave in a pseudo-ductile fashion. As mentioned previously, for multiple cracking to occur, the proportion of fibre must be greater than a critical volume per cent. The critical volume percentages for several fibres in aligned cement composites are given in Table 1. It should be remembered that fibre strength values given in the table are those of pristine fibres. When they are put into cement some of them will suffer sudden and/or progressive losses in strength and therefore the critical volume fractions for these fibres will change with time. It is not impossible that, if the fibre contents are very low, certain types of fibre cement composite may exhibit multiple fracture in the short term but single fracture after several years. When random composites are considered the appropriate efficiency factors should be used in computing the critical fibre volume fractions.

Table 1   **Some typical fibre properties.**

Fibre	Diameter (μm)	Density (10^3 kg/m^3)	Young's modulus (GN/m^2)	Tensile strength (GN/m^2)	Elongation at break (%)	Critical V_f(%) for continuous aligned cement composites
Chrysotile asbestos	0.02-20	2.55	164	4.5	2-3	—
Glass	9-15	~2.6	~80	2-4	2-3.5	0.2 - 0.1
Graphite type I	8	1.90	380	1.8	~0.5	0.3
Graphite type II	9	1.90	230	2.6	~1	0.2
Steel	5-500	7.8	200	1-3	3-4	0.5 - 0.2
Polypropylene	20-200	0.9	5	0.5	~20	1.0
Polycrystalline alumina	500-770	~3.9	245	0.65	—	0.8
PRD - 49 (Kevlar)	~10	1.5	133	2.8	2.6	0.2
Sisal	10-50	1.5	—	0.8	~3	0.6

COMPOSITE CONSTITUENTS

Fibres

The fibres which have attracted most of the attention as reinforcement for cements are listed in Table 1 together with some of their important properties. The figures given should strictly be used as guides since there are different known varieties of some of the fibres listed and the properties that could be obtained could fall outside the range of values shown. Asbestos has been included for comparison and although no published account of the use of the recently developed organic fibre by Dupont, PRD-49, in cement is known to the author, in view of its rather interesting properties it is also included. PRD-49 belongs to a group of fibres appearing under different proprietary names (eg Kevlar) and consists of chains of aromatic rings linked by non-flexible groups such as CO-NH and its specific strength and stiffness values are very large. Table 1 also includes some of the properties of sisal, a natural vegetable fibre. (It is useful to remember that in prehistoric times another fibrous material of vegetable origin, straw, was incorporated in mud bricks to provide integrity before the bricks were sun dried.) Exploratory work has been carried out recently on the feasibility of using thin alumina rods in reinforcing cements. Some of the properties of such rods are also given.

Some of the fibres listed in Table 1 such as steel wire are available as discrete fibres, others such as asbestos, glass PRD-49 and graphite fibres appear as bundles. In the latter case the wetting of the individual filaments in the bundles or tows by the cement poses serious problems and additional porosities are introduced in the composite. Not all the fibres remain unaffected, in a chemical sense, in the presence of the highly alkaline cement matrix. Glass fibres are corroded and lose significant proportions of their pristine strength. It is necessary, therefore, to monitor the properties of fibre cement composites over a long period of time.

Matrix

Different types of cement—Portland, high-alumina, regulated-set, aluminous slag cement etc have been used as the matrix material in the production of fibre cement composites. Of these the use of high-alumina cement is not to be recommended since several structures made from this cement have failed recently in the UK. Fillers such as sand or pulverised fuel ash (pfa) can be gainfully used to replace a portion of the cement. The dimensional stability of the matrix is much improved as a result of such additions and the weatherability of the fibre composite is also likely to improve. In certain cases, for instance with glass fibres which are susceptible to attack by hydrating Portland cements, sand or pfa may remove some of the $Ca(OH)_2$ liberated by forming calcium silicate hydrates and thereby increase the durability of the composite.

Another type of matrix addition, which is becoming increasingly popular, is that of polymeric materials. The effect of incorporating various types of polymer dispersions in cement or concrete has been the subject of numerous investigations (6) and it is now recognised that in relatively dry environments polymer modified cement pastes show considerable improvements, when compared with neat cements, in properties such as strength, toughness and weatherability. In wet environments, however, most of the advantages of polymer addition are lost (7). This does not appear to be the case when the cement or concrete is 'impregnated' with polymers. However, this procedure (8) is much more expensive.

The external additions to cements described above, of silicious as well as polymeric materials, have the effect of reducing the stiffness of the matrix. If there is an increase in the failure strain of the matrix as a consequence, experience has shown that this is of some benefit to fibre reinforcement. It is also well known that the properties of cements are strongly dependent on the water/cement (w/c) ratio used in preparing the pastes. In fibre cement composites, higher strengths are obtained if the w/c ratio is kept at a minimum, say 0.3–0.35 in the case of ordinary Portland cements (OPC). Polymer dispersions can be used to achieve adequate flowability at low w/c ratios.

Some typical strength properties of the neat OPC paste, OPC plus filler and OPC plus a polymer additive given in Table 2 are taken from current work at the Building Research Station. All specimens were initially kept for the first 7 days in air at 100% RH and subsequently in air at 60% RH for 21 days before they were transferred to their long-term curing environments. No precaution was taken to prevent carbonation.

Because of the particulate nature of the cement matrix, the amount of discontinuous fibres which can be usefully incorporated in it is relatively small and this depends on the surface properties and the aspect ratio (length/diameter) of the fibre. The improvements in the mechanical properties of cements which can be expected from fibre addition are limited by this factor. Aligned composites which use continuous fibres offer the most effective exploitation of fibre properties but because of anisotropy such composites are not suited for many applications.

Table 2   **Some typical matrix flexural strengths at 20 °C***

Matrix	28 days		2 years in air (60% RH)		2 years in water	
	MOR (N/mm^2)	Deflection (mm)	MOR (N/mm^2)	Deflection (mm)	MOR (N/mm^2)	Deflection (mm)
OPC (w/c = 0.30)	8.5	0.22	13.5	0.40	—	—
OPC + 40% pfa (w/c = 0.40)	5.5	0.38	—	—	7	0.27
OPC + 10% styrene-acrylic polymer (w/c = 0.30)	14	0.70	21	1.03	15	0.41

* 150 x 50 x 6 mm samples were tested in four-point bending using a span of 135 mm.

Interface

The interface plays a most vital role in the development of properties in fibre cement composites. By interfacial properties is commonly meant the strength of the 'bond' between the fibre and the matrix. The physiochemical nature of this bond is difficult to determine with some fibres. In the case of polymer (eg polypropylene or nylon) and graphite fibres, the fibre/matrix bond is perhaps entirely frictional in character, its strength depending on the surface roughness of the fibre. With glass fibres, it is to be expected (9) that the bond is partly chemical since the nucleophyllic attack of OH^- ions on the Si-O bond cannot be eliminated by altering the composition of silicate glasses. In the case of steel wire, there is also evidence (10) to suggest that diffusion of ions takes place across the interface. For such examples it is best to describe the interface as a zone across which transition of properties occurs from that of the fibre to that of the matrix.

Bond strength of several types of fibres embedded in cementitous matrices are conventionally determined by 'pull-out' techniques. Single filaments are used in such studies and a method suitable for steel wire or thin glass rods has been described (11). The lowest practical depth of embedment of the fibre needs to be used in such experiments if the highest fibre/matrix bond strength is to be measured. In cases where the reinforcement is in the form of bundles, for instance in the case of asbestos, glass or graphite fibres or where its shape is not well-defined, simple pull-out methods are less satisfactory. To deal with fibre bundles, Laws, Lawrence and Nurse (12) have described a method where several strands are pulled out simultaneously and the 'depth of embedment' of the reinforcement can be varied at will. A typical multiple strand pull-out curve is shown in Fig 2a. At short embedment depths the (average) debonding load is the peak load reached, and this increases uniformly with increasing embedment depth. As the embedment depth increases further, some strands break and the relationship between peak load and embedment depth is no longer linear (Fig 2b). The bond strength is obtained from the initial slope of the peak load/embedment depth curve.

It is interesting to note that Allen (13), in his strength theory for thin laminates, has used (single strand) pull-out load/embedment depth curves to find the average stress developed in the fibres. This procedure neatly avoids problems (for example perimeter estimation) associated with the calculation of bond strength.

Aveston, Cooper and Kelly (1) have measured bond strength by using an entirely

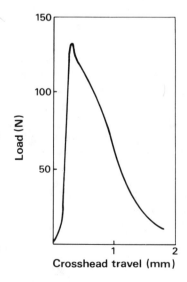

Figure 2a

Multiple fibre-strand pull-out curve; depth of fibre embedment 2 mm; w/c = 0.3; specimens were cured in air and tested at 28 days.

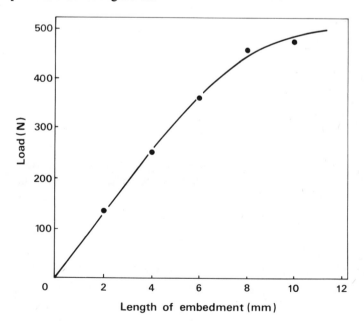

Figure 2b

Peak load vs length of embedment. Preparation and testing of specimens as in Figure 2a.

different technique. According to their theory, during multiple cracking the cement matrix is broken into a series of blocks of length between X' and $2X'$ where

$$X' = \frac{V_m}{V_f} \cdot \frac{\sigma_{mu} r}{2\tau}$$

From the measured values of crack spacing, it is thus possible to calculate τ. For steel fibre reinforced cement a value of 6.8 N/mm² is obtained and Aveston (14) has pointed out that there is little to be gained by increasing this value. He estimates that in order to exploit the strength of a steel fibre in a perfectly bonded brittle matrix composite, the shear strength of the matrix must be of the order of 200 N/mm². This is many times more than is likely to be attained with cement pastes. Bond strength values measured for various fibre/cement combinations are given in Table 3.

Table 3 Fibre/matrix bond strength

Fibre	Matrix	w/c	Age at test (days)	Bond strength (N/mm²)	Reference
Asbestos	Cement paste	–	–	0.88-3	11
Alkali resistant glass (thin) rod	″	0.3	3	5	BRE work
″	″	″	90	>10	″
Cem-FIL glass fibre strand	″	0.3-0.4	28	3-2	″
Steel	″	0.35-0.4	14	8.3 - 6.8	26
″	Mortar	0.5	28	5.4	26
Polypropylene monofilament	Cement paste	0.3-0.4	180	1.2 - 0.7	54
Polycrystalline alumina (thin) rods	″	0.3	20	11.7	25

In a recent comprehensive study Tattersall and Urbanowicz (15) have investigated the effects of various chemical and physical treatments of the wire surface upon the bond between steel wire and cement paste or mortar. The effect of the application of pressure during casting on the bond strength was also studied. The authors present their results in terms of pull-out loads and bond improvement factors. These values are not strictly comparable to bond strength values of other investigators and for this reason they are omitted from Table 3. No simple mechanical or chemical treatment was found very efficient in improving the 28-day bond strength although at 7 days significant increases were measured in several cases. Because of the difficulty in measuring the effective diameter of graphite fibre tows no value can be given for this fibre. However Aveston et al (26) have reported a figure of 5.2×10^{11} N/m³ for τ/r at 14 days; w/c used was 0.5.

The interfacial bond strength in fibre cement composites is influenced by the curing conditions and in some cases, notably with glass fibres, the bond improves with age. The changes occurring in the cement paste with time as well as its potent reactivity can account for most of these variations.

Since a low bond strength is associated with large impact resistance but poor tensile strength of the composite and for a good bond the converse applies, there is some advantage in trying to reinforce cement by a mixture of fibres forming good and poor bonds respectively with the matrix. From this point of view mixtures of graphite or asbestos or glass with either polypropylene or nylon are considered to be suitable combinations.

COMPOSITE FABRICATION

The mechanical properties of a fibre reinforced composite material depend very strongly on the fabrication method employed. The usual variables associated with the fibres, eg its concentration and aspect ratio, its orientation and distribution, affect these properties in a way which can be predicted from theory. In fibre cement composites additional factors such as porosity and the degree of penetration of the reinforcement (particularly if it is present in the form of a bundle) by the cement play important roles in the development of properties. Furthermore, commercially produced composites rarely attain the properties obtained with specimens carefully made in the laboratory.

In a cementitious matrix continuous fibres have been placed either unidirectionally or orthogonally. By a winding process, the rovings or tows (of say glass or graphite fibres) are first impregnated with cement by passing them in an opened-up state through a bath of cement slurry and then applied to a suitable formwork. Developed by Biryukovich and co-workers (16) in the USSR for glass fibre cement, the winding process promises to become the most elegant method of producing composites with exact properties. The 'tape process' developed recently in Canada (17) also places the reinforcing fibre in the matrix in one direction although absorbent fibres such as cotton are woven into the tape at right angles to the reinforcing fibre to aid the 'pick-up' of the cement slurry. Fibre mats having orthogonal arrays of the reinforcement are also becoming available and they can be introduced into the matrix by the lay-up process.

Short fibres are incorporated in the matrix by a variety of techniques. They can be mixed with the cement in a mixer prior to fabrication of the components. In the mixer, a random three-dimensional distribution of the fibre obtains but subsequent processing (eg spraying or extrusion) may influence the fibre orientation significantly.

For introducing short fibres, the spray-up process initially developed at the Building Research Station (18) has proved to be efficient and versatile. In this method, the fibre roving or tow is chopped continuously into predetermined lengths and streams of short fibre meet a spray of cement slurry on a forming surface. The surface is a filter under vacuum which reduces the water content of the matrix. The product, usually a fibre cement sheet, has adequate tear strength to be bent round sharp corners and objects having complicated shapes can be produced. In this method the fibres have been found to lie in the plane of the mould in an approximately random manner. Premixed fibre/cement slurries can also by sprayed but the orientation of the fibre in this case is less specific. The addition of suitable admixtures such as butadiene styrene rubber latex reduces the water requirement of the slurry for flow and in this case the suction operation can be dispensed with.

When using a mixture of fibres slight variations in the fabrication procedures are required which depend on the nature of the particular fibres used. For mixing polypropylene or nylon with glass fibres, a viable system incorporating the principles of the spray suction process has been developed (19,54).

PROPERTIES AND TESTING

The ease with which fibre cement composites can be compacted during manufacture depends on the nature of the fibre used, its amount and most importantly for short fibres, on their aspect ratio. There is a dearth of knowledge in this area for most fibres listed in Table 1. For cement paste and mortar reinforced with chopped steel wire Edgington, Hannant and Williams (20) found that the V-B test gave a realistic assessment of the workability of such composites. The magnitude of the effect of fibre additions on the V-B time for a mortar reinforced with chopped steel fibre is shown in

Figure 3   Effect of fibre aspect ratio on V-B time of fibre reinforced mortar (20).

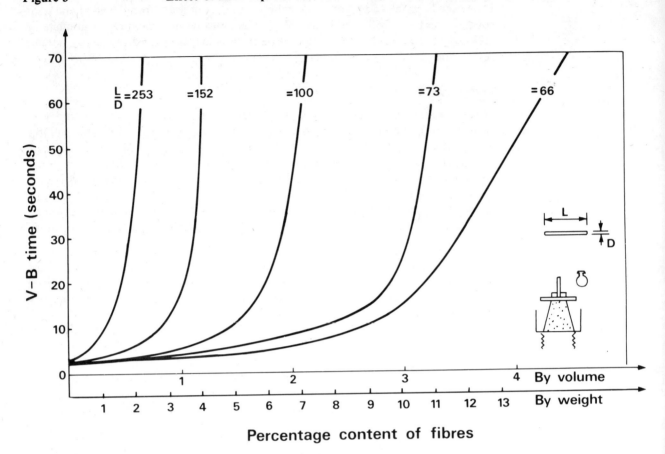

Fig 3. At each aspect ratio, a critical fibre concentration was identified beyond which the rate of increase in the V-B time with respect to increased fibre content approached infinity and the presence of aggregate particles less then 5 mm in size did not have any effect on this critical fibre concentration. For fibres which are not very stiff, for example polymer fibres, the flexural stiffness may also be an important parameter controlling workability.

As the properties of fibre composites are critically dependent on fibre orientation, it is imperative that suitable non-destructive methods be developed for assessing the distribution of the fibre in the matrix for on-line quality control of the product. A method applicable for glass fibres has been developed recently (21) which appears to have considerable promise. It has been found that glass fibre embedded in an opaque matrix is capable of transmitting light, by total internal reflection, over a distance of several centimetres. It is thus possible to use a technique of thick section transmission microscopy provided the depth of the section is sufficiently less than the fibre length. Fig 4a is a photograph of a section 1 cm thick taken from an aligned glass fibre reinforced cement composite with continuous reinforcement in the form of strands of approximately 200 filaments. The distribution and orientation of the strands is clearly visible. Fig 4b shows the individual filaments in one of the strands in the same section.

In cases where the X-ray density of the fibre phase is sufficiently different from that of cement paste so that useful contrast can be obtained, X-ray radiographic techniques may be suitable for unravelling fibre distribution. Such a method is being developed for steel fibres in cement (22). Also, using standard petrographic and metallographic techniques valuable information about the distribution and orientation of the reinforcement in the matrix can be obtained. A three-dimensional picture can be built up from examinations of several sections. The scanning electron microscope can also

Figure 4a   Section of an aligned glass fibre reinforced cement composite taken perpendicular to fibre direction (x5).

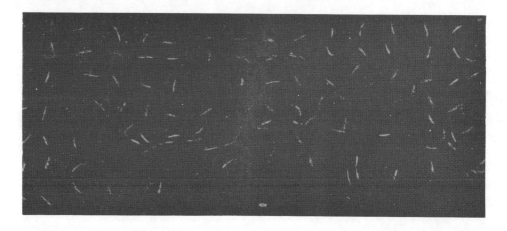

Figure 4b   Photograph of a single fibre strand in Figure 4a (21).

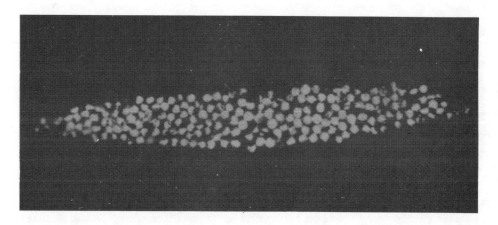

be used profitably (23) utilising its large depth of focus and the facility to obtain 'stereo pairs'. There is scope for developing suitable stereological techniques in this area of research.

Strength

Fibre reinforced materials offer substantial improvements over those of the matrix in various strength properties, the degree of improvement depending on the fabrication variables used. For established fibre composites such as those based on resin matrices some standard test methods have already been agreed upon. By comparison fibre cement composites are relatively new and testing procedures are still evolving. The details of the experimental procedures (including the curing of the composite) adopted by different investigators are, therefore, not given in this review and the readers are referred to the original papers.

Tensile strength: The strength of various types of fibre cement composites in direct tension has been measured using universal testing machines such as the Instron. Rectangular strips of the sheet-like composite materials have been tested most commonly but sometimes waisted (eg dog's bone) specimens have also been used. Edgington et al (20) used prism shaped specimens in their study of cementitious composites containing steel fibres and these were tested employing specially designed scissor-type grips. Others (24,25) have used flat, serrated or wedge-shaped grips. It must be remembered that the determination of the ultimate tensile strength (UTS) of a brittle-matrix composite such as fibre cement poses serious experimental problems. Slight misalignment of the specimen arising from non-parallel surfaces introduces bending stresses whereby the UTS is underestimated. The failure of exceedingly brittle specimens inside the jaw of the testing machine is another common occurrence.

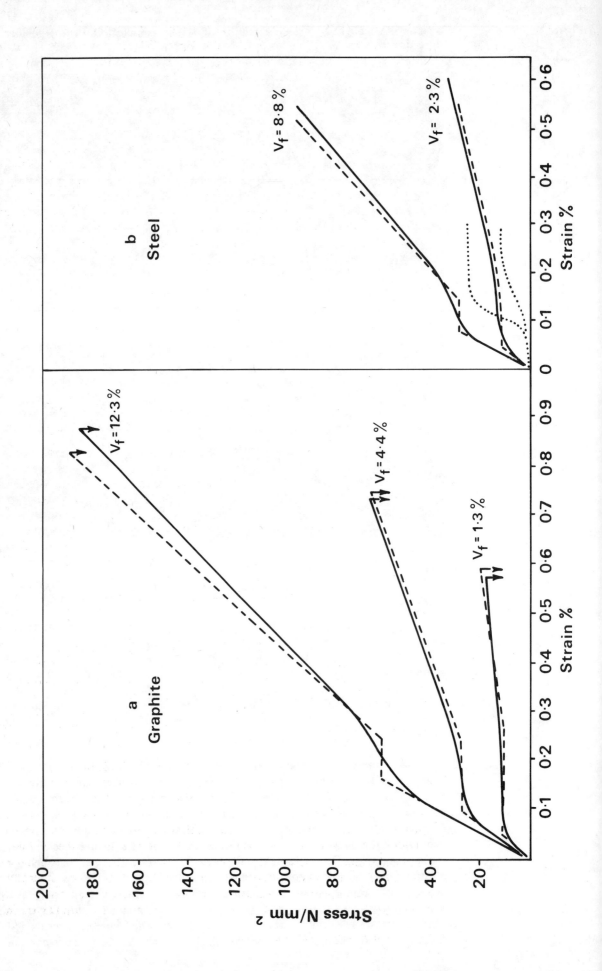

Figure 5 Tensile stress-strain diagrams for continuous carbon and steel wire reinforced cement: ———— experimental; – – – – theoretical; ·········· integrated acoustic emission in arbitrary units (26).

Strains suffered by the specimen in a tensile test have been measured by strain gauges or extensometers of a suitable gauge length. An increasing use is being made of linear variable differential transformer (LVDT) type transducers (20,26). The stress-strain curves of fibre cement composites are obtained by feeding the average output from a pair of LVDT's to the X-axis and the load output from the Instron to the Y-axis of an X-Y recorder.

The tensile stress-strain diagrams for continuous graphite and steel wire reinforced cement as determined by Aveston, Mercer and Sillwood (26) are reproduced in Fig 5. Full lines are the experimental curves, broken lines are the curves predicted by the ACK theory (1) and the dotted lines represent integrated acoustic emission plotted using arbitrary units. The experimental curves clearly show that the strain as well as the stress at the LOP increase with V_f. The failure strain of unreinforced cement paste cannot be determined very accurately—Aveston et al obtained a value of 0.02%—but even if one accepts the failure strain of 0.06% (typical in flexure) as being the upper bound, results depicted in Fig 5 leave no doubt that fibre incorporation increases the cracking strain of unreinforced cement pastes. A good agreement with theory is also noticeable. Laws (27) has confirmed a similar trend for aligned glass fibre cement composites and similar results were reported by Biryukovich et al (16) more than a decade ago.

Some idea of the degree of improvement in the UTS of cement pastes which can be effected by the addition of graphite or steel fibres can also be obtained from Fig 5 remembering that the UTS of neat cement pastes of w/c ratios used is probably not greater than 5 N/mm². Using glass fibres as the reinforcement and a special type of cement, Biryukovich et al (16) reported tensile strengths of the order of 60 N/mm² for composites containing 9 wt % fibre and cured under ambient conditions for two years.

These improvements, of course, refer to aligned composites containing appreciably larger quantities of fibre than is thought practical for real composites. In materials containing 2–3 vol % of chopped steel wire arranged in a three-dimensional random array, Edgington et al (20) did not obtain any significant improvement due to the fibre. The 28-day strength values of glass fibre cement composites containing different proportions of short alkali-resistant Cem-FIL* fibres in a random two-dimensional distribution are shown in Fig 6 (28). An increase in the stress and strain at the LOP with increasing V_f is indicated in this study also. Because of the uncertainty regarding the most appropriate orientation efficiency factor applicable to these composites it has not been possible to compare the experimental results with theoretical predictions in a rigorous manner.

Bending strength: The bending strength of fibre composites has been determined using either three-point or four-point bend tests carried out on a universal test machine such as the Instron. Different investigators have used different specimen sizes and spans and nearly all have expressed bending strength in terms of the modulus of rupture (MOR) calculated from simple homogenous-beam theory. In general, the effect of span depth ratio on MOR has not been studied. It is also worth pointing out that in the case of composites made from cement the shear strength increases with time and such a study needs to be conducted with both young and old samples.

For elastic materials, the MOR is ideally equal to the UTS. In practice, the MOR is usually greater than the UTS and for fibre cement composites its magnitude may be 2–3 times that of the UTS. Allen (29) and Aveston et al (26) have examined the flexural behaviour of such a material from a theoretical point of view. They explain the observed departure from ideal behaviour in the case of fibre cement composites on the ground of the progressive movement of the neutral axis towards the face in compression during a flexural test which results in an increase in the bending moment. Allen considers the bending test to be very insensitive to the way the material first cracks in tension. Although the tensile stress-strain curve exhibits a very sharp bend when the cement starts to crack, the curve of bending moment vs bending strain shows,

---

* Registered trade mark of Fibreglass Limited

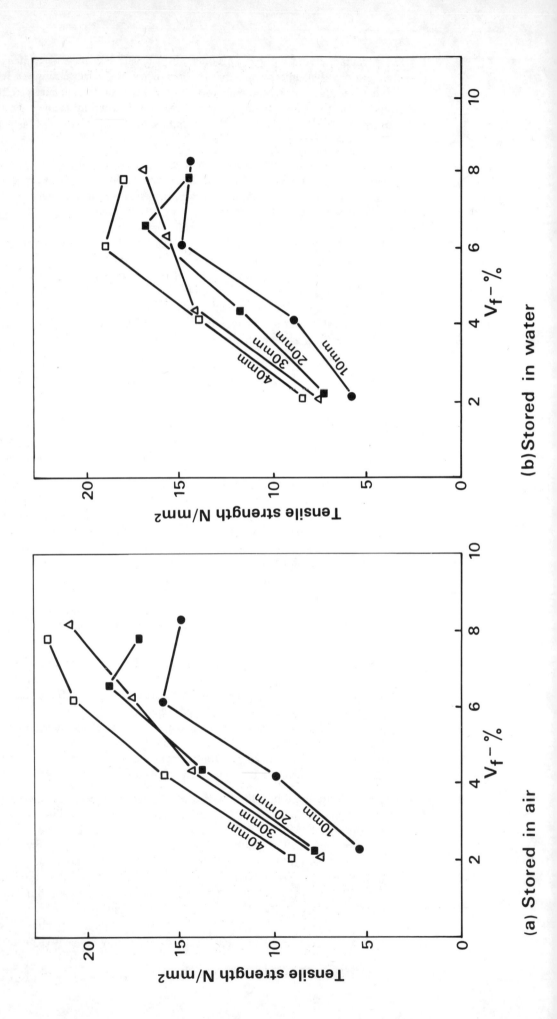

Figure 6 Relation between fibre volume fraction and tensile strength of glass fibre cement at 28 days for different fibre lengths (28).

in many cases, only the beginning of a deviation from linearity, which cannot be located with precision.

There is agreement amongst various investigators that the MOR of cement pastes are substantially improved by fibre addition. Even low-modulus fibres such as those derived from organic polymers are reported to increase the MOR of cement, albeit only by a small amount. At the other extreme, MOR values greater than 150 N/mm² have been obtained with aligned composites having approximately 8 vol % graphite fibres (30) and at 12.5 vol %, values as high as 295 N/mm² have been claimed (31). The Russian authors (16) obtained bending strengths of about 90 N/mm² by reinforcing special cements with 9 wt % glass fibres. These improvements are very considerable indeed.

As mentioned previously, when discontinuous short fibres are used there is a limit beyond which fibre addition does not improve the strength of the composites. This limit depends on fibre characteristics as well as the method of fabrication used in the preparation of the composite. Ali and co-workers (28) have investigated these optimum limits with glass fibres of different lengths. Their findings with 28-day old specimens prepared by the spray-suction method are shown in Fig 7. The apparent loss in the strength of the composite beyond a certain V_f is ascribable to an increase in its porosity. Fibre bundles are porous themselves and the introduction of large quantities of such reinforcements also requires greater amounts of mixing water; the two effects combine to reduce the compressive strength of the material.

Compressive strength: Unlike tensile and bending strengths the compressive strength of fibre reinforced cements has not been measured routinely, perhaps because they have been fabricated mainly as laminates. In composites fibres act as aggregates of a special shape and in view of their small percentages in practical materials they are not expected to have a predominant influence, the compressive strength depending largely on the consolidation of the composite. Briggs et al (30) have reported that the compressive strength of aligned graphite fibre composites consolidated at 7 N/mm² and tested parallel to the fibres decreases progressively as a function of increase in fibre content. 10 mm cubes were used in these tests which might have overestimated the compressive strength of the material. Using prisms and cylinders, Edgington and co-workers (20) have determined the compressive strength of steel fibre reinforced mortar. At fibre additions up to 4 vol % no large differences from the value of the unreinforced mortar were recorded although the results suggest that in composites with three-dimensional arrays of short fibre, small increases in the compressive strength can be expected.

Torsional strength: Edgington et al (20) have also measured the torsional strength of steel fibre mortar using an Avery torsion machine and loading at an angular twist of 1 degree/min. The maximum diagonal tensile stress at failure was calculated using elastic, plastic and semi-plastic analyses and the conclusion was reached that for fibre contents up to 4 vol %, the reinforced material had a slightly higher torsional strength than its unreinforced counterpart.

Shear strength: Information on the shear strength of a composite material and in particular its relation with the orientation of the fibre is very important in design. Fibre cement composites being relatively new products such information is not readily available. Briggs and co-workers (30) have reported on the shear properties of aligned graphite fibre composites and some preliminary results are also available for glass fibre cement.

(i) Interlaminar shear: This was determined for graphite fibre composites by three-point bend tests using short beams with span to depth ratios ranging from 2 to 5. Specimens with different proportions of high-modulus graphite fibre were investigated and it was concluded that the interlaminar shear strengths of composites containing 2 and 8 vol % fibres were at least 10 N/mm² and 16 N/mm² respectively. It is accepted that such a test does not achieve a state of pure shear and it is noticeable that the results obtained are much higher than that expected from the matrix.

Using a different method, described in ASTM(D 2733-70) in which direct inter-

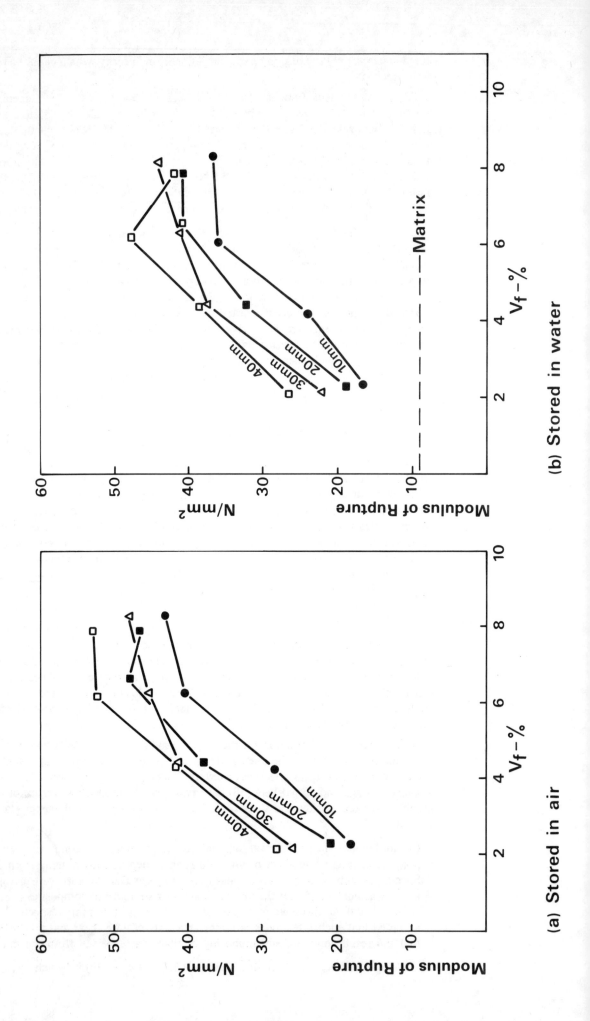

Figure 7 Relation between fibre volume fraction and modulus of rupture of glass fibre cement at 28 days for different fibre lengths (28).

laminar shear is produced at specimen failure, a much lower value, comparable to that of the matrix, has been obtained with glass fibre cement ($V_f = 0.04$) (32).

(ii) Transverse shear: This was determined by Briggs et al (30) on I-section beams in four point bending. The dimensions of the specimen were so chosen as to cause shear failure in the web. This did not happen always and the conclusion reached was that for high-modulus graphite fibre composites ($V_f = 0.05$) the transverse shear strength was not less than 17 N/mm². For a high-strength fibre composite the corresponding value ($V_f = 0.04$) was greater than 31 N/mm². This led the authors to conclude that the transverse shear strength of the composite is dependent on fibre strength.

Using the procedure described in the specification US Fed 406 B, a 'punch-through' shear strength of ~ 30 N/mm² has been obtained for the glass fibre cement composite described above (32).

Effect of polymer addition: It is well known that the addition of certain polymer emulsions increases the cracking strain of a cement-based matrix but reduces its stiffness. Kubota and Sakane (33) reported on the improvement in the tensile strength of polypropylene reinforced cement brought about by the addition of polyethyl acrylate to the matrix. Te'eni and Scales (34) have recently extended such studies. The effect of incorporating several types of polymer dispersions on the properties of glass fibre cement has been studied in detail recently at the Building Research Station (35) and the results will be presented in a separate paper at a later date. The main conclusion of this study is that the addition of polymers produces substantial improvements in the strength properties of the composite but the long-term retention of these benefits depends on the conditions of use. In wet environments polymer addition has little or no effect on the strength of the composite.

Polymer impregnation of fibre cement composites has also attracted the attention of several researchers recently (8). Initially a monomer is added when the composite is fabricated and polymerisation is effected in-situ subsequently by various means. Flajsman, Cahn and Phillips (36) have investigated the properties of polymer impregnated steel fibre reinforced mortars in some detail. From their experimental results there appears to be little doubt that, with steel fibres, polymer impregnation can produce a fibre cement composite which is strong, flexible, tough, impermeable and corrosion-resistant. Similar studies are currently in progress in several laboratories with composites containing other types of fibres.

Young's modulus

The static Young's modulus can be easily obtained from the tensile stress-strain curves (Fig 5). With small fibre additions which are practical in cements, only minor increases in stiffness are obtained in the composite. This is not a serious limitation since the matrix phase is reasonably stiff on its own. For aligned composites containing long fibres, the modulus is given, for all practical purposes, by the mixture rule which predicts that at the same V_f, the higher the fibre modulus the higher will be the modulus of the composite. Aligned composites containing 5 vol % high-modulus graphite fibres have given (30), from bend tests, Young's modulus values of the order of 27 GN/m², which is nearly twice that of the matrix used.

In glass fibre cement, the environment has a marked effect on the Young's modulus of the composite. For two-dimensional random composites containing approximately 4 vol % of 34 mm long alkali-resistant Cem-FIL glass fibres, typical values of the modulus at various ages under different storage conditions are given in Table 4.

It is clearly seen that storage conditions have a profound effect on the Young's modulus and some of the values have changed with time also. Understandably the composites cured in water have the highest stiffness, as this is related to the degree of hydration of the cement paste.

Dimensional stability

Portland and other hydraulic cements expand if allowed to set and harden in water. On drying they undergo a shrinkage which is only in part reversible if the cement paste is subsequently re-wetted. Irreversible drying shrinkage is from half to one third of the

Table 4   Tensile Young's modulus of GRC* in different environments at various ages.

Composite	w/c	Environment	Young's modulus (GN/m^2)			
			28 days	1 year	2 years	5 years
GRC	0.28	Air, 40% RH, 20 °C	–	22	21	20
GRC	"	Water, 20 °C	31	33	34	33
GRC	"	Weathering at Garston	–	28	28	27
GRC + 40% pfa	0.30	Air, 60% RH 20 °C	–	18	16	–
"	"	Water, 20 °C	–	26	17	–
"	"	Weathering at Garston	–	22	17	–
GRC + 10% styrene-acrylic polymer	0.25	Air, 60% RH 20 °C	27	20	17	–
"	"	Water, 20 °C	–	24	22	–
"	"	Weathering at Garston	–	21	20	–

* GRC stands for OPC composites containing Cem-FIL glass fibres and prepared by the spray suction method.

total shrinkage, which may amount to 0.2–0.3%. The effect of fibre addition on shrinkage and moisture movement has not been examined very thoroughly in most cases. Edgington et al (20) found that the presence of 2–4 vol % of short steel fibres did not reduce the drying shrinkage of mortars. Grimer and Ali (24) reported that although the drying shrinkage of glass fibre cement is reduced significantly with increasing glass content, even at 10 wt % fibre addition this amounted to only 20%, suggesting that the fibres modify the shrinkage of the matrix very little. On the other hand Briggs et al (30) have reported marked reductions (about ten fold) in both expansion and shrinkage of cements brought about by the incorporation of 5.6 vol % of high modulus graphite fibres. The shrinkage of the matrix in air at 60% RH was reduced by the same margin by which expansion in water storage, also at 20 °C, was reduced.

Dimensional stability is of critical importance to the performance of fibre cement composites in many applications and the subject will call for much closer attention in future.

Impact strength and fracture energy

An examination of stress-strain curves in tension and compression, load-deflection curves in bending or torque-twist curves in torsion obtained with cement composites reinforced with various types of fibre reveals that a considerable amount of energy is expended before such a material finally fails. This manifests itself in a very marked improvement in impact resistance and this feature perhaps constitutes the principal reason for adding fibres to a brittle matrix.

The resistance to impact can be measured empirically by a dropping ball test for example, or with a little more meaning by the conventional pendulum test methods, Izod or Charpy. Even in the latter methods the energy expended in the complete separation of the specimen is measured and this does not provide a valid assessment of the proneness of the material to crack development at any earlier stage of failure. Such information can be obtained by determining the critical stress intensity factor k_{1c} and estimating the work of fracture. Both can be derived from the load-deflection traces of slow bend tests of notched samples.

Figure 8a Impact fracture energy of composites as a function of volume fractions of graphite fibres.

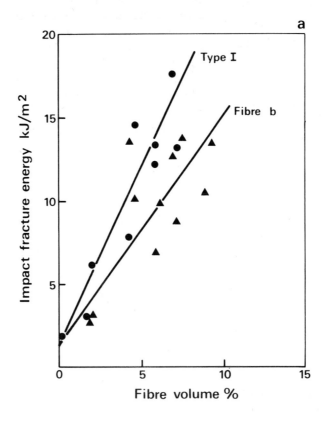

Figure 8b Work of fracture as a function of volume fraction of fibres.

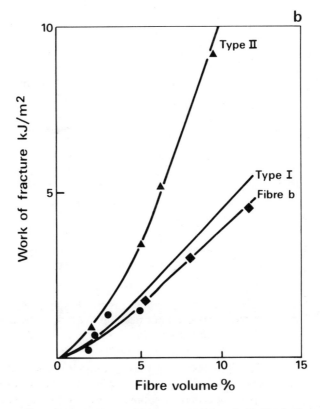

Fibre b is an experimental fibre with low modulus (110 GN/m^2) and low strength (270 N/mm^2) (30).

The Izod or Charpy impact strength of fibre cement composites increases linearly with fibre volume fractions and this does not seem to depend strongly on fibre orientations. The results obtained with aligned graphite fibre composites are shown in Fig 8a. Glass fibre cement, in which the distribution of fibre was random two-dimensional, also indicates a similar linear dependence (28). The estimates of work of fracture of aligned graphite fibre composites are plotted in Fig 8b as a function of V_f. These values were derived from load-deflection traces of three-point bend tests by electromechanical integration of areas under the curves. Comparison of Figs 8a and 8b shows that the energy absorbed during impact failure was much greater than the corresponding energy in the slow bend failure. The authors (30) ascribe this to the differences in failure modes operative in the two tests.

Brown (37) has pointed out that for non-homogenous solids such as fibre composites the critical stress intensity factor is not a material constant since the effective k_{1c} increases as the crack opens and fibre fracture or pull-out occurs. He has measured the 'apparent' fracture toughness k_a of glass fibre cement using 250 × 38 × 38 mm specimens having a 14 mm deep notch sawn in the centre of the tension face. The specimens were loaded in four-point bending. Computed values of k_a could be expressed as $k_a = 0.4 + 0.025\, C\, V_f$ where C is the crack growth in mm. When C is zero, $k_a = 0.4\, NM/m^{3/2}$, the value for the unreinforced matrix. This means that the toughness of the fibre reinforced material is no higher than that of the matrix at the initiation of crack growth, but thereafter increases with crack growth. A mechanism for arresting crack growth and avoiding catastrophic failure in fibre cement composites is thus provided.

Brown's observation that the apparent fracture toughness of fibre cement composites at crack initiation is not materially influenced by the presence of fibres is supported by the work of Harris, Varlow and Ellis (38) who used sand/cement mixtures containing random three-dimensional dispersions of 2 vol % steel and glass fibres. It must be stressed, however, that in these two studies fibre volume fractions were relatively small and it would be necessary to verify the conclusions for composites having larger fibre contents. From integrated load/deflection curves of notched specimens in bending Harris et al obtained estimates of the work of fracture and in the case of mortar reinforced with steel wires, an increase by at least two orders of magnitude over the work of fracture of the matrix was recorded. Improvements attained with glass fibres were less spectacular. These authors also measured the impact strength of some of their composites by the Charpy method using notched samples. In agreement with Briggs et al (30), they found that the impact strength values were higher than the corresponding work of fracture estimates but suggested that the Charpy test may be quite suitable for on-site testing of fibre cement composites.

Kelly (39) has recently summarised the current views on the fracture resistance of fibre composites. Obviously, the fibre parameters such as its tensile strength and aspect ratio are very important but equally important are the factors which control the interfacial bond strength in the composite which determines the degree of fibre pull-out. Bond strength can be improved by compacting the composite and Briggs et al (30) observed with graphite fibre composites that an increase in the compaction pressure brought about a reduction in the work of fracture. A similar reduction in the impact strength of glass fibre reinforced gypsum was reported by Ali (40).

Improvements in the impact resistance of cements can be realised by the addition of cheap natural vegetable fibres such as sisal or jute. Organic polymer fibres such as polypropylene do not develop a strong interfacial bond ($\tau \sim 1\, N/mm^2$) when placed in the cement matrix and their effectiveness in increasing the impact strength of cementitious materials has been known for some time (41). They, however, do not greatly improve the tensile strength. But if a combination of such a low-modulus fibre and a stiff fibre such as graphite or asbestos or even glass is used as reinforcement the resulting cement composites may be strong and at the same time have a high resistance to impact. Some studies along these lines are in progress in several laboratories. In this context it should be remembered that there is overwhelming evidence that addition of

Behaviour under load

polymers to the cement matrix also increases the impact strength of fibre reinforced composites quite substantially.

Creep: As the proportions of fibres in practical composites are usually very small, the creep behaviour of the reinforced material is likely to be controlled by the viscoelastic cement phase. Experiments with glass fibre cement composite ($V_f = 0.04$) made by the spray-suction process have shown (32) that at working stresses below the LOP, the bending creep is identical to that of the cement paste specimens prepared in the same way and is roughly proportional to the magnitude of the applied stress. It was also observed that the age and the curing history of the specimen had a marked effect on creep strain and creep rate, young and dry-stored materials creeping more rapidly. After several weeks under load these effects largely disappeared but the bending modulus was reduced appreciably. Preliminary creep measurements in tension support these observations.

The compressive creep data on steel fibre concrete gathered by Edgington and coworkers (20) over a period of 12 months indicate that the addition of steel fibres in concrete does not reduce the creep strains of the composite. Working with aligned graphite fibre composites Briggs et al (30), on the other hand, claimed that even in concentrations as low as 2 vol % fibres had a profound effect on flexural creep reducing the deflection by a factor of six. At higher fibre volume fractions the effect was even more pronounced. On removal of the load, creep recovery increased with V_f, again suggesting a contribution from the fibre. Obviously, much further work is required in this area before definite assessments regarding the influence of fibres on the creep properties of cement composites can be made.

Static fatigue: Under load, practical fibre cement composites crack at very nearly the matrix failure strain which is of the order of 0.02–0.06%, although their ultimate failure strength may be several times that of the matrix. When such a load is sustained over a long period of time, it is to be expected that for stresses in excess of that at the LOP deterioration of strength will take place. This is confirmed in the study by Briggs, Bowen and Kollek (30) with aligned graphite fibre composites. Several levels of stress were used, between 0.08 and 0.27 of the ultimate strength of the composite, and experiments were carried out in different environments—air, wet/dry and freeze/thaw and some of the tests lasted for 26 weeks. All specimens ($V_f = 0.07$) showed a consistent deterioration in strength while in the loaded condition and this loss ranged from 17% to 27% with respect to the unloaded condition.

Specimens of glass fibre cement ($V_f = 0.04$) are being weathered on the exposure site at Garston under a bending load. At stress levels below that at the LOP no significant reduction in the bending strength of the composite has been observed after one year although permanent deformation has taken place.

These results pose a serious question regarding the suitability of brittle matrix composites in truly structural applications.

Cyclic tensile tests: Allen (29) has presented stress-strain curves for glass fibre cement laminates loaded cyclically in tension (Fig 9). The envelope of the cyclic stress-strain curves is roughly comparable with the simple stress-strain curve (Fig 5) of such a material. It is clear that the tensile strain consists of a permanent deformation and a recoverable elastic strain, their respective magnitudes being related to the maximum total strain ϵ_{np} to which the specimen has been subjected. As ϵ_{np} increases, the permanent residual strain rises and the stiffness of the specimen E_n decreases (Fig 9c). At very small values of ϵ_{np}, the matrix is presumably uncracked and in such circumstances E_n should be independent of ϵ_{np} (idealised as AB in Fig 9c (inset)). There is little doubt that the sharp drop in E_n is associated with the development of transverse cracks. The limiting value of the asymptote corresponds roughly to the stiffness of the fibre array.

Similar information on cement composites made from other fibres is not available yet. Obviously the subject matter is of great importance in the design of components and further work in this area is urgently necessary.

Figure 9a   Tensile stress-strain curves for glass fibre cement under cyclic loading. For clarity, the vertical scale is halved.

Figure 9b   Method of analysing cyclic tensile stress-strain curve.

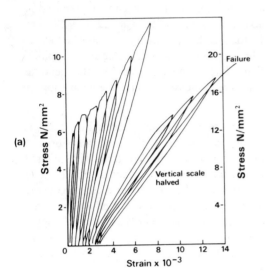

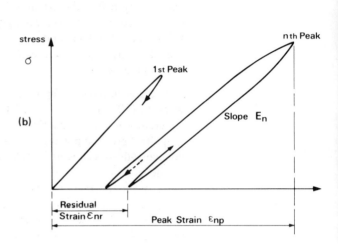

Figure 9c   Variation of stiffness (E_n) with previous peak strain (ε_{np}). Circles and crosses represent two different materials. DE corresponds roughly to the stiffness of the fibre array.

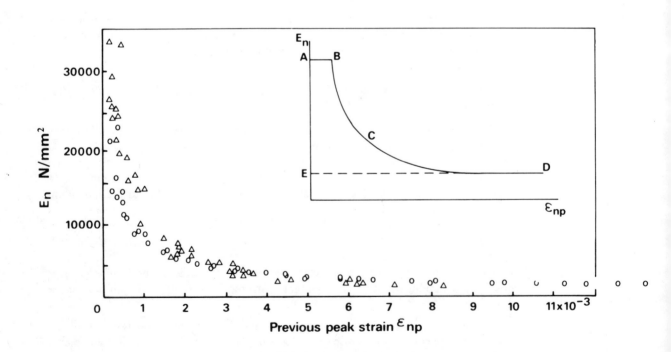

Figure 10a   Fatigue life of graphite fibre cement (30).

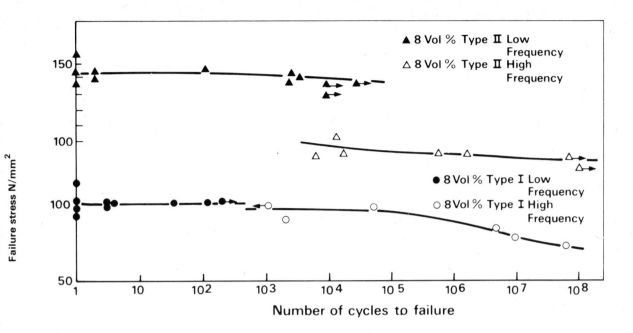

Figure 10b   Fatigue life of glass fibre cement (42).

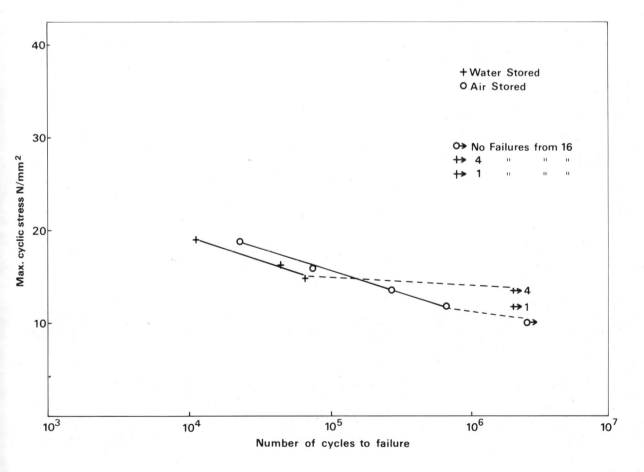

Dynamic fatigue: Briggs, Bowen and Kollek (30) have reported on the effect of repeated loading on the bending strength of aligned graphite fibre composites. Two types of fibre were used; for both V_f was 0.08, and the specimens were subjected to both low-frequency (30 cycles/min) and high-frequency (2000 cycles/min) loading. The results are shown in Fig 10a. Data points carrying arrows pointing to the right represent specimens that had not failed when a test was stopped. The points with an arrow pointing left represent specimens which failed in an unknown number of cycles. The tendency of the curve to level off around 10^8 cycles suggests that the material had fatigue limits in the region of 70 N/mm² for Type I fibre composite and 80 N/mm² for composites containing Type II fibres. These limits occurred at stresses much higher than the matrix cracking stress. This suggests that matrix cracking at relatively low stresses may not be very harmful to the composite performing at stresses below its fatigue limit.

Hibbert and Grimer (42) have determined the fatigue life of glass fibre cement (V_f=0.04) in four point bending using a frequency of 3 Hz with the stress varying between zero and a predetermined maximum. They developed a special rig capable of testing up to 16 specimens simultaneously and the results were presented in the form of Probability-Stress-Number of Cycles to failure (PSN) diagrams. It was not possible to determine the endurance limit but at stresses up to 90% of the stress at the LOP in static bending (10–13 N/mm²) a fatigue life of 10^6 cycles was recorded for specimens stored for one year under various environmental conditions including natural weathering. Some of these results are shown in Fig 10b. For a given stress the fatigue life increased with increase in V_f but at stress levels above the LOP, the fatigue life was reduced appreciably, its magnitude depending on the applied stress.

Crack detection

It is useful to have some idea of the onset and progression of cracking as fibre cement composite specimens are stressed. One method which is currently finding favour is based on the application of stress wave (acoustic) emission. Such emissions are produced when the stored elastic energy is released suddenly by a material undergoing deformation or fracture. A part of this energy propagates through the material as elastic wave and can be detected at the surfaces by sensitive transducers. In the work on wire reinforced cement described by Aveston, Mercer and Sillwood (26), the maximum integrated acoustic emission coincided with the minimum slope of the stress-strain curve. The authors concluded that the stress value at this point corresponded to the mean cracking stress of the matrix.

Edgington, Hannant and Williams (20) and others have used ultrasonic pulse time along the longitudinal axis of steel fibre reinforced specimens as an indicator of cracking. The flexure specimens are instrumented so that continuous recordings of the pulse time can be made at increasing loads. Changes in pulse time are considered to be indicative of cracking. Edgington et al observed that a change in pulse time occurred at a load lower than those which marked the change in slopes of either the load deflection or load-strain curves. They concluded that the presence of fibres did not have a significant influence in increasing the strain at which microcracking was initiated in the matrix. It is worth pointing out, however, that these specimens contained only small (V_f=0.03) amounts of fibre in approximately three-dimensional random arrangements and therefore the matrix failure strain is expected to be affected only marginally due to the presence of the fibres.

Application of X-ray radiography in detecting cracks is well known to concrete technologists and so is the method of crack decoration by fluorescent dyes. For identifying surface cracks the new technique of holographic interferometry using a laser beam (43) offers much promise.

Durability

Durability is defined here as the degree of retention of the initial mechanical properties of the hardened composite at long ages. Since civil engineering materials are expected to last for fifty years or more, often under severe environmental conditions, information on the long-term properties of fibre cement composites is essential if they are to be used widely. With new materials, this requires the development of accelerated testing procedures which, in the field of concrete technology, has had rather controversial history. It is self evident that we have to await the results of long-term field studies.

Among the fibres listed in Table 1 it is well known that natural vegetable fibres absorb water and are also attacked by the alkalinity of cement pastes. Synthetic organic fibres are supposedly immune from such attacks at ambient temperatures. Graphite fibres are likely to be unaffected even at relatively high temperatures. Steel is not attacked by the cement as such but if the alkalinity of the matrix is reduced by carbonation, and this is likely to happen if the composite is cracked, steel fibres may deteriorate on rusting. Silicate glasses are notoriously unstable in an alkaline medium and the durability studies on glass fibre cement composites are particularly important.

Using an accelerated form of curing, in water at 50°C, Ali, Majumdar and Rayment (44) have shown that the strength of graphite fibre composites is reduced only slightly with time under these conditions. Briggs et al (30) have extended such studies to other environments including wet/dry and freeze/thaw cycling. Under all conditions investigated composites containing graphite fibres of good quality did not show strength losses on ageing.

To overcome the lack of alkali resistance shown by borosilicate E glass fibres (16,24), new glass fibres containing Z_rO_2 have been developed (45) and some are now commercially produced (46,47). Glass wool has been produced from basalt in Czechoslavakia which also contains a small amount of additional Z_rO_2. Others have experimented with various types of resinous coatings on conventional glass fibres. Although the zirconia containing fibres are more resistant to attack by OH^- ions than E glass fibres, they are not totally immune from corrosion when placed in a cementitious matrix. The loss in strength suffered by this type of fibre in solutions of two cement extracts (9) at 20°C up to 2 years are shown in Fig 11. In these graphs measurements carried out on single filaments (dia $\sim 10\mu$m) are presented and although each experimental point is the mean of about twenty individual determinations, the data shown might not represent the strength of fibre bundles in aged composite specimens. Fig 11 shows a very sharp fall in the strength of the glass filaments in initial stages but the rate of deterioration has decreased substantially after 180 days and after 2 years strength values of the order of 1100 N/mm^2 are retained by the fibre. During this two-year reaction period, the Young's modulus of the fibre has not changed appreciably.

Attempts have also been made to extract fibres from composite boards cured in various environments. The results corresponding to dry and wet storage conditions are shown in Fig 12. It is seen that wet curing of the composite has a more deleterious effect on the strength of the reinforcement but even in this case strength in excess of 1100 N/mm^2 has been obtained for the fibres extracted from the board after 3 years. Tensile strengths of this order have also been measured with fibres removed from composite specimens which were exposed to natural weather at Garston, England for five years. It is well to remember that even this reduced strength of glass fibres is higher than that of typical organic fibres and is similar to that of some mild steel wires.

Experiments are being carried out (48) to ascertain whether asbestos fibres suffer any long-term loss in strength when placed in the matrix of ordinary Portland cement. The results to date indicate no large deterioration in the UTS of the fibre in the long-term. Chrysotile fibre bundles (dia 40–15 μm) extracted from a 7 year old asbestos cement sheet used outdoors have given a mean tensile strength of 415 N/mm^2 compared with the strength of 550 N/mm^2 obtained with similar bundles taken from fresh boards. It must be admitted that the results in these experiments have shown a large variability.

These results roughly correspond to the degree of long-term durability observed with glass and asbestos reinforced cement composites. The tensile and flexural strength of asbestos cement do not show any loss in strength on weathering whereas glass fibre cement strength decreases under similar conditions. The change in the flexural strength of cement composites reinforced with 4 vol % of a type of zirconia containing alkali-

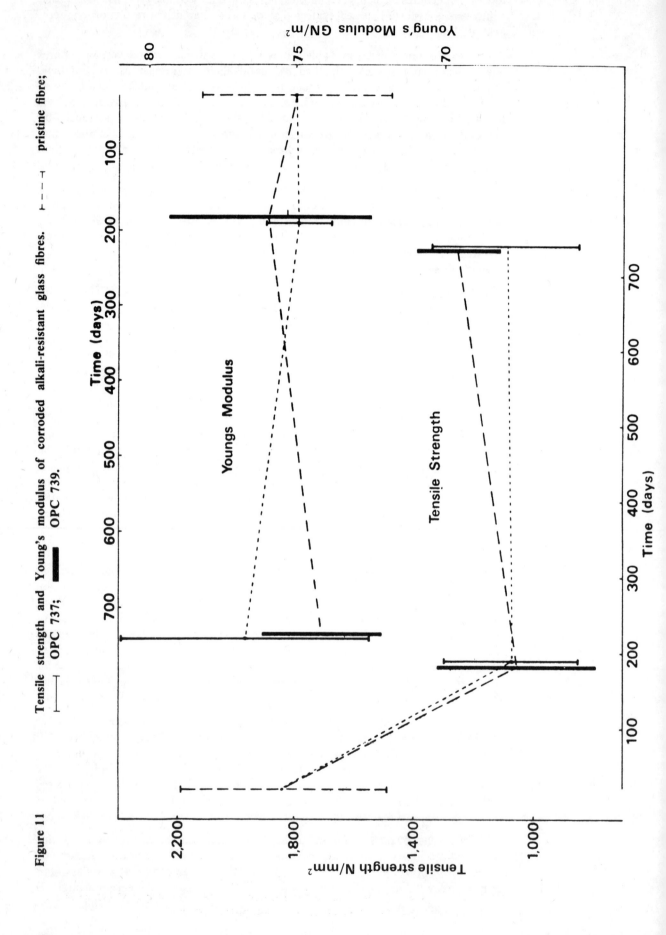

Figure 11 Tensile strength and Young's modulus of corroded alkali-resistant glass fibres. ⊢——⊣ pristine fibre; ⊢—⊣ OPC 737; ▬▬ OPC 739.

Figure 12a

Tensile strength of alkali-resistant glass fibres extracted from cement composites. ├──┤ control (chopped fibre); ▭ fibres from air stored composite, ▬ fibres from water stored composite.

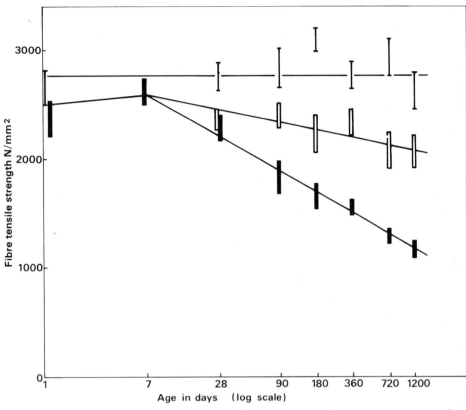

Figure 12b

Tensile strength of alkali-resistant glass fibres extracted from cement composites. ├──┤ control (chopped fibre); ▭ fibres from air stored composite; ▬ fibres from water stored composite.

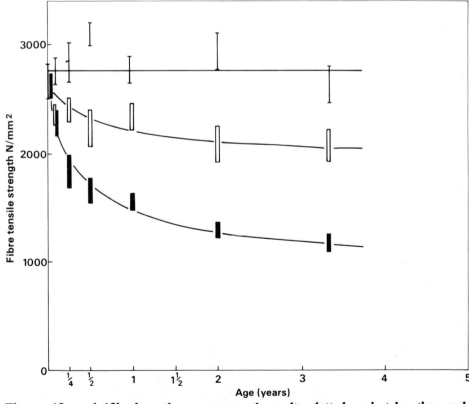

Figures 12a and 12b show the same strength results plotted against log time and time respectively. The bars represent 90% confidence limits.

Figure 13a   Age-strength relationship of cement composites made from Cem-FIL glass fibres in various environments: ▭ stored in air; 40% RH, 20°C; ■ stored under water; ▨ weathered.

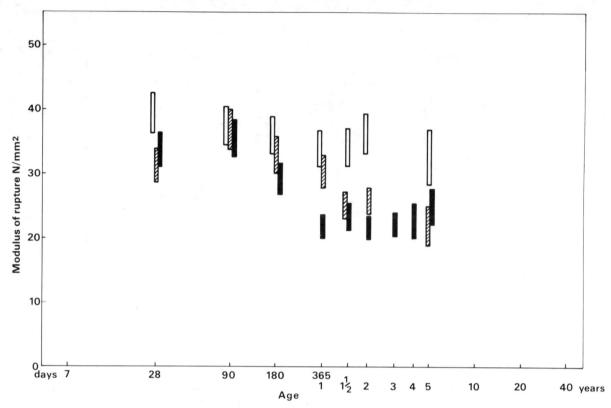

Figure 13b   Age-strength relationship of cement composites made from Cem-FIL glass fibres in various environments: ▭ stored in air, 40% RH, 20°C; ■ stored under water, 20°C; ▨ weathered.

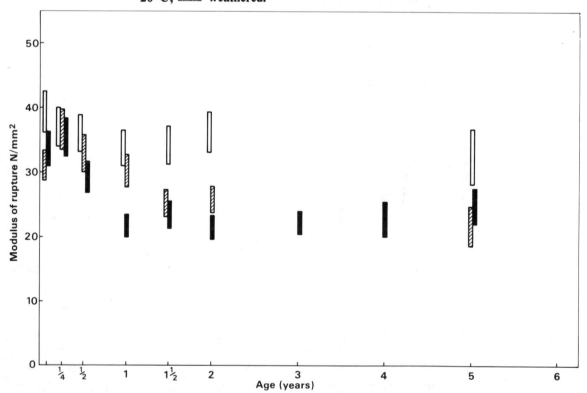

Figures 13a and 13b show the same strength results plotted against log time and time respectively. The bars represent 90% confidence limits.

Figure 14 The effect of environment on the stress-strain behaviour of glass fibre high-alumina cement:
(A) E-glass: (a) water stored, (b) air stored;
(B) Alkali-resistant glass: (a) water stored, (b) air stored (49).

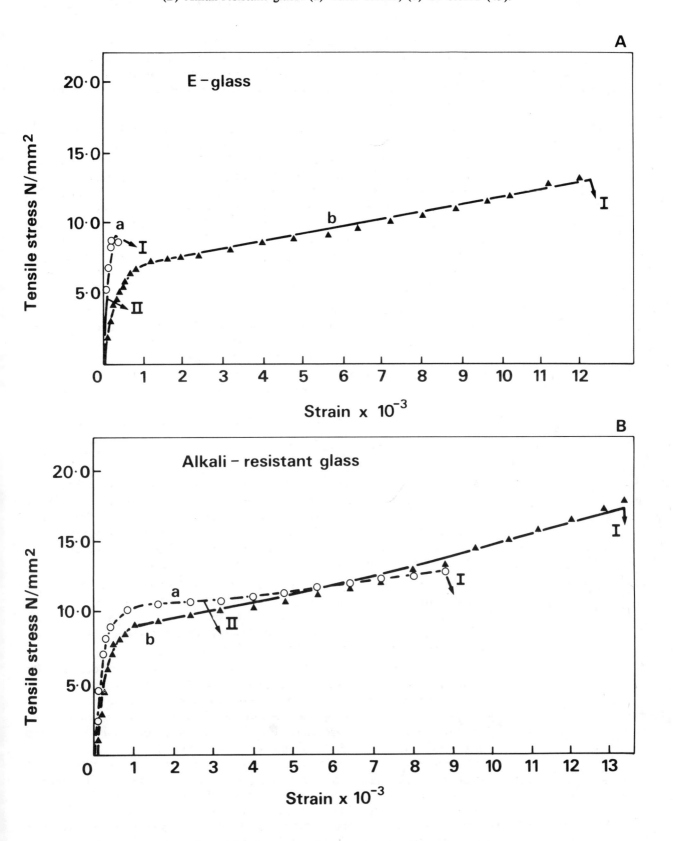

resistant fibre is shown in Fig 13 for three different environments. The bars represent 90% confidence limits. It is noticeable that in continuous water storage, there has not been any reduction in the strength after one year, but under natural weathering conditions prevailing in the south east of England, there probably has been a small reduction in strength. In dry air storage the durability of these composites is adequate for long-term applications.

It is becoming clear that the loss in the strength of fibre cement composites is not solely due to the OH⁻ attack on the glass. Equally important is the possibility of the development of flaws on the surface of the fibre resulting from crystal growth on or around it. And in this respect the microstructure of the paste in the vicinity of the fibre might play a crucial role.

For the reasons mentioned above it is not possible to predict the long-term strength properties of glass fibre cement from results obtained with single filaments such as those shown in Fig 11. The results presented in Fig 12 are more relevant but even here fibres attached to the cement (and therefore corroded to the greatest extent) cannot be removed from the composite. However, the current view assumes a residual fibre strength of the order of ~ 1000 N/mm^2 obtaining in cement composites after several years of weathering in a temperate climate. For reinforcing cementitious matrices fibre tensile strengths of this magnitude are adequate for many purposes providing, of course, there is no further reduction in the longer term. As the UTS of the fibre largely determines the critical fibre volume fraction, it is essential that reduced and not pristine fibre strength values are used in glass fibre cement mix designs.

A far more serious problem is posed by the progressive loss of the ultimate failure strain of glass fibre cement composites with time. The results obtained with high-alumina cement composites having two different glass fibres are shown in Fig 14 (49). In each case V_f was approximately 0.04 and 34 mm long fibres were distributed uniformly in the matrix in a two-dimensional random array. Points I and II refer to stress-strain properties at failure obtained with composite specimens after approximately one and four years respectively. In air storage, point I remains virtually constant up to four years but under wet conditions the composites suffer a serious loss in pseudo-ductility with time which is significantly less pronounced in the case of the alkali-resistant glass. Obviously, the loss in fibre UTS and the increase in the interfacial bond strength with time (which reduces the pull-out of the fibres) are important factors here. Furthermore, the deposition of the products of cement hydration inside the fibre bundle by solution-precipitation processes may also cause a loss of flexibility in the reinforcement. The composite will then fail in a single-fracture mode. This clearly happens in the case of E glass high-alumina cement composites cured in water for more than a year. With OPC, which is more corrosive than HAC, the gradual loss of ductility of the composite is likely to occur at an enhanced rate.

It is known that in the case of asbestos cement, where the fibres are not corroded to any great extent by the matrix, the impact strength of the material is reduced by about 50% of its initial value after 10–15 years exposure to natural weather (34). The explanation for this reduction may be the same as described above since asbestos reinforcement is exploited also in the form of bundles.

As mentioned previously, in order to overcome the loss in composite ductility with time and also to increase their alkali-resistance, attempts have been made to coat the surface of conventional borosilicate glass fibres with various types of resins. Several years ago the potentiality of specially coated glass fibre rods in prestressed concrete was explored (50) and later Klink (51) produced short thin rods by coating conventional glass fibre rovings with epoxy resins and used them in place of chopped steel wire in reinforcing concrete. A more recent study in this field is due to Roper, Stitt and Lawrence (52) who have tried various coating formulations on E glass fibre bundles in an attempt to make the composite more durable. One of the resin formulations (composite A) tried by these authors gave very promising results. This is shown in Fig 15 in terms of impact strength. One of the advantages claimed for this particular type of reinforcement was that it could be produced with a range of diameters. Those

Figure 15   Impact energy vs time for various cement products (52).

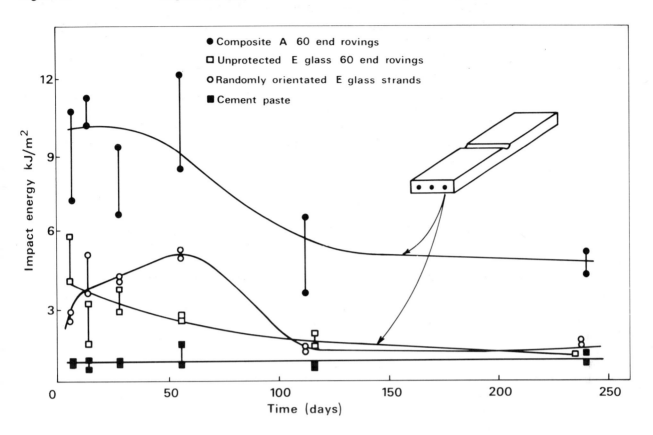

with small diameters could be chopped and sprayed as is the case with ordinary glass rovings. A fair amount of flexibility in composite manufacturing is thus assured.

It must be pointed out here that there is very little information in the current literature on the long-term durability of cement composites made with resin coated glass fibres. In one such study (53) where S glass fibre bundles coated with epoxy-phenol resins in the form of a tape were used to reinforce Portland cement mortars, the composite showed a pronounced loss in strength under static loading in a moist environment. Even in the unloaded condition, a reduction in bending strength of the order of 15% was recorded after 2 years. The authors concluded that even very small pinholes or flaws in the coating adversely affect the durability of the reinforcement and in this respect the ends of the fibre bundles are particularly vulnerable.

An alternative approach to ensure the long-term retention of the pseudo-ductility in glass fibre cement is to use a very small amount of a low-modulus fibre such as nylon or polypropylene with the glass reinforcement. A study of cement composites having mixtures of various types of fibre is in progress (54) in the author's laboratory and the tensile stress-strain curve of a composite having a two-dimensional random dispersion of a mixture of glass and polypropylene fibre is shown in Fig 16 (49). The composite in this case had received an accelerated form of curing. In this environment the composite failed initially at the LOP (point 1) but the polypropylene fibres were able to support the cracked composite at reduced loads. Even at very high strains the specimens were not physically separated into two halves. The stress-strain graph of a composite containing 4 vol % of glass fibre only and cured in air at 20°C is also included in Fig 16 for comparison.

CONCLUSIONS

In this review, an attempt has been made to present the current state-of-the-art picture of the possibilities of fibre cement composites as materials of construction by

Figure 16   Tensile stress-strain curves: ———— glass fibre cement; ·—·—· glass fibre cement with 1 vol % polypropylene monofilaments.

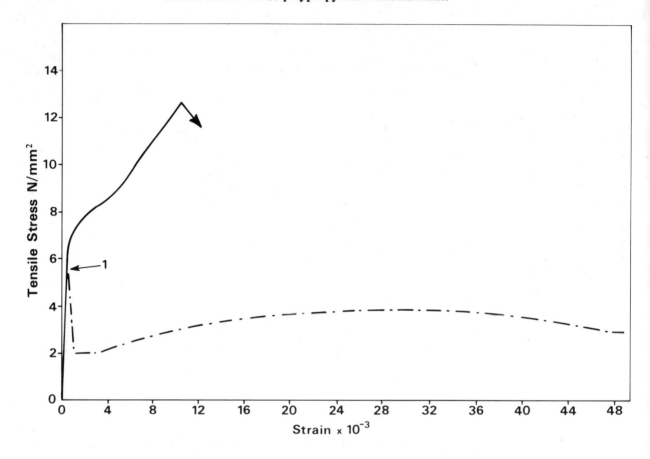

describing the properties which such composites have been found to possess. The long-term durability of these composites has been discussed in terms of glass fibre cement because (a) this type of fibre is particularly prone to attack by cements and (b) a considerable amount of systematic information is available on the long-term properties of such a composite. Similar data should be collected on cement composites containing other fibres which are being considered at present as reinforcements. The case of steel wire is particularly important as it is known to rust when the alkalinity of the cement is reduced by carbonation. Such a condition might prevail after the composite has cracked.

Careful attention should be given to the loss with time of the shatter resistance of those composites which are reinforced with fibre bundles and the mechanisms for such a degradation unravelled. It is more than likely that large scale use of fibre cement composites will depend on the retention of good impact strength they show at young ages. The shear properties of these materials should also be studied in more detail than heretofore.

In the final analysis, it will be the weathering characteristics which will determine the usefulness of fibre cement as a construction material. Acknowledging the fact that cement pastes, on their own, break up when weathered while asbestos-cement sheets do not, there is a high probability that many fibre cement composites will be reasonably stable over a period of many years when exposed to weather. In this respect, the design of the composite would be very important. The fibres should be finely divided, they should be evenly distributed in the matrix and in this condition they must retain a significant proportion of their pristine strength over a long period of time. Their aspect ratio should be so selected as to optimise ease of composite manufacture and the development of interfacial bond of appropriate strength. Some of these properties may be more easily obtained if the cement matrix is modified by external additions. The main design objective should be to ensure the multiple cracking of the matrix and fibre pull-outs in composites after many years of use.

ACKNOWLEDGMENT

The author wishes to thank Miss V Laws for reading the manuscript and offering valuable comments. The work described has been carried out as part of the research programme of the Building Research Establishment of the Department of the Environment and this paper is published by permission of the Director.

REFERENCES

1. Aveston, J, Cooper, G A and Kelly, A, Single and multiple fracture. Proceedings of a Conference on 'The properties of fibre composites', National Physical Laboratory, November 1971, published by IPC Science and Technology Press Ltd, pp 15-26.
2. Aveston, J and Kelly, A, Theory of multiple fracture of fibrous composites. J. Mater. Sci., Vol 8, 1973, pp 352-362.
3. Kelly, A, Some scientific points concerning the mechanics of fibrous composites. Proceedings of a Conference on 'Composites—standards, testing and design', National Physical Laboratory, April 1974, published by IPC Science and Technology Press Ltd, pp 9-16.
4. Romualdi, J P and Batson, G B, Mechanics of crack arrest in concrete. J. Eng. Mech. Div., Proc. Amer. Soc. Civ. Engineers, Vol 89, No EM3, 1963, pp 147-168.
5. Krenchel, H, Fibre reinforcement. Akademisk Forlag, Copenhagen, 1964.
6. Synthetic Resins in Building Construction, RILEM Symposium, September 1967, Editions Eyrolles, Paris, 1970.
7. de Vekey, R C and Majumdar, A J, Durability of cement pastes modified by polymer dispersions. Supplementary paper, VI International Congress, Chemistry of Cement, Moscow, September 1974.
8. Steinberg, M, Concrete polymer materials and its worldwide development. Polymers in Concrete, American Concrete Institute, SP-40, 1973, pp 1-13.
9. Larner, L J, Speakman, K and Majumdar, A J, Chemical interactions between glass fibres and cement. BRE PD 43/74 (in press).
10. Martin, H, The adhesion of the reinforcement in (steel) reinforced concrete. Radex Rundschau Vol 2, 1967, pp 486-509.
11. de Vekey, R C and Majumdar, A J, Determining bond strength in fibre-reinforced composites. Mag. Concrete Res. Vol 20, No 65, December 1968, pp 229-234.
12. Laws, V, Lawrence, P and Nurse, R W, Reinforcement of brittle matrices by glass fibres. J. Phys. D. Appl. Phys. Vol 6, 1973, pp 523-537.
13. Allen, H G, The strength of thin composites of finite width with brittle matrices and random discontinuous reinforcing fibres. J. Phys. D. Appl. Phys. Vol 5, 1972, pp 331-343.
14. Aveston, A, Fibre reinforced materials. Paper presented at the Spring Meeting of the Institute of Metallurgists, Practical Metallic Composites, Palma, Majorca, March 1974.
15. Tattersall, G H and Urbanowicz, C R, Bond strength in steel fibre reinforced concrete. Mag. Concrete Res. Vol 26, No 87, June 1974, pp 105-113.
16. Biryukovich, K L, Biryukovich, Yu L, and Biryukovich, O L, Glass fibre reinforced cement. Kiev, Budi velnik, 1964, trans. from Russian by G L Cairns, Civil Engineering Research Association, CERA translation No 12, London (1965), p 41.
17. Dempster, D P, Cement-coated tape and its possibilities. Concrete, Vol 7, No 12, December 1973, pp 34-35.
18. Steele, B R, Proceedings International Building Exhibition Conference on 'Prospects for fibre reinforced construction materials', London, November 1971, published by BRE, Watford, 1972, pp 29-39.
19. National Research and Development Corporation, London, Mixed fibre reinforcement of cement. UK Patent Application No 9537/73, February 1973.

20 Edgington, J, Hanant, D J and Williams, R I T, Steel fibre reinforced concrete. BRE Current Paper No 69/74, July 1974, p 17.
21 Hibbert, A P, A method for assessing the quantity and distribution of glass fibre in an opaque matrix. J. Mater. Sci. Vol 9, 1974, pp 512-514.
22 Swamy, R N, Sheffield University Private Communication, 1974.
23 Majumdar, A J, The application of scanning electron microscopy to textural studies. Proceedings No 20, Brit. Ceram. Soc., June 1972, pp 43-69.
24 Grimer, F J and Ali, M A, The strengths of cements reinforced with glass fibres. Mag. Concrete Res. Vol 21, No 66, March 1969, pp 23-30.
25 Bailey, J E, Barker, H A and Urbanowicz, C, Alumina filament reinforced cement paste. Trans. Brit. Ceram. Soc. Vol 71, No 7, 1972, pp 203-210.
26 Aveston, J, Mercer, R A and Sillwood, J M, Fibre reinforced cements—scientific foundations for specifications. Proceedings of a Conference on 'Composites—standards, testing and design', National Physical Laboratory, April 1974, published by IPC Science and Technology Press Ltd, pp 93-103.
27 Laws, V, ibid, pp 102-103.
28 Ali, M A, Majumdar, A J and Singh, B, Properties of glass fibre cement—the effect of fibre length and content. BRE PD 93/74 (to be published).
29 Allen, H G, Stiffness and strength of two glass fibre reinforced cement laminates. J. Composite Materials, Vol 5, 1971, pp 194-207.
30 Briggs, A, Bowen, D H and Kollek, J, Mechanical properties and durability of carbon-fibre-reinforced cement composites. Proceedings International Conference on 'Carbon fibres their place in modern technology', II International Carbon Fibre Conference, The Plastics Institute, London, February 1974, Paper No 17, p 8.
31 Waller, J A, Carbon fibre cement composites. Fibre Reinforced Concrete, American Concrete Institute SP-44, 1974, pp 143-161.
32 Oakley, D R, Pilkington Bros Ltd, Private Communication, 1974.
33 Kubota, H and Sakane, K, A study of the improvement of cement mortars by admixing polymer emulsion and synthetic fibre. Proceedings of the RILEM Symposium on 'Synthetic resins in building construction', 1967, Editions Eyrolles, 1970, pp 115-126.
34 Te'eni, M and Scales, R, Fibre reinforced cement composites. Technical Report 51–067, Materials Technology Division, Concrete Society, London, 1973, pp 30-31.
35 de Vekey, R C and Majumdar, A J, Polymer modified glass fibre cement. Unpublished Report, Building Research Station, 1974.
36 Flajsman, F, Cahn, D S and Phillips, J C, Polymer impregnated fibre-reinforced mortars. J. Amer. Ceram. Soc. Vol 54, No 3, 1971, pp 129-130.
37 Brown, J H, The failure of glass fibre-reinforced notched beams in flexure. Mag. Concrete Res. Vol 25, No 86, March 1973, pp 31-38.
38 Harris, B, Varlow, J and Ellis, C D, The fracture behaviour of fibre reinforced concrete. Cement and Concrete Res. Vol 2, No 4, 1972, pp 447-461.
39 Kelly, A, Microstructural parameters of an aligned fibrous composite. Proceedings of a Conference on 'The properties of fibre composites', National Physical Laboratory, November 1971, published by IPC Science and Technology Press Ltd, pp 5-14.
40 Ali, M A, Study of brittle fibre-brittle matrix composite system. M. Phil. Thesis, University of Surrey, 1972.
41 Goldfein, S, Fibrous reinforcement for Portland Cement. Modern Plastics, Vol 42, No 8, April 1965, pp 156-160.
42 Hibert, A P and Grimer, F J, Flexural fatigue of glass fibre reinforced cement. BRE, PD 51/74 (to be published).
43 Stroeven, P and De Hass, H M, Detection of cracks in concrete by holographic interferometry. Paper presented at the RILEM Symposium on 'New developments in non-destructive testing of non-metallic materials', Constantza, Rumania, September 1974, (to be published).
44 Ali, M A, Majumdar, A J and Rayment, D L, Carbon fibre reinforcement of

cement. Cement and Concrete Research, Vol 2, No 2, 1972, pp 201-212.
45 Majumdar, A J and Ryder, J F, Glass fibre reinforcement of cement products. Glass Technology, Vol 9, No 3, June 1968, pp 78-84.
46 Thomas, J A G, Glass fibre reinforced cement composites. Vol 2, No 2, June 1971, pp 95-97.
47 (a) Marsh, H N and Clarke, L L, Glass fibre reinforced cement-based materials. Fibre reinforced concrete, American Concrete Institute SP-44, 1974, pp 247-264.

(b) Majumdar, A J and Tallentire, A G, ibid, pp 351-362.
48 Larner, L J, BRE, Private Communication, 1974.
49 Majumdar, A J, Modification of grc properties. Proceedings of a Conference on 'Composites—standards, testing and design', National Physical Laboratory, April 1974, published by the IPC Science and Technology Press Ltd, pp 108-110.
50 Soames, N F, Resin bonded glass fibre tendons for prestressed concrete. Mag. Concrete Res. Vol 15, No 45, November 1963, pp 151-158.
51 Klink, S A, Fycrete, a new material for structures. PhD Dissertation, Renselaer Polytechnic Institute, Troy, New York, 1967.
52 Roper, H, Stitt, D and Lawrence, P, Properties of resin coated glass fibres as reinforcement in concretes, mortar and paste. Fibre Reinforced Concrete, American Concrete Institute SP-44, 1974, pp 221-245.
53 Cahn, D S, Phillips, J C, Ishai, O and Aroni, S, Durability of fibre glass—Portland cement composites. American Concrete Inst. J., Vol 70, No 3, March 1973, pp 187-189.
54 Walton, P L and Majumdar, A J, Cement-based composites with mixtures of different types of fibres. BRE PD 48/74 (to be published).

7.2 Validity of flexural strength reduction as an indication of alkali attack on glass in fibre reinforced cement composites

Eleanor B Cohen
Owens-Corning Fiberglas Corporation
and
Sidney Diamond
Purdue University

Summary

This paper examines the assumption that the characteristic drop in flexural strength of fibre cement composites starting at about four weeks reflects the effect of alkaline attack on the glass. Hot alkaline solution tests are considered to be misleading as indicators of alkaline attack on fibres in cement, both because of temperature effects and because of reduced availability of fluid to the glass within the cement composites. Fibres removed from cast AR-glass fibre composites which had suffered loss of strength starting after several weeks appeared to have retained the same level of strength as they had at one day of ageing. Microstructures within these composites are examined and show no direct evidence of attack. The chemistry of cement pore solutions suggests the possibility of long term alkali-silica reaction, but symptoms of such attack have not been observed nor are they predicted for alkali-resistant glass by standard susceptibility tests. Decreases in flexural strength have also been reported in composites having a low concentration of hydroxyl ions (high alumina and supersulphated cements). It is concluded, therefore, that the reasons for change in flexural strength of the cement composites at early and intermediate ages (four weeks to about six months) cannot be specifically attributed to alkaline attack on glass fibres. However, provision of alkali-resistant fibres remains potentially important in connection with the long-term durability of the composites.

Résumé

On examine, dans cette étude, l'hypothèse suivante: la perte de résistance à la flexion, caractéristique des composés de ciment F R (renforcé en fibre de verre), constatée à partir de quatre semaines après la prise environ, serait dûe à une attaque alcaline du verre.

On considère que les expériences menées avec des solutions alcalines chaudes sont des indicateurs trompeurs de l'attaque alcaline sur les fibres de verre incluses dans le ciment, tant à cause de l'effet de la température que du moindre contact fluide-verre dans les composés du ciment. Les fibres prélevées sur des composés en fibres de verre AR (résistant aux alcalins) qui ont subi une réduction de résistance commencant plusieurs semaines après la coulée, semblent avoir conservé leur résistance à 24 h. L'examen des microstructures de ces composés ne permet pas de conclure directement qu'il y ait eu attaque. La chimie des solutions aqueuses contenues dans les pores du ciment laisse croire à la possibilité de réactions lentes alcalins-silice, mais les symptômes d'une corrosion de ce genre n'ont pas été observés, et on ne saurait les prévoir à partir des tests standards de sensibilité pour les verres résistants aux alcalins.

Des pertes de résistance à la flexion ont également été rapportées dans des composés à faible concentration d'ions hydroxyles (ciments à forte teneur d'alumine et supersulfatés).

La conclusion est donc que l'on ne saurait attribuer à la corrosion alcaline du verre la perte de résistance des composés cimenteux d'âge jeune ou intermédiaire (4 à 6 mois). Cependant il reste important de fournir des fibres résistants aux alcalins afin d'assurer la durabilité à long terme de ces composés.

INTRODUCTION

The marked improvements in the mechanical properties of cementitious materials arising from the addition of relatively small volumes of glass fibres have stimulated considerable development effort in recent years in various countries. One concern limiting practical application of these developments is the potential durability of such composites, especially those made with Portland cement. According to Biryukovich et al (1), results of early tests carried out by Chinese workers indicated that strengths of composites made from Portland cement and "high alkali glass" fibres increased with age only up to about 28 days and subsequently decreased; at ages of the order of 5 months they were typically only 40 to 50 percent of the 28-day strengths. Apparently after this period "there is no further decrease in strength"

Work in the Soviet Union reported by Biryukovich et al was largely restricted to high alumina cement systems, with glass usually incorporated as continuous filaments, mats, or in ways other than randomly oriented chopped strands, which these workers considered to make relatively unsatisfactory composites. Under these circumstances no reductions of strength on ageing seem to have occurred.

In 1968 Majumdar and Ryder (2) confirmed that flexural strengths of composites based on Portland cement incorporating E-glass fibres were considerably lower at 90 days than at 28 days. A-glass and Pyrex glass composites showed similar trends, as did those made with several varieties of experimental alkali-resistant glasses, but the loss of strength for the latter was less marked. Similar tests were reported by Grimer and Ali (3) for E-glass bearing composites made with a variety of cement types, including high alumina and supersulphated cements as well as ordinary Portland cement. All appeared to show loss of strength with time. Further tests of composites made with Portland cement aged for longer periods were reported by Majumdar (4), Steele (5), Majumdar and Tallantire (6), and Majumdar (7). The pattern of loss of flexural strength with time varied somewhat with different test series depending on type of glass used, exposure, and presence or absence of supplementary additives such as pulverized fuel ash (fly ash).

Data for a large number of test series similar to those reported above have been accumulated by workers at the Owens-Corning Fiberglas (OCF) Technical Center, with substantially similar results. Figure 1 provides a sampling of some of these previously unpublished results, in the form of flexural strength ("modulus of rupture") values plotted against exposure time. The composites concerned were cast using 1.3 cm chopped strands of an alkali-resistant glass (OCF AR-glass) and Type I Portland cement at various water:cement ratios. The composites were hand-compacted into moulds, demoulded at 1 day, and exposed at 22 °C and 50 percent RH for various periods before testing. The general level of flexural strength attained is related to the fibre content and orientation of the strands. Essentially similar trends have been recorded for composites made by spray-suction techniques, and by composites aged under various environmental conditions different from air-curing.

The data of Figure 1 conform to the following pattern:

1) Increase in flexural strength with hydration of the cement, up to about 4 weeks, followed by

2) Irregular (and not necessarily reproducible) progressive decrease in flexural strength for a number of months, gradually changing to

3) Apparent long term stability, with significant changes not detected for prolonged periods.

This time trend is similar to that attributed by Biryukovich et al to Chinese experience with composites of chopped fibres and Portland cement, and is substantially similar to the published results, if not necessarily the interpretations, of the workers at the Building Research Establishment.

It has been generally considered (2,7) that this reduction of flexural strength is due to alkaline attack on the glass fibres as a consequence of the high pH developed in the

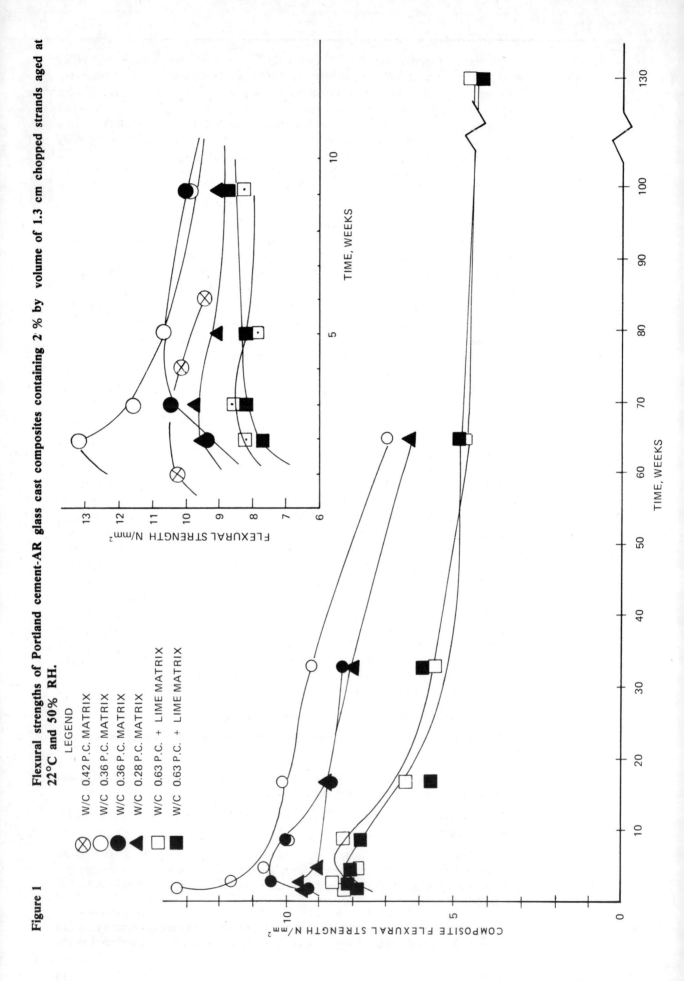

Figure 1 Flexural strengths of Portland cement-AR glass cast composites containing 2 % by volume of 1.3 cm chopped strands aged at 22°C and 50% RH.

internal pores of the hydrated cement paste matrix. This supposition spurred the development of alkali-resistant glasses, and commercial application of cement-based products containing these alkali-resistant glass fibres is underway.

Nevertheless, the evidence for the basic premise that the reduction in flexural strength which occurs after several weeks to several months of exposure is in fact due to alkaline attack on the glass seems to the present writers to be open to question. It appears to us that while durability of some glasses in Portland cement composites can be a potential problem (especially with high alkali cements), the specific phenomenon of the loss in flexural strength occurring after a few weeks or months may be due to some other cause.

DEGRADATION OF GLASS IN ALKALINE SOLUTIONS

Studies documenting the attack of hot alkaline solutions on various glasses have been reported by Majumdar and Ryder (2), and additional results appear in the patent literature (8,9). Three different conditions of exposure have been reported by the workers at the Building Research Establishment: immersion of glass fibres in saturated $Ca(OH)_2$ solution at 100 °C for 4 hours; immersion in 1 N NaOH solution at 100 °C for $1\frac{1}{2}$ hours; and immersion in a so-called "cement effluent solution" made up to resemble the composition of the solution separated from early-age cement slurries as reported by Lawrence (10).

The pattern of degradation of glass found to occur with these hot solutions was quite variable (8,9). For example, in saturated lime solution at 100 °C, E-glass fibres were reduced only slightly in diameter but very significantly in strength, while an alkali-resistant glass was hardly affected. In the hot NaOH solution, E-glass fibres were reduced in diameter by 59 percent in $1\frac{1}{2}$ hours (corresponding to an 83 percent loss of cross-sectional area), yet the remaining core apparently suffered no intrinsic loss in strength. In contrast, the alkali-resistant glass lost only 5 percent of its diameter (corresponding to about 10 percent of its area), but the strength of the undissolved core dropped significantly. It appears that the nature of the attack is different in the two tests, and the response of the glasses different in each instance.

It has been suggested that "cement effluent solution" might be more representative of the kind of attack to be expected in actual composites. Results at 80 °C indicated that most glasses tested in such solutions lost considerable strength over a period of 3 or 4 days (2). Trials reported in the patent literature (8) indicated that E-glass lost most of its strength after 24 hours of such exposure while alkali-resistant glasses suffered much less degradation. However the strength loss cited for a 16 percent zirconia glass was progressive with time, being 18, 31, and 57 percent, respectively, after 24, 48, and 72 hours. Experimental alkali-resistant glasses containing tin were shown to retain about 700 N/mm^2 tensile strengths after as long as 4 weeks in the hot cement effluent solution (9).

However, the applicability of all these results as indicators of what might actually be occurring in cement composites at ordinary ambient temperatures is questionable. The increase in temperature to 80 °C or 100 °C would not only be expected to increase reaction rates, but might activate reaction processes that could have effectively zero rate constants at the lower temperatures. In this connection, it is unfortunate that while Majumdar and Ryder (2) state that cement effluent solution tests were carried out at 50 °C and at 25 °C, the results of these tests at normal temperatures were not published.

Results of recent tests carried out at Owens-Corning Fiberglas Corporation are given in Figure 2. These results describe the load-bearing characteristics in tension of strands (not individual fibres) exposed to the "cement effluent solution" at 24 °C for many weeks. The initial strength levels are 1900 N/mm^2 for E-glass and 940 N/mm^2 for the AR-glass. The data are plotted in terms of percentage of the initial load-carrying capacity retained by the strand, rather than absolute strength, since reductions in cross-sectional area were difficult to measure reliably. The results indicate that E-glass strands at 24 °C undergo prolonged but gradual weakening over a period of about 9

Figure 2   Reduction of tensile breaking load with time for glass strands exposed to cement effluent solution at 24° C.

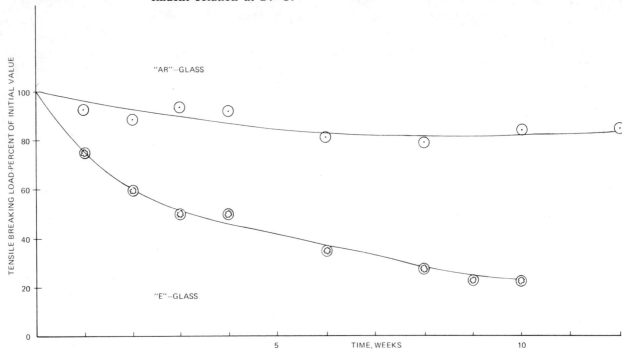

weeks to a level only about 20 percent of their initial value; AR-glass is weakened only slightly over a period of some weeks, and is thereafter stable against further strength loss, the long-term level being about 80 percent of the original value.

However, it is our opinion that the significance of these results should not be overestimated, since close correspondence between even room temperature exposure of glass fibres in solution and in cement paste is not likely. The physical exposure of the glass to the solution within the composite is significantly restricted. In the solution tests described above, the volume of solution is essentially infinite compared to the fibre volumes; in contrast, in the usual composite the total volume of pore solution present is only a few times that of the fibres even before set, and most of the water is withdrawn by incorporation into solid C–S–H gel and other reaction products as hydration proceeds. Furthermore, this limited volume of solution is largely constrained in small diameter pores within the set cement, and is not particularly mobile except for slow diffusion.

Perhaps of greater importance, it has been shown by Hadley (11) that during initial contact of fresh cement paste with a glass surface, a complete or virtually complete film of calcium hydroxide deposits next to the glass, backed with a single layer of fine fibrous C–S–H gel particles. Such layers are also seen when AR-glass fibres are in contact with tricalcium silicate paste (12). This duplex layer would interfere with access of the pore solution to the glass surface, the calcium hydroxide layer itself acting as a barrier toward further contact. Even if a layer of early solution is trapped between the glass and the calcium hydroxide layer, such a layer is probably much less alkaline than bulk solution developed on further hydration according to the results of Longuet et al (15), in the bulk of the pore solution.

In view of these factors, it is important to base our estimates of the severity of alkali attack on evidence derived not from attack on the fibres exposed to alkaline solutions, but on evidence from fibres taken from the cement composites themselves.

RETENTION OF STRENGTH OF AR-GLASS IN THE FIBRE CEMENT COMPOSITES

Evidence is now available indicating that the loss in flexural strength in fibre cement composites at intermediate ages is not due to reduction in strength of glass fibres as a

Figure 3   **Tensile strengths of AR-glass fibres removed from Portland cement composites after various periods of ageing at 22°C and 50% RH.**

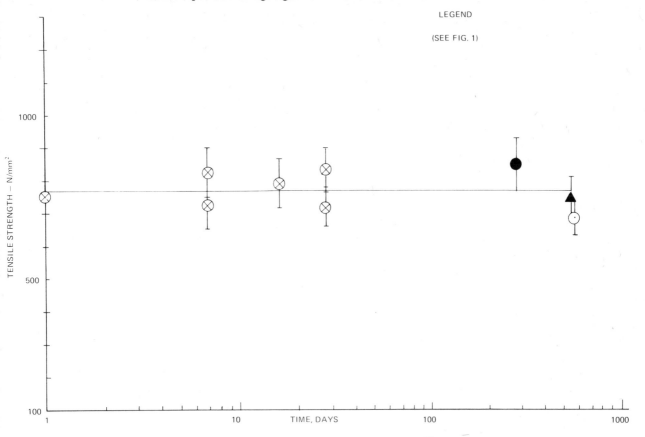

consequence of alkali attack on them.

The results of tensile strength tests on individual AR-glass fibres removed from several cast fibre cement composites at various ages are given in Figure 3. The series from which the fibres were removed is among those whose flexural strength versus time behaviour are illustrated in Figure 1. All showed the characteristic intermediate term flexural strength reduction.

The tensile strength measurements were taken using a Tecam$^{(R)}$ Micro Tensile Testing Machine, manufactured by Techne Ltd, Duxford, Cambridge, England. Each point illustrated in Figure 3 represents the average of individual tensile strength measurements for 10 to 15 different fibres removed from the same specimen. An indication of the spread of the data is given by the range shown about each point, the limits illustrated representing $\pm$ 1 standard error of the mean.

Unfortunately, all of the data do not represent results from a single series, data for such a series over the whole range of age not being available. The early points (up to 28 days) represent one series; each of the last three data points represent "flexural" specimens of three additional series. However, all of the specimens were fabricated and cured in the same manner, and individual details have been discussed in connection with Figure 1.

Initially it was found that the tensile strengths of the AR-glass batches used ranged from 1050 N/mm^2 to 1380 N/mm^2, these figures reflecting the condition after chopping but before the mixing process. The earliest strength value after mixing was that recorded after one day of curing 750 N/mm^2. As indicated in Figure 3, all of the tensile strengths of the fibres from all of the composite series shown did not vary significantly from this one-day value, even after more than 500 days of exposure.

The data suggest that the observed loss in flexural strength suffered by the composites starting after about four weeks is apparently not associated with a reduction in strength of the glass during this period since no such reduction seems to take place. Rather, the glass strength seems to drop at a very early age to a long-term consistent value.

ENVIRONMENT OF FIBRES IN CEMENT COMPOSITES

It was recently suggested by Majumdar (7) that the flexural strength of composites may be associated in some fashion with unspecified changes in the interfacial region between the glass fibres and the cement paste. Pull-out tests reported earlier by deVekey and Majumdar (13) had indicated that the interfacial bond strength increases with time, for all varieties of glass tested. It is presumed that Majumdar had in mind changes leading to such increased bond strength; at least such is the impression given in a recent survey of the current state of glass-fibre concrete development by Walgate (14). While it is obvious that increased bond strength might be detrimental to impact resistance, one would normally consider that its effect on flexural strength would be positive.

While the true cause of the degradation in strength of composites is not apparent to us, we feel strongly that it rests with the details of the peculiar physical and chemical environment surrounding the fibres. Existing publications in our opinion have not given an adequate picture of the situation, details of importance being omitted from the brief descriptions usually provided. In consequence, we now attempt to provide a description of some important factors which may affect the behaviour of the "typical" chopped strand fibre cement composites. A survey of the physical features of such interfacial regions for C_3S-based glass fibre composites has recently been provided (12).

Glass fibre strands when mixed with cement paste will retain their strand form, or will separate into smaller fibre groups or single filaments depending on the cohesiveness of the strand and the process used for making the composite. The fibre bundles that retain their strand form are surrounded by cement paste in such a manner that only the outer fibres of the strand are surrounded by cement paste. If water penetrates the strand, calcium hydroxide, C–S–H gel and other precipitates may deposit on the surfaces of the inner fibres; however, even if this occurs, the pore spaces between these inner fibres remain largely unfilled. Often little or no evidence of penetration of solution into these spaces can be detected. Where fibre bundles disperse into single filaments, cement paste usually surrounds each filament. These features are illustrated in Figure 4A and 4B, where part of the strand shown has separated into smaller groups of fibres and a few isolated single fibres. The interior space within the fibre groups has remained unfilled after 80 weeks of ageing. Neither the single fibres nor the fibres still in bundles observed microscopically show deterioration.

The composite shown in Figure 4 was a Portland cement–AR-glass cast composite of the series shown in Figure 1 as open circles. Further details of the

Figure 4 (A) Optical photomicrograph from a section of the composite indicated by open circle in Figure 3, illustrating the physical relationship of a partly dispersed strand to the surrounding matrix; (B) A higher magnification view, illustrating the contact of the cement paste matrix to the outside of the fibre groups and to isolated single fibres, and the lack of contact to interior fibres.

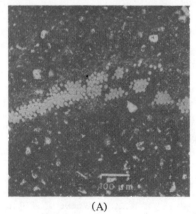
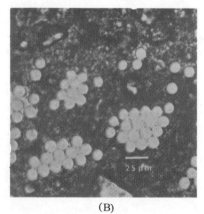

(A)   (B)

Figure 5   (A) Scanning electron photomicrograph of a fracture surface on the OCF AR-glass fibre Portland cement composite shown in Figure 4. The photomicrograph shows the outer portion of a glass fibre strand which had been in contact with the cement matrix for 80 weeks; (B) Interfacial area surrounding some of the fibres, showing the incomplete filling of hydrated cement between the outer layer of fibres, a film of Ca(OH)$_2$ backed with C-S-H gel surrounding the fibre itself and part of this film removed by fracture; (C) High magnification view of an exposed fibre from the same composite showing detail of the fibre surface itself.

(A)

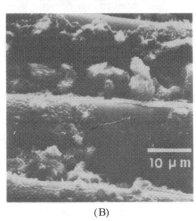
(B)

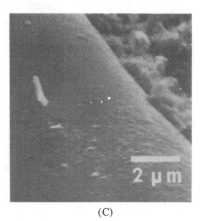
(C)

environment surrounding the fibres are provided in Figure 5. Figure 5A is a scanning electron micrograph of a group of fibres constituting part of the outer rim of an undispersed strand. It is apparent that the upper surfaces of these fibres, at least, have been in contact with the cement paste. As seen in Figure 5B, part of the surface of the fibres is still coated with what appears to be the duplex film of oriented calcium hydroxide backed by a layer of short C–S–H gel fibres as described by Hadley (11). In some areas the film has been removed in fracturing the specimen, exposing the underlying glass surface. Figure 5C is an enlarged view of the surface of another fibre from the same composite, and indicates the lack of visible damage at the glass surface. As indicated in Figure 3, the strength level of fibres removed from this composite is about 700 N/mm² (100 000 psi).

The influence of the chemistry of the pore solution may be quite important on the behaviour of the composite. Unfortunately, brief discussions of this factor in the literature on glass fibre composites have tended to be misleading. It is true that as has been indicated in the literature, the aqueous phase in Portland cement systems rapidly becomes saturated or slightly supersaturated with calcium hydroxide, and that in the absence of other alkali constituents such solutions have hydroxyl concentrations of the order of 0.04 N and pH levels of about 12.5. However, all Portland cements contain some sodium and potassium, and some have contents of these in excess of 1 percent Na$_2$O equivalent. In consequence, the aqueous phase commonly develops concentrations of 0.1 to 0.2 N in these cations at early ages (10,15) and corresponding hydroxyl concentrations usually also develop rapidly. The long term status of such fluids becomes even more alkaline; recently Longuet et al (15) expressed fluid from cement pastes up to a year old and found concentration levels as high as 0.7 N in hydroxyl ions, with corresponding pH levels as high as 13.7. Concurrently, the concentrations of calcium ion in solution decreased markedly (the "common ion" effect) and became so low as to be undetectable after several months. Thus the long-term status of cement fluids seems to be more nearly that of strong alkali hydroxide solutions than saturated calcium hydroxide solutions. In consequence, the potential for alkali attack on the glass certainly exists.

The suspected long-term alkaline attack on glass, especially in systems where high alkali Portland cements have been used, should be marked by characteristic symptoms of distress such as are found with conventional "alkali-aggregate" or "alkali-silica" reactions in concrete. These symptoms include softening of the affected glass, expansion, cracking of the matrix, and appearance of characteristic reaction product sol which permeates the cracks, and dries to a whitish powder on the exposed surfaces.

Such symptoms are normally developed when Pyrex glass is incorporated in high alkali cement systems, and measurement of the expansion that occurs forms the basis of a standardized test method (ASTM Designation: C 441–67) to check the efficacy of additives meant to overcome the effects of such reaction. Without such additives, Pyrex glass containing cement systems made with high alkali cements behave as typical alkali-silica reaction systems. Recently similar effects occurring when recycled waste glass was used as concrete aggregates have been documented by Johnson (16).

However, to the knowledge of the writers, no composite containing Portland cement and any of the recently developed alkali-resistant glasses has ever displayed any such symptoms of alkali attack.

In a recent trial carried out through the courtesy of Dr D V Braddon and Dr L J Stryker of the Westvaco Charleston Research Center, Charleston, S.C., OCF AR alkali-resistant glass has been used instead of Pyrex glass in the ASTM test mentioned above. Mortar bars made with the glass as aggregate and a high alkali cement were exposed for eight weeks in a humid cabinet at a temperature of 38 °C. No expansion was recorded and no visible symptoms of alkali-aggregate attack were displayed by the mortar bars.

Thus it appears that such attack does not occur in the normal sequence of events with alkali resistant glass fibre composites, and such attack is clearly not the source of the strength drops detected at early ages.

RESULTS WITH CALCIUM ALUMINATE AND SUPERSULPHATED CEMENTS

As previously mentioned, Grimer and Ali (3) showed that the characteristic strength reductions observed with Portland cement matrices also occurred when either high alumina cement or supersulphated cement was substituted for Portland cement, and further confirmation of such strength losses has been provided in the patent literature (17). It is well known that the pH levels of the aqueous fluid generated in such systems are very much lower than those generated with Portland cements, being of the order of 11.5 or less for the calcium aluminate systems and only about 10.5 for supersulphated cements. The hydroxyl ion concentrations corresponding to these values would approximate only about 0.003 N for the former and as little as 0.0003 N for the latter. It is extremely unlikely that solutions of such low hydroxyl concentrations would have any serious effect even on E-glass, and certainly none would be expected on alkali-resistant glass.

It is of considerable interest that the characteristic strength reductions were not found for calcium aluminate cement composites by the Soviet workers, as indicated by Biryukovich et al (1). However, these workers normally used continuous fibres or mats rather than chopped strands. Chan and Patterson (18) also failed to report such losses in tensile tests with calcium aluminate based composites in which chopped strand glass mats were used. One might speculate that the reduction in flexural strength is associated with some characteristic feature of composites prepared using individually dispersed chopped strands.

CONCLUSIONS

On the basis of a number of arguments and experimental results, it is concluded that the characteristic drop in flexural strength commonly observed to begin after about four weeks in glass fibre cement composites is not a result of alkaline attack on the glass. Experimental data from AR-glass composites appear to indicate that glass strengths remain essentially unchanged from the strength level attained after one day in the composite, and that glass fibres in cast fibre cement composites do not follow the progressive strength degradation behaviour exhibited in alkaline test solutions at high temperatures. No symptoms of alkali-silica reaction have been reported or observed. Furthermore, drops in flexural strengths are reported with chopped strand composites made with matrices of such lower alkalinity that alkaline attack is highly unlikely.

Despite the conclusion that alkali attack is not responsible for the intermediate-term flexural strength reduction, the very high pH levels attained in many cement pore fluids suggest that the alkali resistance of the glass fibre remains highly important to the potential long-term durability of glass cement composites.

ACKNOWLEDGEMENTS

The authors acknowledge the contribution of the Materials Analysis Laboratory and the Product Testing Laboratory of Owens-Corning Fiberglas for measurements made and to various members of the Research and Development Group for valuable discussions.

REFERENCES

1. Biryukovich, KL, Biryukovich, YL and Biryukovich, DL, Glass Fibre Reinforced Cement (in Russian), Budivel'nik Publishing House, Kiev, USSR, 1964.
2. Majumdar, AJ and Ryder, JF, Glass Fibre Reinforcement of Cement Products, Glass Technology, Vol 9, No 3, June 1968.
3. Grimer, FJ and Ali, MA, The Strengths of Cements Reinforced with Glass Fibres, Mag. Con. Res., Vol 21, No 66, 1969, pp 23–30.
4. Majumdar, AJ, Glass Fibre Reinforced Cement and Gypsum Products, Proc. Royal Soc. (A), Vol 319, 1970, pp 69–78.
5. Steele, BR, Glass Fibre Reinforced Cement in Prospects for Reinforced Construction Materials, Building Res. St., London, 1971, pp 29–36.
6. Majumdar, AJ and Tallantire, AG, Glass Fiber Reinforced Cement Based Materials, Fiber Reinforced Concrete Publication SP-44, Amer. Concr. Inst., Detroit, 1974, pp 351–362.
7. Majumdar, AJ, The Role of the Interface in Glass Fibre Reinforced Cement, Cement and Concr. Res., Vol 4, No 2, 1974, pp 247–268.
8. United Kingdom Patent Specification No 1 243 972, filed 4 Aug 1967, issued to National Research Development Corp.
9. United Kingdom Patent Specification No 1 307 357, filed 3 Apr 1969, issued to National Research Development Corp.
10. Lawrence, CD, Changes in Composition of the Aqueous Phase During Hydration of Cement Pastes and Suspensions, Special Report 90, Highway Res. Bd., Washington, DC, 1966, pp 378–391.
11. Hadley, DW, The Nature of the Paste-Aggregate Interface, Ph.D. Thesis, School of Civil Engineering, Purdue University, 1972.
12. Cohen, EB and Cohen, CI, Hydration Morphology of Tricalcium Silicate at Fiber-Paste Interfaces, paper presented at 76th Annual Meeting, American Ceramic Society, Chicago, May 1974.
13. deVekey, RC and Majumdar, AJ, Interfacial Bond Strength of Glass Fibre Reinforced Cement Composites, J. Mater. Sci., Vol 15, 1970, pp 183–185.
14. Walgate, D, Bright Future for Glassy Concrete, New Scientist, 24 Oct 1974, pp 276–278.
15. Longuet, P, Burglen, L and Zelwer, A, The Aqueous Phase of Hydrated Cement (in French), Rev. des Mater. de Constr., No 676, 1973, pp 35–41.
16. Johnston, CD, Waste Glass as Aggregate for Concrete, J. Testing and Evaluation, Vol 2, No 5, 1974, pp 344–350.
17. United States Patent No 3 783 092, issued 1 Jan 1974 to National Research Development Corp.
18. Chan, HC and Patterson, WA, Effects of Ageing and Weathering on the Tensile Strength of Glass Fibre Reinforced High Alumina Cement, J. Mater. Sci., Vol 6, 1971, pp 342–346.

7.3 Microstructural features in glass fibre reinforced cement composites

A C Jaras and K L Litherland
Pilkington Brothers Ltd, Lathom, Ormskirk, Lancashire

Summary   *A description is given of the general microstructural features of glass fibre reinforced cement (GRC). Cement matrix types considered are rapid hardening Portland cement and Portland cement-fly ash mixes. The techniques employed include optical microscopy, scanning electron microscopy and electron microprobe analysis.*

The course of hydrolysis of the cement clinker close to the glass fibres and the build up of hydration products around the fibres and in voids are discussed in relation to the matrix hardening conditions.

The effect of the cement environment on the corrosion of the glass fibre surface is discussed with special reference to a comparison between fibres of E-glass and high zirconia compositions.

Résumé   *Étude au microscope optique, au microscope électronique et à la microsonde électronique des microstructures des ciments armés à la fibre de verre. Les matrices traitées se composent du ciment de Portland au durcissement rapide et des mélanges ciment de Portland-cendres volantes.*

On commente le cours d'hydrolyse du ciment auprès des fibres de verre et l'accumulation des produits hydratés autour des fibres et dans les vides par rapport aux conditions de durcissement.

L'influence du milieu cimentaire sur la corrosion de la surface des fibres est traitée et on compare à cet égard les fibres de verre E et celles de verres à haute teneur en ZrO_2.

INTRODUCTION

As with any material an understanding of glass fibre reinforced cement (GRC) composite behaviour must be based on an appreciation of its structure. This paper describes some of the observations made on the microstructure of GRC and attempts to relate these observations to other properties. Some microstructural observations have been reported previously (1).

With GRC composites, in addition to the inherent characteristic structural features of the glass fibre strands and the hardening cement paste, there are also interaction features. These vary from those produced by the method of incorporation of the glass fibres into the cement matrix to those caused by chemical changes at the molecular level during the curing and life of the material. Each of these features has a particular size scale and in describing the structure it is convenient to start with the coarser and then proceed to the finer features.

EXPERIMENTAL

GRC composites were made using ordinary or rapid hardening Portland cement at a water/solids ratio of 0.3, reinforced with about 4% by volume of 38 mm long bundles or strands of 204 glass filaments, each fibre being of about 13 μm diameter and of either E or an alkali-resistant glass composition.* The glass strands were incorporated by the method of spray suction or direct spray as described previously (2). The details of setting and hardening the matrix are given in the test for each particular sample discussed.

Polished sections were prepared by vacuum impregnation of composite specimens with epoxy resin. These were then mounted in blocks of resin and polished using conventional petrographic and metallurgical techniques. Microscopic observation was made using reflected Nomarski light.

Samples of composite for stereoscan observation were prepared by cleavage parallel to the plane of orientation of the fibre and coating with gold palladium. Fibres for stereoscan examination were initially released from the cement matrix by ultrasonic agitation in ice cold 20% HCl solution. The separated fibres were then further cleaned in the same way, washed and deionised water and dried.

GENERAL FEATURES OF THE STRUCTURE OF GRC COMPOSITES

Figure 1 shows a fracture surface produced by cleavage in the plane of a sheet of GRC material which had been continuously hardened and cured in water at 22 °C for 2 months. This shows the glass fibre reinforcement arranged in flat taped bundles (of up to 204 separate parallel fibres) and indicates that there are substantial regions of matrix in which there are no reinforcing fibres. The arrangement of the reinforcing fibres is in direct contrast to the filamentised and dispersed asbestos fibres in asbestos cement and has important consequences with regard to the mechanical behaviour of the composite (3).

Cleavage of specimens parallel to the bundle A demonstrates the 2-dimensional arrangement of the reinforcement in the composite. With this structure it is important in composite manufacture to obtain a uniform deposition of fibre bundles throughout the thickness of the composite in order to provide the maximum contact of cement with the glass fibre bundles.

The gross voids seen here, of 0.2–0.4 mm diameter, are caused by air trapped in the wet cement paste. These tend to reduce the strength of the hardened cement and may reduce the efficiency of fibre bonding (see below).

CEMENT ENVIRONMENT AROUND AND WITHIN THE FIBRE BUNDLE

Figure 2 is an optical micrograph of a polished section perpendicular to the glass fibre bundle and demonstrates features at a higher magnification which are related to the

---

*Alkali-resistant fibre is sold under the trade mark Cem-FIL by Fibreglass Limited.

Figure 1　　Stereoscan photograph of cleavage surface in GRC composite after two months in water at 22 °C (bar = 500 μm).

Figure 2　　Optical photo-micrograph of fibre bundle in cement paste after 3 days at 100% RH (bar = 150 μm).

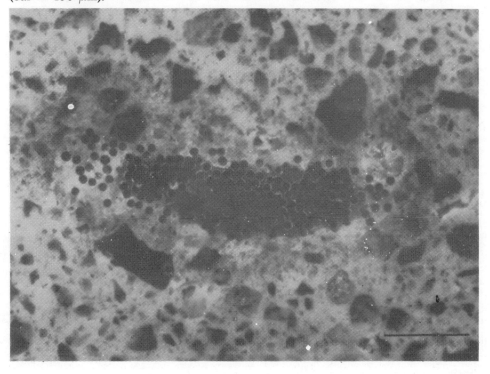

329

microstructure of the glass fibre bundle and the cement clinker. The composite specimen in this case had been hardened for 3 days at 100% RH and ambient temperature.

The lens or tape shaped arrangement of fibres within the bundle (of approximate dimensions 0.6 x 0.15 mm) contains variable packing of the fibres and leaves cement free voids within the bundle. The generally close-packed nature of the fibres precludes the relatively large (2–85 μm) cement clinker particles from entering the bundle and it is only in those regions where the fibres are more dispersed, i.e. at the points of the lens, that the whole circumference of the fibres is incorporated into the fine 'gel' products of cement hydration.

A study of a sequence of polished cross-sections of composite taken as cement hydration proceeds shows that with time, especially in water storage conditions at

Figure 3 (a) Stereoscan photograph of void in five-year old, air stored GRC sample (bar = 15 μm); (b) Stereoscan photograph of calcium hydroxide filled void in five year old, water stored GRC sample (bar = 50 μm).

elevated temperatures, the voids within the bundle (interstitial voids) become filled with crystalline material (independently confirmed in unpublished BRE studies). With an OPC matrix, electron probe analysis has shown that this interstitial material is substantially pure calcium hydroxide.

The filling of voids with calcium hydroxide crystals in products based on Portland cement has been reported previously (4). A supersaturated solution of calcium hydroxide resulting from hydration of the calcium silicate phases fills the voids. Crystals can then nucleate either on the walls and flow inwards towards the centre of the void or they can nucleate on the surface of the glass fibre and form outwards from this surface. The rate of development of crystals in the solution filled interstitial fibre voids is increased at higher temperatures and for continuous water immersion of the composite. However, for a variable hardening environment such as natural weathering the rate depends on the proportion of time the interstitial voids are liquid filled. This is illustrated in Figures 3 (a) and (b) showing stereoscan views of trapped air voids in GRC samples which were stored in air and water respectively for 5 years.

The void in the air stored sample shows no evidence of filling with calcium hydroxide whereas the void in the water stored sample has been filled with well developed crystals. Natural weathering conditions give filled-void characteristics closer to the water stored sample than the air stored sample after the same period of time.

Observations have also been made on the crystalline material which is deposited in the bundle voids from hydrating OPC-Fly ash (60:40 blend) matrix. After 2 years of hydration by water immersion, electron probe analysis of the crystals showed that they contained 45% by weight of CaO, 15% Al_2O_3 and 10% SiO_2 demonstrating that the alumina and silica components of the fly-ash were being taken into solution. This may well be the reason for the reduced alkalinity of the aqueous phase in a hydrating OPC—Fly ash matrix.

FIBRE-CEMENT BOND FORMATION IN GRC

The nature and strength of the bond between fibre reinforcement and matrix are of undoubted importance in determining the mechanical properties of GRC composites (5). In the previous section some of the gross features of the interaction between fibre bundle and matrix were described. Some additional details are reported below.

Regions such as that marked A in Figure 1 are the imprints left by a stripped-out fibre bundle. These permit examination of the areas where hydrating cement has come into closest contact with the outer fibres in the bundle, and reveal the finer features of the interaction between cement and fibres.

Figure 4 shows higher magnification views of two such regions. In (a) the cement matrix has set and hardened for three days at 100% RH at ambient temperature and in (b) for 7 days at 100% RH followed by 5 years in water at 18 °C.

Figure 4(a) shows that, in the early stages of cement hydration, the area of actual contact between the matrix and the outer fibres is considerably less than the geometrical area of the fibre surface. This arises because of the discrete nature of the calcium silicate hydrate structures (A). At this stage the hydrates have not developed fully and at the same time the voids between them have not yet been filled with calcium hydroxide. In Figure 4(a) it can be seen also that the hydrate structures have flattened surfaces where they accommodate to the fibre surface during growth. Another interesting feature is marked B in Figure 4(a). This is a large calcium hydroxide crystal which has formed during the initial hydration and illustrates the inhomogeneity of the matrix when observed at this scale of magnification.

The region marked A in Figure 4(b) illustrates the extent to which, after 5 years hardening in water, the cement hydration products have formed an intimate association with the fibre surface. At the same time, as described previously, the voids in the fibre bundle have also been filled with calcium hydroxide.

Under wet storage conditions, the reinforcing element in the cement matrix has thus changed from (originally) a bundle of fibres with the outer fibres only in poor contact

Figure 4   Stereoscan photographs of fibre/cement contact zones (bar = 5μm):
(a) After hardening and setting for three days at 100% RH;
(b) After seven days at 100% RH and five years in water at 18 °C.

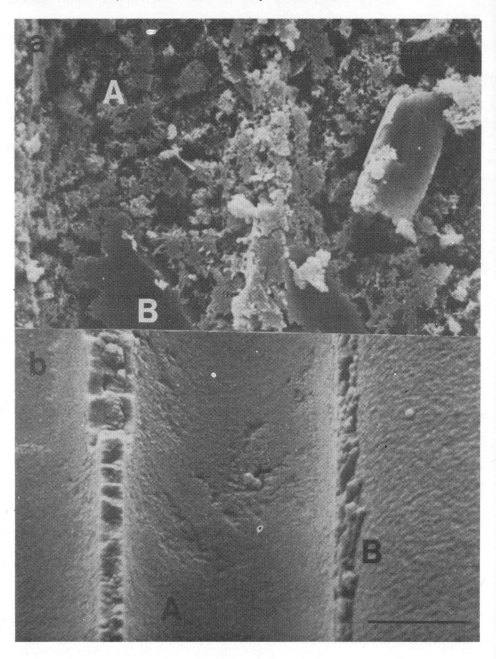

with the hydrating cement to (finally) individual fibre elements in intimate contact with cement hydration products and grouped in clusters throughout the matrix. These features have to be incorporated into any realistic model describing the mechanical behaviour of GRC.

The changes in the bond microstructure and reinforcement element type, which are reflections of the matrix changes, occur at different rates and to different extents depending on the conditions under which the composite is exposed. For example in natural weather exposure of GRC under temperate climatic conditions the transformation from the structure in Figure 4(a) to that in (b) occurs over a longer time span than water immersion and it is not complete at 5 years. After dry air storage for the same period the structure is still much more closely related to that shown in 4(a).

MICROSTRUCTURAL FEATURES IN THE GLASS FIBRE SURFACE

Microstructural features in the glass fibre surface are important from two points of view. Firstly surface corrosion will normally reduce the strength of the fibre since this is almost always determined by the size and number of flaws in the fibre surface. The type and extent of surface corrosion will therefore clearly determine the magnitude of fibre strength loss. Secondly the chemical continuity of the bond between the fibre and the hydration products will increase if corrosion takes place at the same time as, or subsequent to, the deposition of these products around the fibres. Increased bond continuity between glass fibre and cement matrix will reduce the crack stopping property of the interface, leading to a decrease in composite toughness.

Both these features can be observed in GRC most clearly and severely when E-glass composition strands are used as reinforcing elements.

Figure 5   **Stereoscan photographs of the surfaces of extracted and cleaned fibres (bar = 15 μm):
(a) E–glass fibre after one year indoor storage of the composite;
(b) Cem-FIL fibres after five years indoor storage of the composite.**

Figure 5 shows fibres extracted and prepared as described where the surface of the fibres shown is that remaining after cleaning off all the reaction products which are soluble in cold hydrochloric acid.

Figure 5(a) shows fibres of E-glass composition taken from a composite stored in indoor ambient conditions for 1 year: 5(b) shows alkali resistant fibres from a composite stored under the same conditions for 5 years. The E-glass fibres have suffered considerable corrosion leading to substantial fibre stength loss, but the alkali-resistant fibres show only very slight corrosion correspondingly giving a barely significant change in fibre strength.

The changes in composite mechanical properties reflect these observations. With E-glass fibres both the composite ultimate strength and the toughness decrease markedly, due to the fibre strength loss and the increase in bond continuity with the corrosion; on the other hand the Cem-FIL composite reinforced with alkali-resistant fibres retains substantially its original ultimate properties under these conditions.

Under more aggressive conditions of attack by the hydrating cement (eg in wet environments) etch features can be seen on the surface of alkali-resistant fibres. However, the degree of etching on E-glass fibres after one year dry storage in cement has not been seen on alkali-resistant fibres after five years ambient water or natural weathering storage. Region B in Figure 4(b) is a fibre adhering to the matrix and gives some idea of the surface condition remaining after five years in water.

CONCLUSIONS

In GRC composites the glass fibres are arranged in bundles or strands in a matrix of considerable structural variability containing inhomogeneities of the order of size approaching the fibre bundles. The microstructure of the matrix changes considerably in wet storage conditions, voids in the matrix and the fibre bundles are filled with calcium hydroxide crystals and the individual fibres then become bonded into the matrix. Relatively few such changes have been observed in dry storage of the composite.

Surface corrosion of the fibres may also take place to varying degrees dependent on the glass composition and composite storage conditions. This may both reduce fibre strength and increase fibre-matrix bonding. With E-glass compositions this occurs rapidly and to a considerable extent under dry storage conditions but with alkali-resistant glass compositions more aggressive conditions are necessary for a significant effect.

ACKNOWLEDGEMENTS

Our thanks are due to the Building Research Establishment for the provision of the five year old samples of GRC.

The authors also wish to thank the Directors of Pilkington Brothers Limited and Dr D S Oliver, Director of Group Research and Development, for permission to publish this paper.

REFERENCES

1 Majumdar, A J, The role of the interface in glass fibre reinforced cement, Cement and Concrete Research, Vol 4, No 2, March 1974, pp 247-268.
2 Majumdar, A J and Ryder, J F, Glass-fibre reinforcement of cement products, Glass Technology, Vol 9, No 3, June 1968, pp 78-84.
3 Nair, G, Mechanics of glass fibre reinforced cement, this publication.
4 Bache, H H, Morphology of calcium hydroxide in cement paste, Symposium on Structure of Portland Cement Paste and Concrete, Highway Research Board Special Report 90, pp 154-174.
5 Oakley, D R and Proctor, B A, Tensile stress-strain behaviour of glass fibre reinforced cement composites, this publication.

7.4 Can asbestos be completely replaced one day?

Herbert Krenchel and Ole Hejgaard
Structural Research Laboratory, Technical University of Denmark

Summary The annual consumption of asbestos by the asbestos-cement industry throughout the world is enormous. For this and other reasons it is essential to find a suitable replacement fibre, but this is by no means a simple task and the problem has several different aspects. Blown or spun mineral wool forms a group of fibre materials with a great potential for the purpose—high fibre strength, low production cost, ease of production in huge quantities, thin fibres giving high specific fibre surface in the composite, resulting in an excellent crack distribution etc. But the question is how to put these fibres into the matrix so that the reinforcement becomes fully effective.

A special technique is described in which the composite is built up as a laminate using a high wet-strength tissue paper of mineral wool fibres as reinforcement (paper weight 40 g/m², fibre diameter about 5 microns). Cement paste is scraped through the paper, whereby it enters the pores, displacing all the air in each layer of the reinforcement, so that a completely uniform distribution of these fine fibres in 2-d random orientation is achieved throughout the material.

The thickness of the individual layers of reinforcement in the laminate is approximately 0.3 mm (0.01 inch) and the fibre content in the composite is about 6% by volume. The aspect ratio of the fibres—and thus the overall efficiency of the reinforcement—is high, and due to the well dispersed fibres and very high specific fibre surface, tensile cracking in the matrix takes the form of invisible micro-cracking only all the way up to the ultimate point.

Résumé La consommation annuelle d'asbeste par l'industrie asbeste-ciment est énorme dans le monde entier. C'est entre autres pour cette raison qu'il est essentiel de trouver une fibre de remplacement qui convienne. Mais la tâche n'est pas des plus simples car le problèm présente différents aspects. Le coton minèral, embouti ou soufflé, est un groupe de fibres qui présente à cet effet de vastes possibilitiés – grande résistance de la fibre, prix modéré de la fabrication, facilité de produire en quantités énormes, le fait que les fibres minces donnent une grande surface spécifique du renforcement dans le composé, ce qui a pour résultat une excellente distribution des fissures, etc. Mais le probléme est de placer ces fibres dans la matrice de façon à obtenir une armature optimale.

Description est faite d'une technique spéciale par laquelle le composé est construit comme un matériau laminé dont l'armature consiste en un papier pelure de fibres de coton minéral à forte résistance humide (poids de la feuille 40 g/m², diamètre de la fibre, environ 5 microns). La pâte de ciment est raclée sur la feuille de façon à pénétrer dans les pores, déplacant l'air dans chaque *couche de l'armature, si bien qu'une distribution complètement uniforme de ces fibres fines est obtenue dans tout le matériau, avec une orientation de hasard de 2-dimension.*

L'épaisseur des couches individuelles du renforcement dans le matériau laminé est environ de 0.3 mm et la quantité des fibres dans le composé est environ de 6 vol-%. Le rapport ℓ/d de ces fibres—et par conséquent l'efficacité générale de l'armature—est élevé, et grâce à la dispersion adéquate des fibres et à la haute surface spécifique du renforcement, la fissuration de traction dans la matrice prend seulement la forme d'invisible micro-fissuration jusqu'au point de rupture.

INTRODUCTION

A very considerable part of the world's asbestos consumption goes to the production of fibre-reinforced materials—plane and corrugated roofing materials, side panels, asbestos-cement pipes, fireproof and insulation materials of various types. In 1950, the world asbestos consumption was about 1 million tons, in 1958, 2 millions tons, and in 1970, 4 millions tons, so it can be foreseen that the world's deposits of asbestos may well be exhausted before the turn of the century (1,2).

Furthermore, it has been realized in recent years that there may be a considerable health risk for persons working with asbestos (the so-called asbestosis), and the public health authorities in several countries have already banned the use of asbestos altogether for the manufacture of certain of the above-mentioned products. We are thus forced to seek alternative solutions.

It has at any rate been known for a long time that the world resources of asbestos would one day be exhausted, and there has been no lack of suggestions as to how we can manage with other types of fibres instead (3, 4). However, the matter is not as simple as it may seem. Several problems arise when we try other types of fibres, and in most cases, it is very difficult to solve them completely satisfactorily. Asbestos seems, from Nature's hand, to have been richly endowed with very valuable properties, which make it easy to work with in the mixing and construction of all the types of materials mentioned above. In addition, asbestos has a large number of very useful mechanical properties: high strength, natural buoyancy for the particles of the matrix material, resistance to high temperatures and great resistance to various chemicals—acid-proof, alkali-proof, etc. When to all this we add the fact that the types of asbestos of interest have—at any rate up to the present time—been available at a very reasonable price in relation to that of most synthetic fibres, it will be easy to understand the reluctance to desert asbestos for other types of fibres.

Reference (5) mentions various types of fibres that might prove suitable as a substitute for asbestos, and examines their advantages and drawbacks for the different products. From the point of view of price in relation to strength, interest concentrates particularly on the synthetic mineral-wool fibres, glass-wool, rockwool and basalt-wool; however, the technical problems involved in fiberizing and incorporating these fibres effectively in the brittle matrix materials have not yet been solved.

In recent years we have been working at our research laboratory on a special process for the production of fibre-reinforced materials with synthetic mineral-wool fibres as the only reinforcing material. There are two prime problems to be solved if we are to be able to utilize the comparatively high tensile strength of these fibres and to achieve as good a compatibility between this type of reinforcement and the brittle matrix as is the case in asbestos-cement: firstly, there is the problem of opening up the mineral-wool and defiberizing the material completely—without any reduction in the fibre length, or only a limited and strictly controlled reduction—and secondly, the problem of placing and fixing the fibres in correct positions in relation to each other during the building-up of the composite, so that the variation in fibre orientation and in fibre concentration from point to point in the finished product is kept to an absolute minimum.

Once these two problems have been solved—and, as we shall see, they can be solved in a relatively easy manner—we shall be able to make a fibre-reinforced brittle matrix material with mechanical properties that are equal to those of normal good quality asbestos-cement. With a reasonably high fibre concentration ($V_f = 0.05$–0.06) and purely randomized, 2-dimensional fibre-orientation, we thus get tensile and bending strengths, impact strengths and elongations at rupture that are just as good as—if not better than—those of asbestos-cement, and the behaviour of the material under loading beyond the critical point at which the matrix material begins to fail is ideal. Due to the uniform fibre distribution and the very high specific surface of the reinforcement, failure of the matrix material, in the form of discrete, visible cracks, does not occur until the stress-strain curve has reached its ultimate point.

Over the entire range, from the time the matrix material begins to crack until it reaches final failure, the composite is riddled with a finely distributed network of invisible micro-cracks throughout the tensile zone, just as in a high quality asbestos-cement.

The last major problem, that of the alkali-resistance of the fibres, ie their resistance to attack by calcium-hydroxide and related substances formed during the hydration of the cement, has been left for the time being as we have found the solution of the problems described above to be of greater importance.

However, we have recently succeeded in producing a steam-hardened calcium-silicate product with pure glass fibres of ordinary glass-wool as the only reinforcement (fibre diameter d $\sim 5\mu$m), and even these very thin, single fibres are fully effective in the end product, without any signs of attack on the fibre surface, even after the harsh treatment received in the autoclave.

SHORTENING AND DEFIBERIZATION

To obtain a reasonably high efficiency of the reinforcement the wool material must be pre-treated and defiberized in order to lay bare the entire fibre surface and so that, in the next stage, the matrix material can envelop and anchor each separate fibre. Seen with the naked eye, only partially defiberized tufts of mineral-wool, with a diameter of, say, 0.5–1 mm, may seem insignificant, but each of them may contain hundreds of tightly tangled individual fibres, which prevent penetration of the matrix material. Such inadequately defiberized reinforcement may do more damage than good because the individual tangles form notches in the finished material from which early failure may originate.

In (6, p 44), a method is described for systematically shortening a mineral-wool material to a desired maximum fibre length. The material is spread out in a thin layer and cut, by means of special knife rollers, into squares in which the diagonal represents the maximum fibre length of the shortened material.

In tests carried out along these lines from 1951 to 1955 (6, p 134), after shortening, the fibres were just suspended in water and then mixed with the cement, after which a plate material was built up by drawing off the excess water in much the same way as in the asbestos-cement industry. In order to ensure complete defiberization it was at that time necessary to work with a very limited maximum fibre length, knife-roller spacing 2.0 mm, maximum fibre length about 2.8 mm. As many of the fibres in the mineral-wool material cut in this way are naturally considerably shorter than the maximum length, the poor overall efficiency of the reinforcement obtained at that time can undoubtedly be attributed to inadequate anchorage of a large part of the reinforcement. Another factor that must also have played a part was an insufficiently uniform distribution of the reinforcement within the final material. Once the fibres, cement and water have been mixed together in a muddy suspension, it is impossible to see how the reinforcement is distributed during the subsequent stages of the process.

REINFORCING MATERIAL

With the new method, developed during the last few years, these difficulties have been overcome by separating the process into two parts. First, a thin and open tissue paper is produced from the mineral-wool fibres, and then the composite material is built up as a laminate of many layers of this reinforcing paper, each separate layer being impregnated with a suitable suspension of cement and water.

It is possible to ensure a very uniform distribution of the fibres during production of the reinforcing paper, and if this is waterproofed by means of a light impregnation (eg 4 per cent by weight of a phenol-formaldehyde binder), the fibres will all be completely fixed and will not shift when the matrix material is scraped in. In this way we get the same, uniform fibre orientation and fibre concentration throughout the composite.

In order to be able to work with fibres as long as possible, a suitable dispersing agent is added to the mixing water during production of the paper. Different types of dispersing agents have been tried with varying success; however, this is a separate problem, on which a great deal more research is required. With the addition of, for instance, 0.5% *Polyox* (Union Carbide, WSR N 3000) and a fibre concentration of 0.04% by weight in the mixing water, a very smooth sheet of paper is obtained, with a dry weight of about 40 g/m² at a maximum fibre length of approximately 7 mm (max $\ell_f/d \sim 1400$), and it is even possible to achieve a suitable fibre dispersion at max $\ell_f = 12$ mm (max $\ell_f/d \sim 2400$). With other types of dispersing agents (e g polyvinyl-pyrrolidone, as proposed by Callinan (7)), it is possible to work with even longer fibres and, at the same time, to obtain a more uniform fibre distribution in the reinforcing paper.

BUILDING-UP OF THE MATERIAL

When the sheets of paper are dry and the phenol-formaldehyde dressing has hardened, the matrix material, in the form of a suspension of paste-consistency, is scraped into each sheet, and the material is built up from a suitable number of sheets, normally about three sheets per 1 mm plate thickness. During the hardening of the cement these sheets are kept lightly compressed ($Q \sim 0.45$ N/mm² or less); the material thereby grows smoothly together and there is no risk of the finished material delaminating, even under loading with high shear stresses. An increase in the pressure during hardening of the plate material will result in the removal of some of the excess water and presumably in a certain improvement of the properties of the composite, but there is also a risk of the fibres becoming broken. This problem has not yet been investigated systematically.

Figure 1   **Tensile stress-strain curve of a three-component epoxy material specially composed for indirect tensile tests on staple fibres of mineral-wool.**

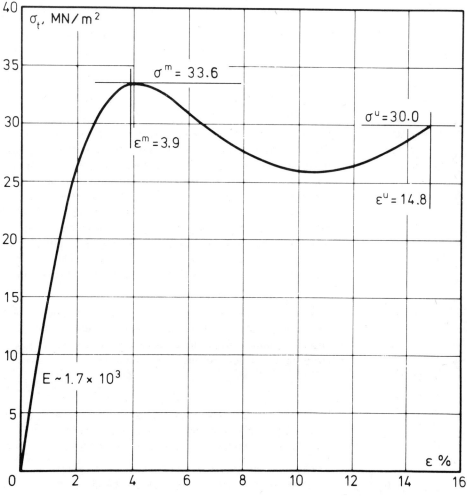

Figure 2   Tensile test specimen of the unreinforced epoxy materials (t = 5 mm) and the epoxy/glass–fibre laminates (t = 3–4 mm).

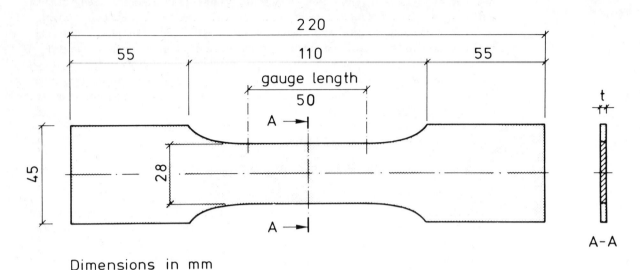

Dimensions in mm

If ordinary Portland cement is used as matrix material, it is normally pre-mixed with a total of 50% by weight of water, and a dispersing agent, eg polyvinyl-acetate, is also added in order to obtain the ideal consistency for scraping-in. However, further research is also required on this. If pure calcium-silicate is used as matrix material, no dispersing agent is normally necessary, especially if the SiO_2 material is very finely grained.

MECHANICAL PROPERTIES OF THE REINFORCEMENT

Direct measurement of the tensile strength of the individual fibres is possible, but difficult, and will not represent random testing because such investigations can only be carried out on the longest fibres; furthermore, it will seldom be possible to extract samples of the material with an equal representation of both strong and weak fibres. A reliable determination of the elastic properties of the fibres by direct measurement is even more difficult.

We have therefore found it preferable to carry out an indirect determination of these properties on the basis of the reinforced material produced, and for this purpose we have built up laminates of the reinforcing paper with an epoxy-resin matrix. It is, of course, necessary to use a matrix material that has an elongation at rupture in excess of that of the fibres. We have achieved this by a specially composed, 3-component epoxy-resin (*Araldite**: GY 257, 72.2% + CY 208, 18.1% + HY 951, 9.7% by weight). The resultant material had a stress-strain curve as shown in Fig 1. An ultimate strain of about 4% should provide ample insurance that failure of the laminate will be dictated by the reinforcement and not by the matrix. The tensile test specimens of both the unreinforced epoxy plates and the epoxy/glass-fibre laminates were cut as shown in Fig 2.

The laminates were made using traditional production methods. The plates were compressed during hardening, $Q = 0.45$ N/mm². The mechanical properties of the test specimens appear from the tensile stress-strain curve in Fig 3, from which the properties of the fibres, calculated in accordance with (6) are:

Ultimate strength: $\sigma_f^u = 945$ N/mm²,   Ultimate strain: $\epsilon_f^u = 1.15\%$

Modulus of elasticity:
Upper limit: $E_{f.u} = 106$ kN/mm², Lower limit: $E_{f.1} = 82$ kN/mm².

* Produced and supplied by CIBA, Switzerland

Figure 3   Tensile stress-strain curve of epoxy/glass-fibre laminate with 2-d random fibre orientation. Reinforcement: Staple fibres of glass wool, $d \sim 5\mu$, max $\ell_f = 12$ mm, $V_f = 0.135$. Matrix: Three-component epoxy material, see Figure 1.

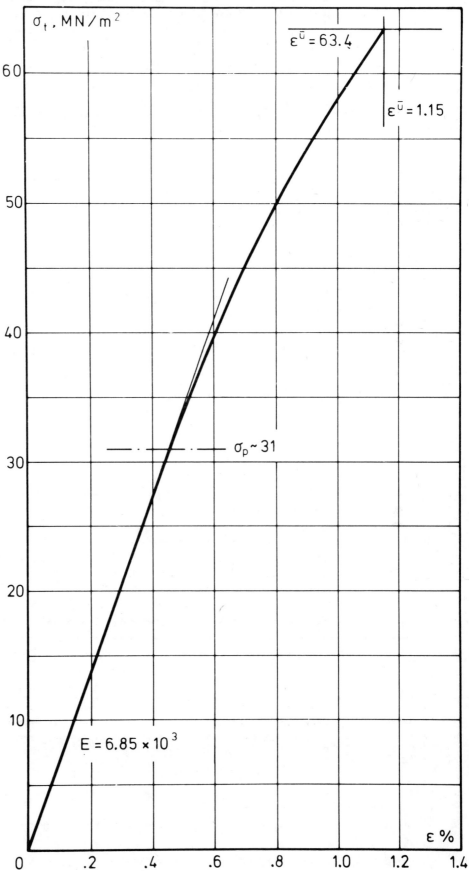

MECHANICAL PROPERTIES OF THE FRC-MATERIAL

For the FRC-material produced in the laboratory today by the process described above, we normally use an ordinary, Danish, rapid-hardening Portland cement as matrix material (Blaine fineness 3800 cm^2/g). The test specimens are moist-cured for a total of 5 days, after which they are placed in laboratory atmosphere (20-22 °C, 35-55% RH) for 9 days until tested at an age of 14 days.

On the test day, 3 bending tests were carried out on 50 x 220 x 6 mm specimens, and 6 impact tests on 20 x 70 x 6 mm specimens. The impact test plates were loaded to bending failure in a pendulum testing machine with a 3-point load over a span of 50 mm (Alpha machine, capacity 1.0 Nm, impact speed 3.4 m/sec, see (6), Fig. 47). The bending plates were tested (100 kN Instron, 5 kN load cell, 200 N range), with a 4-point loading as shown in Fig 4. In these tests, the edge strains at mid-span in the tensile and compressive zones were measured up to final rupture by means of 17 mm electric strain-gauges, glued directly on the faces of the plates at the centreline crossings. The signals from each strain-gauge were recorded directly by a 2-pen x/y-plotter. The rate of loading was kept constant throughout the test, crosshead speed: 0.05 mm per minute (total test time for one bending plate from zero to rupture: about 15 min).

Typical stress-strain curves from these tests are shown in Fig 5, where the ordinate is the edge bending stress (σ_b = M/Z), and the abscissa represents the edge strains in the tensile and the compressive zone (both shown as positive). The curves represent the mean values for the three tests from one series.

It will be seen that the two stress-strain curves follow the same course up to a bending stress of about 12 N/mm^2 and a strain of 0.1%. After this, greater strain occurs in the tensile zone than in the compression zone, indicating that the matrix material is cracking and that the reinforcement in the tensile zone is taking the load. The ultimate strain in the tensile zone is about 0.55%. At this point the distance from the compressed edge to the neutral axis is about 40% of the total height of the plate.

Comparison with Fig 3, where an elongation at rupture of the fibres of about 1.15% was obtained, shows that we so far only have a utilization of the reinforcement of about

Figure 4   **Arrangement for bending tests with FRC-plate material.**

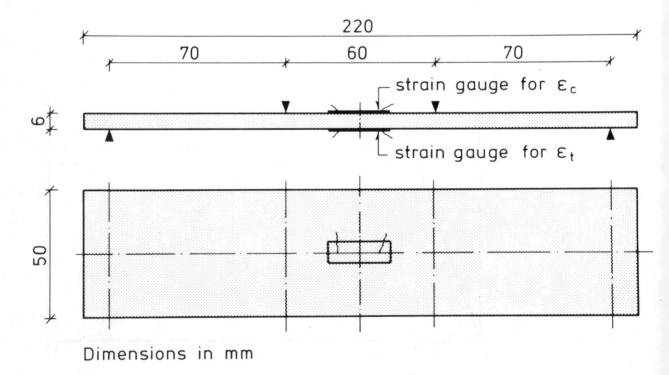

Dimensions in mm

Figure 5 Stress-strain curves from bending tests with special FRC-plate material built-up as a laminate with 2-d random fibre orientation. Reinforcement: Staple fibres of glass wool, $d \sim 5\mu$, max $\ell_f = 12$ mm, $V_f = 0.058$. Matrix: Portland cement, w/c-ratio ~ 0.24.

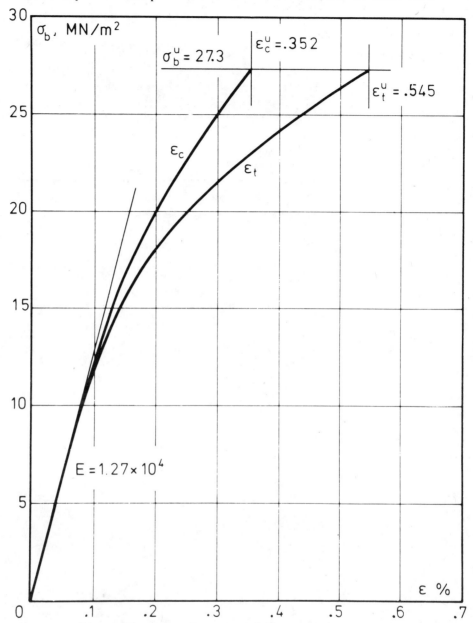

48% in our FRC-material. This is presumably due to a relatively high content of rather short fibres in the reinforcing material. This can naturally be improved by using a greater maximum fibre length and a more suitable dispersing agent in the production of the paper, and possibly also by using a cement that gives a better adhesion to the surface of the fibres. All the same, it should be emphasized that the strength properties and elongations at rupture achieved here are already better than those normally obtained in asbestos-cement production today.

The impact strength of the material is also somewhat better, averaging about 3.3 kNm/m², against about 1.8 to 2.8 kNm/m² for a medium quality asbestos-cement, depending on whether the material is tested in the weakest or the strongest principal direction.

Another very important point is the way in which the material cracks when it is loaded beyond the elongation at rupture of the matrix. Due to the very high specific fibre surface (SFS ~ 50 mm²/mm³, see (8)), all cracks in the tensile zone occur as

Figure 6   Rupture face in impact prism of fibre-reinforced, steam hardened calcium-silicate with 3-d random fibre orientation of glass wool, $d \sim 5\mu$, max $\ell_f = 7$ mm, $V_f = 0.013$. Matrix $Ca(Oh)_2$, 40% + SiO_2, 60% by weight. (Magnifications, 6a: 100 x, 6b: 2,000 x).

6a

6b

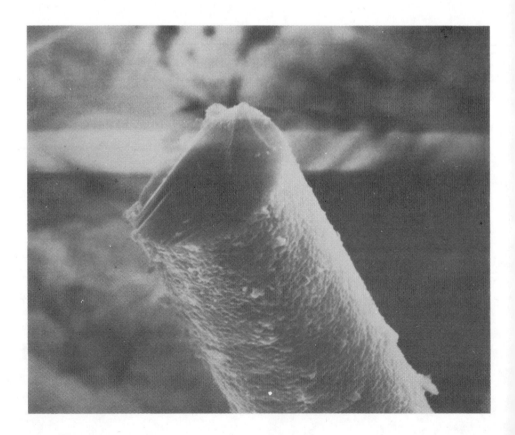

finely distributed, invisible micro-cracks, which are hardly likely to have any deleterious effect on the material in practice.

ALKALI-RESISTANCE OF THE MATERIAL

The glass fibres used in our investigations are not specifically alkali-resistant—quite the reverse. As mentioned earlier, they are taken from ordinary insulating wool—a special glass of the A-type (alkali content: $Na_2O + K_2O \sim 15\%$).

The prime reason for using this fibre material for our investigations is that it can be produced at reasonable cost and in the very large quantities that will be needed if we are one day to arrive at a fully acceptable and feasible alternative to asbestos for ordinary FRC-materials. As mentioned, we are also interested in this particular fibre because of its very high specific surface and the influence of this factor on the crack distribution (the specific surface is presumably of the same order of magnitude as that of the asbestos fibres in a high quality asbestos-cement). However, at the same time, a big surface increases the risk of attack on the individual fibres by the free calcium-hydroxide and alkali content of the matrix.

Up to the present time the prime purpose of our tests has not been to achieve a high alkali-resistance. We take the view that this problem must be solved in the next round—and it will undoubtedly be solved once we have an FRC-material with the right mechanical properties, including a correct crack distribution, and at a price that stands comparison with that of ordinary asbestos-cement.

The problem of the alkali-resistance might be solved in a number of ways: by changing the analysis of the fibre material, coating the fibres, changing the matrix material, or using additives to the normal Portland cement matrix, or by combinations of these measures. It is as yet too early to say which procedure will be preferred.

We have tried some of these alternatives during the past year, and the results are, in fact, promising, even though the question has not yet been investigated exhaustively.

As an interesting example of how far we can go in this direction, even with these quite ordinary glass-wool fibres, we can mention a special, fibre-reinforced oven-insulating block material of very lightweight, steam-hardened calcium-silicate, which has been developed in connection with this research. Various versions of such material, normally reinforced with asbestos, are on the market, but we know that, here too, there is very great interest in finding a suitable substitute fibre.

After steam-hardening (200 °C and 100% RH for 8 hours), followed by drying, the material has a specific weight of about 0.3, and its strength properties compare favourably with those of corresponding, asbestos-reinforced materials. In particular, a high bending strength and very high impact strength show that even these thin glass-wool fibres can survive the extremely harsh treatment received in the heavily alkaline and highly reactive environment of the autoclave, as will also be seen directly from Fig 6, which shows a rupture surface in such a test specimen.

It seems probable that we shall one day be able to proceed along these lines with the ordinary FRC-plate materials as well.

REFERENCES

1. "Fibro-Cement Composite", United Nations Industrial Development Organization, United Nations Publication, Sales No E 71 II B1, New York, 1970.
2. "World Asbestos Output—1973", Asbestos, (Stover, Pennsylvania), July 1974, p 22.
3. Hayden, R, "Grundsätzliche Fragen der Herstellung von Asbestzement" ("Basic Problems Regarding the Production of Asbestos-Cement"), Zementverlag, Berlin, 1942.
4. Hatschek, L, French Patent 303 663, Paris, 12 Sept 1900.
5. Krenchel, H, "Fibre Reinforced Brittle Matrix Materials", ACI-Symposium: Fibre Reinforced Concrete, Ottawa, Oct 1973, Publication SP-44, American Concrete Institute, Detroit, 1974, pp 45-77.

6 Krenchel, H, "Fibre Reinforcement", Dissertation, Technical University, Copenhagen, Akademisk Forlag, 1964.
7 Callinan, T D, "Powdered Resin Improves Mechanical Properties of Inorganic Specialty Papers", Report of NRL Progress, Naval Research Laboratories, Washington DC, Sept 1955, pp 12-17.
8 Krenchel, H. "Fibre Spacing and Specific Fibre Surface", RILEM Symposium: Fibre Reinforced Cement and Concrete, London, Sept 1975, (see chapter 3.3).

7.5 Tensile stress-strain behaviour of glass fibre reinforced cement composites

D R Oakley and B A Proctor
*GRC Research Department, Pilkington Brothers Limited,
Ormskirk, Lancashire, United Kingdom*

Summary

Tensile stress-strain curves have been obtained for a number of different types of glass fibre reinforced cement composites with cement paste and cement-sand mortars as matrices. The composite properties have been related to measurements of crack spacing and to separate measurements of fibre strength and stiffness.

The effects of reinforcement configuration and orientation, matrix type, cure and ageing are discussed in some detail with regard to the bonding between fibre and matrix, and the effectiveness of the reinforcement in preventing cracking and providing strength and stiffness at various stages of the composite's stress-strain behaviour.

Résumé

Des courbes de tension-élongation ont été obtenues pour un nombre de matériaux composites armés aux fibres de verre et ayant pour matrice d'une part la pâte de ciment, et d'autre part des mortiers mixtes au sable et au ciment. Les propriétés des matériaux composites ont été reliées aux mesures d'entre-fissures et aux mesures obtenues séparément de la résistance et de la rigidité des fibres.

On commente les aspects de la configuration et de l'orientation du renforcement, le genre de matrice, la maturation et le vieillissement par rapport au degré de cohésion entre la fibre et la matrice aussi bien qu'à l'efficacité du renforcement d'empêcher la fissuration et de fournir la résistance et la rigidité aux divers stades du comportement du matériau composite sous tension-élongation.

INTRODUCTION

The addition of a few percent of glass (or other) fibres to cement paste enables quite large and thin sheets to be produced and results in a composite material which can have a tough and apparently ductile behaviour. Such materials are increasingly being used in the construction industry for cladding and shells. They form part of a relatively new class of composite material in which the matrix is brittle and fails at a much lower strain than the reinforcement. Clearly it is important to understand the way in which fibres and matrix interact to control failure.

Aveston et al (1,2) have published theoretical models describing the behaviour of brittle matrix composites and previous work in these laboratories (3) has shown that practical materials manufactured by the spray/suction-dewatering method (4,5) follow their predictions quite closely. The present work extends investigations of tensile stress-strain behaviour and multiple cracking to materials using sand mortars as a matrix. Measurements of the strength of glass strands in a cement environment, of loads needed to pull glass strands out of cement blocks and of the elastic modulus of glass strands are used in a comparison of observed composite behaviour with theoretical predictions.

EXPERIMENTAL

Composite manufacture and testing

Spray/suction-dewatered sheets 2 m × 1 m × 8 mm with glass contents between 4 and 5 volume % were manufactured. The glass was in the form of 38 mm long chopped strands of Cem-FIL alkali resistant glass fibre and the strands consisted of about 200 fibres, each 12.5 μm in diameter, in a flat bundle held together by a size coating. The cement used was a rapid hardening Portland cement and the sand was a high silica sand with an even size grading between 1mm and 150 μm.

One of the cement paste matrix boards (LA 77) was cured under polythene for one day, then in a fog room at 95% RH and 21 °C for six days and thereafter kept in the laboratory at 40-60% RH. The remaining boards were made as one series at varying sand contents and were cured under polythene for one day and then in water at 21 °C for six days before laboratory storage.

They were made to a nominally constant water/solids ratio of 0.26, but at the high sand contents this level could not be maintained. Complete specifications of all boards are given in Table 1A, numbers LA 177 to 189 being the water cured mortar matrix set.

Tensile tests were carried out on 150 × 25 × 8 mm strips, cut from these boards by means of a water lubricated diamond saw and then pulled to fracture in an Instron testing machine at strain rates of about 10^{-3}/minute using wedge action grips. Strain was measured with a clip on an extensometer with a 25 mm gauge length, and in the initial elastic region with 12 mm strain gauges.

Tested samples were prepared for crack spacing measurements by painting with a penetrant dye and then polishing to remove the surface layer. Crack spacing measurements were made using a travelling microscope and values for some of the samples at one month and one year are included in Table 1B.

Fibre and strand measurements

The Young's Modulus of the glass strands used for board manufacture was determined directly using long gauge length samples (8.5 metres) and static loading. The strand cross-section was determined by weighing known lengths of fibre.

The relevant strengths of the glass fibres, both initially and in the composite, were estimated from direct tensile tests on strands in a cement environment. Samples were stored for various times in both wet and dry conditions prior to testing.

In order to estimate the bond strength between fibre and matrix, cement paste discs 10 mm in diameter of varying thickness were cast round one end of lengths of glass strand which were again protected outside the embedded lengths. The load was measured as the strand was pulled from the disc which was supported by its flat face. This test is similar to that used by Bartos (6). The water/cement ratio of the cement block was 0.4, the minimum that could be reliably achieved. These samples were given a cure history similar to that of cement paste board LA 77.

Table 1　Specific of cement paste matrix, glass reinforced boards

	Board number		LA 77		LA 179		LA 177		LA 181		LA 183		LA 185		LA 186		LA 189	
	Sand/cement		0		0		0.11		0.25		0.48		0.67		1.0		1.5	
	Glass content (%)		4.55		4.4		4.9		4.2		4.2		4.1		3.9		4.4	
A	Water/solids		0.32		0.28		0.26		0.25		0.26		0.18		0.23		0.18	
	Water/cement		0.32		0.28		0.29		0.32		0.39		0.30		0.45		0.44	
	Density (kg/m^3)		2.05		2.1		2.1		2.15		2.15		2.1		2.05		1.85	
	Sample orientation		L	T	L	T	L	T	L	T	L	T	L	T	L	T	L	T
	Ultimate tensile	1 m	17.5	8.3*	15.3	8.5	14.1	9.8	13.7	9.2	15.3	9.8	12.6	9.9	11.2	8.8	8.2	5.8
	strength (N/mm^2)	1 yr	12.4	8.1	13.5	8.6	12.6	8.4	11.7	8.6	13.2	9.4	12.5	8.5	11.3	7.9	8.6	6.0
	Failure strain	1 m	1.37	1.06*	1.15	0.82	1.05	1.34	1.23	1.12	1.28	1.01	1.16	0.70	1.11	1.11	1.05	1.11
	%	1 yr	0.9	0.96	1.12	0.83	1.20	0.96	0.94	1.10	1.00	1.06	1.09	0.93	0.97	1.02	0.92	0.92
	Bend-over point	1 m	8.5	5.4*	7.9	6.2	7.9	6.3	7.7	6.9	8.0	6.4	6.6	5.6	5.1	4.7	3.0	2.9
B	(N/mm^2)	1 yr	6.8	5.7	6.5	5.6	6.2	5.4	6.9	6.2	7.3	6.4	5.7	5.7	4.1	4.4	3.6	3.1
	Post-cracking	1 m	0.91	0.43*	0.84	0.48	0.80	0.50	0.71	0.44	0.84	0.52	0.73	0.49	0.69	0.49	0.71	0.47
	modulus (GN/m^2)	1 yr	0.91	0.54	0.81	0.53	–	–	–	–	0.79	0.48	–	–	–	–	–	–
	Average crack	1 m	1.9	1.2*	1.2	1.0	1.5	1.2	1.6	1.2	–	1.4	1.5	1.6	1.7	1.7	3.2	3.5
	spacing (mm)	1 yr	2.3	1.9	1.9	–	–	–	2.0	1.55	1.9	–	2.5	2.2	2.1	–	–	–
	Calculated	1 m	1.1	1.1*	1.6	1.5	1.2	1.2	1.2	1.4	–	1.2	1.2	1.1	0.8	0.8	0.2	0.2
	bond strength (N/mm^2)	1 yr	0.8	0.8	0.9	–	–	–	0.9	1.0	1.0	–	0.7	0.65	0.5	–	0.3	–
	Crack suppression	1 m	1.3	0.4*	0.75	0.4	0.85	0.4	1.0	0.6	–	0.6	0.7	0.55	0.45	0.4	0.3	0.3
C	efficiency, K_2	1 yr	1.0	0.7	0.8	–	–	–	1.0	0.6	1.05	–	0.9	0.8	0.4	–	0.3	–
	Post-cracking	1 m	0.29	0.13*	0.27	0.16	0.23	0.15	0.24	0.18	0.28	0.18	0.26	0.17	0.25	0.18	0.23	0.15
	stiffness efficiency, K_3	1 yr	0.29	0.17	0.26	0.17	–	–	–	–	0.27	0.16	–	–	–	–	–	–
	Strength efficiency, K_4	1 m	0.31	–	0.28	0.16	0.23	0.16	0.27	0.18	0.30	0.19	0.25	0.20	0.22	0.17	0.15	0.11

* Three month data

RESULTS AND DISCUSSION

Fibre and strand properties

The Young's Modulus of Cem-FIL glass fibres in the form of strands was found to be 70 (± 2) GN/m². No marked changes in strand modulus were observed up to 2/3 of the breaking load.

The as produced strength of glass strands lay between 1450 and 1750 N/mm². However, on incorporation into cement some additional damage took place giving strengths between 1200 and 1300 N/mm² after 24 hours in the cement environment. No further change of strength of strands in cement took place over periods of about a month when stored in either air or water. Detailed data are not available for air storage conditions at times greater than one month, but the results of Majumdar (7) for the strength of extracted single filaments, and strand tests on similar glass compositions indicate only a small drop in strength up to two years. For comparison with composite data on young material the strength results for strands in cement up to one month have been pooled giving a value of 1245 N/mm² (90% confidence limit 50 N/mm²). Tests with sand mortar matrices give identical strength values.

Fibre-matrix bond strengths by pull-out

If the bond between fibre and matrix is frictional in nature the shear stress along the fibre in a pull-out test is constant and the bond strength

$$\tau_b = \frac{F}{t\,p} \quad (1)$$

where F is the pull-out load, t is the embedded length or disc thickness and p is the strand perimeter. The pull-out load increases linearly with embedded length until the fibre failure stress is reached. Fibres then fracture before pull-out occurs and the failure load is constant.

When there is an initial elastic or adhesive bond followed by a frictional bond of similar magnitude the behaviour is indistinguishable from the purely frictional case (8).

While in a pull-out test with a single fibre or rod (9,10) it is possible to obtain a reliable estimate of the glass perimeter in contact with the cement, the problem is more difficult when a multifilament strand is involved. The configuration of the fibres within a composite can vary, depending on the original shape of the bundle and its treatment during incorporation. It can vary from complete dispersion of the fibres in the matrix, so that each fibre is individually in contact with the cement, to retention of a close packed bundle form with only the outer perimeter in contact. Although there is considerable variability within a composite the most common configuration is that of an integral strand (Fig 1).

From a microscopic examination of strands in the pull-out specimens and in composites the average strand has been taken as a tape of 204 12.5 μm fibres about 0.65 mm wide and 0.11 mm thick. The cross-section area of this strand is 0.0735 mm² which means that glass occupies 34% of the strand volume. The perimeter in contact with the matrix is taken as 2.83 mm.

In the present pull-out tests samples tested between 1 and 3 months old did show a linear increase of pull-out load with embedded length for short lengths. Fibre failures occurred for lengths greater than about 13 mm. Between 1 and 3 months the slope of the load length curve gave a constant value of τ_b p of 3 N/mm². Using the measured perimeter length this leads to a bond strength of $\tau_b = 1.1$ N/mm². This is in reasonable agreement with the estimates of Pell (10) and the Building Research Establishment (Laws, private communication) for alkali resistant glass in Portland cement after dry storage, but differs by a factor of at least three from the results of deVekey and Majumdar (9) for wet stored material. Although the absolute value of τ_b is subject to errors in the estimation of the effective perimeter, the product τ_b p for the configuration of strand used in the present work has been determined with more certainty, and it is this product that is used in the analysis of multiple cracking.

After yielding, dynamic frictional forces in the pull-out tests maintain 80–90% of the

Figure 1   Configuration of an integral glass strand within a composite.

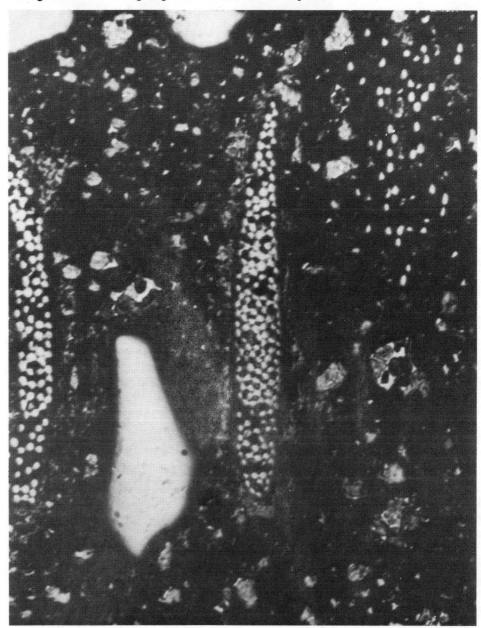

peak load. This value may be compared with estimates of 70% by Pell (10) and deVekey (9).

Using the nomenclature of Laws (16), the critical fibre length is twice the embedded length at onset of fibre failure — or twice the full anchorage length (15). Present experiments indicate a value of about 26 mm, agreeing well with the data of Bartos (6) when the same criteria are used.

The above results refer to cement paste systems only. It has not been possible to make satisfactory specimens for direct pull-out tests with sand mortars.

Stress-strain behaviour of composites

The stress-strain curves of the spray-dewatered glass reinforced cement composites (Fig 2) exhibit four main regions (3);-

(1) An initial elastic region where the Young's Modulus is experimentally indistinguishable from that of the matrix up to a stress approximately equal to the failure stress of the unreinforced matrix.

(2) A transition region with reduced elastic modulus in which some cracking can be detected, but in which crack growth is constrained.

Figure 2   Stress-strain curves of spray-dewatered glass reinforced cement.

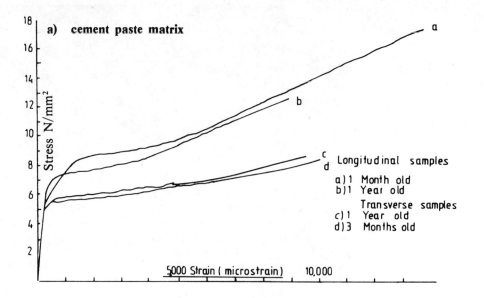

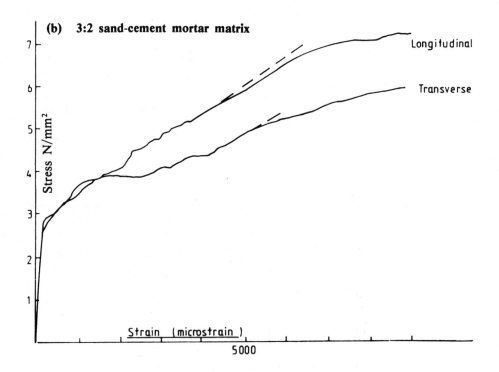

(3) Above a "bend-over point" a region where multiple cracking of the matrix occurs and the matrix is broken down into quite regular blocks.

(4) After completion of multiple cracking, a region where the extra stress is borne entirely by the fibres up to failure.

Values of ultimate tensile strength, failure strain, bend-over point, and post-cracking stiffness at ages of one month and one year are given in Table 1B.

It may been seen from Figure 2 that the stress-strain curves are divided into two groups for samples cut in different directions from the sheet. This reflects a degree of anisotropy introduced by the spray equipment used in the manufacture of the present boards.

Figure 3 The effect of sand content and age on composite and glass properties.
(a) The effect of sand content on tensile properties of composites at one month and one year.

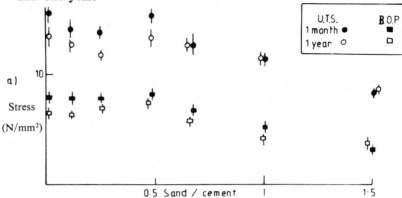

(b) The effect of age on tensile properties of cement paste matrix and 3:2 sand-cement mortar matrix composites.

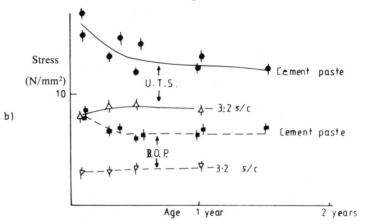

(c) The effect of age on the strength of glass strands within the composites.

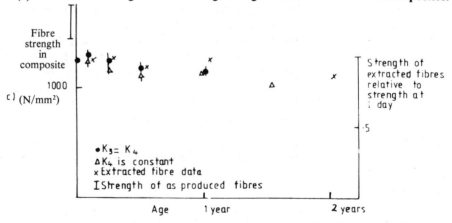

The stress-strain curve of the high sand content board (Fig 2b) is generally similar to that of the paste composites (Fig 2a), but the bend-over point and ultimate tensile stress are both reduced and beyond the multiple-cracking region there is a marked curvature in the stress-strain relationship. Figure 3a summarises the effect of sand content on UTS and bend-over point, and illustrates how these properties change with age.

Multiple cracking behaviour

If the reinforcing elements are sufficiently strong and numerous then multiple cracking occurs and the minimum crack spacing x can be estimated from the force balance at the first crack (1) and in a continuous aligned composite

$$x \, p\tau_b \, N = \sigma_{mc} V_m \qquad (2)$$

where N is the number of reinforcing elements per unit area crossing the crack and σ_{mc} is the stress in the matrix on cracking. For single filament composites N is $V_f/\pi r^2$ and

$$x = \frac{V_m}{V_f} \cdot \frac{\sigma_{mc} r}{2\tau_b} \quad (3)$$

and from statistical considerations (2) the average crack spacing $\bar{x}$ may be taken as 1.364x.

The commercial type of glass reinforced cement examined in this work differs from the above ideal composite in three ways:–
   (a) the reinforcing elements are of finite length,
   (b) they are arranged in a pseudo-random two-dimensional array,
and (c) instead of being single filaments they are multi filament strands.

Aveston et al (11,2) have suggested that for planar, randomly oriented, continuous reinforcement the crack spacing should be $\pi/2$ times that in the aligned case, and that for reinforcement of finite length the crack spacing is little affected, provided that the length is greater than about four times the crack spacing. In these composites strand lengths are some 15 times the observed spacing so that reinforcement, for this purpose, may be regarded as continuous.

An estimate of the effect of arranging the reinforcement in multifilament strands instead of as discrete fibres may be obtained by substituting the total exposed perimeter of the bundle into equation (2) above. It is found that the crack spacing varies with the size and shape of the strand; eg for a 200 filament strand in a rectangular array three fibres thick the spacing would be 2.9 times that in a composite with identical single filaments, and in a rectangular array eight fibres thick the factor would be over six times.

If the strand is not close packed and contains voidage the factor is reduced somewhat since V_m in equation 2 is lowered. In the present work the volume of glass within a strand is typically 34% of the total strand volume, the strand is about five filaments thick and the crack spacing is estimated to be 2.5 times that of single filament reinforcement.

This marked dependence of cracking on the detailed arrangements of fibres within strands means that it is difficult to give a fully quantitative description of cracking behaviour in sprayed glass reinforced cement. The fibre arrangement will depend not only on the form of supply of the fibre, but also on its treatment during the manufacturing process. It should also be noted that the strand-matrix bond strength may also depend on the strand arrangement because of the effect of local voidage (10). Despite these reservations it is possible to use cracking behaviour to study changes in the system when using one type of fibre and one manufacturing process, and this is done below.

Equation (2) may be rearranged to give the average crack spacing

$$\bar{x} = \frac{1.364 \, V_m \, \sigma_{mc} \, A_s}{p \, \tau_b \, K_1 \, V_s} \quad (4)$$

where A_s is the cross-section area of the strand, V_s is the volume fraction of strands in the material and K_1 is a constant indicating the efficiency with which a random planar array of elements controls cracking compared with the aligned case (orientation efficiency factor). Equation (4) reduces to:–

$$\bar{x} = \frac{1.364 \, (1-V_s) \, \sigma_{mc} \, A_f}{p\tau_b \, K_1 \, V_f} \quad (5)$$

with A_f as the cross-sectional area of glass within a strand and V_f the actual glass content in the composite.

The observed value of bend-over point stress, at which multiple cracking begins is taken as σ_{mc}, and then equation (5) can be used with the measured values of crack spacing and τ_b p to estimate the efficiency factor K_1. For board LA 77 for which the bond strength measurements were most applicable a value of $K_1 = 1$ predicted the measured crack spacings exactly (Table 1B) for both the one month longitudinal and three month transverse samples.

Use of this orientation efficiency factor of 1 in equation (5), together with the measurements from the remaining composites in the sand containing series of boards, leads to the calculated effective bond strengths given in the first row of Table 1C. Several points may be noted from these results.

The efficiency factor applied was independent of orientation and gave the same values of calculated bond strength for both longitudinal and transverse samples, ie, at least over this range, the effectiveness of fibres in controlling final crack spacing is independent of orientation.

For the neat cement paste materials, the six day water cured board had a somewhat higher bond strength than the 95% RH cured board when tested at one month. For all materials, dry storage over a year led to some *decrease* in bond strength.

The variation of bond strength with sand content is interesting: a frictional bond might be expected to show a steady decrease in strength with sand content in line with the decrease in drying shrinkage. Instead there is a marked drop at the higher sand contents, associated with high water/cement ratios, and at the highest sand level the bond possibly increases with age suggesting porosity and/or degree of hydration as controlling factors.

Matrix cracking stress level

The analysis of Aveston et al (1) predicts that matrix cracking may be constrained and shows that for aligned fibre composites with $E \simeq E_m$ the matrix cracking stress is given by

$$\sigma_{mc} = \left\{ \frac{12 \, \tau_b \, \gamma_m \, E_f \, V_f^2}{r \, V_m} \right\}^{1/3} \tag{6}$$

where γ_m is the surface work of fracture of the matrix. Combining this with equation (3) for crack spacing gives

$$\sigma_{mc} = \left\{ \frac{6 \gamma_m \, E_f \, V_f}{x} \right\}^{1/2} \tag{7}$$

Once again the problem of taking account of orientation and strand geometry effects is a complex one but an empirical approach of introducing an efficiency factor K_2 to account for these effects and taking $\bar{x} = 1.364x$, leads to

$$\sigma_{mc} = \left\{ \frac{8.18 \, k_2 \, \gamma_m \, E_f \, V_f}{\bar{x}} \right\}^{1/2} \tag{8}$$

If we use a value of 4 J/m² for γ_m (2) and take measured values of strand modulus, crack spacing and bend-over point (for matrix cracking stress σ_{mc}) then we obtain the values of K_2 shown in Table 1C. This efficiency factor for crack suppression *does* depend on the anisotropy of fibre orientation, since longitudinal and transverse values differ. It also depends on the sand content. There is an increase with sand content at low levels, consistent with increases in the work to fracture on addition of sand (12), but this effect is overriden by a marked loss of effectiveness at the high sand levels. In general the efficiency factor does not change greatly with age indicating that equation (8) has some success in describing the changes in bend-over point stress with age.

Although this approach has some success in relating the bend-over point to other composite properties the fit is not entirely satisfactory and further work is needed in this area.

Composite strength and post-cracking stiffness

Once multiple cracking is completed additional load cannot be carried by the matrix; it is all carried by fibres. For the aligned continuous fibre reinforced composite (1)

$$E_{1v} = E_f V_f \qquad (9)$$

and at failure

$$\sigma_c = \sigma_f V_f \qquad (10)$$

where E_{1v} is the slope after multiple cracking is completed.

For the spray-dewatered pseudo-random two-dimensional reinforcement we may again introduce empirical efficiency factors K_3 and K_4 to describe the overall effectiveness of the random short fibres in strand form, then:

$$E_{1v} = K_3 E_f V_f \qquad (11)$$

and $$\sigma_c = K_4 \sigma_f V_f . \qquad (12)$$

Experimental values for K_3 and K_4 have been determined using the post-cracking stiffness, composite strength, fibre stiffness and fibre in cement strength data and are given in table 1C. Where the slope after multiple cracking was not constant the initial value, immediately after cracking was completed, has been used.

Different efficiencies are observed in longitudinal and transverse directions and their ratio corresponds closely with the relative number of fibres oriented in these directions as the microscopic technique of Hibbert (13). It is found that the stiffness efficiency factor K_3 varies very little with age and also is little affected by sand additions, even at high levels. The strength efficiency factor K_4 also varies little with sand additions up to 0.6 sand-cement, but thereafter drops markedly indicating an increasing effect of easy fibre pull-out at failure.

In the cement paste composite and in the mortar composites having low sand contents the efficiency factors K_3 and K_4 are identical. It is only at high sand contents that these differ and for these composites the final region shows a marked curvature as shown in Figure 2b, which is again consistent with sample failure by progressive pull-out of fibres.

Average values determined for these efficiency factors, for all of the composites except those with the two highest sand contents are for stiffness (K_3), 0.26 (± 0.02) and 0.16 (± 0.01) for longitudinal and transverse directions and for strength (K_4), 0.27 (± 0.03) and 0.17 (± 0.02). These values may be compared with other experimental estimates of 0.26 for strength efficiency by Allen (14) in a glass reinforced sample and 0.24 and 0.21 for stiffness and strength respectively by Aveston (2) in a carbon fibre reinforced material.

A number of theoretical estimates have been made of the efficiency of strength reinforcement in random two-dimensional short fibre composites (2,14,15,16,17). These are shown in Figure 4, as a function of fibre length. Allen (14) has also pointed out how the composite strength depends on sample dimensions. If the critical length ℓ_c is estimated from the single strand pull-out test ($\ell_c \approx 26$ mm) the strength of 25 mm wide samples with 38 mm long strands would be expected to be only 81% of that of a wide strip. Taking account of this the efficiency factors become 0.32 and 0.21 for longitudinal and transverse specimens respectively. The average value agrees well with the predictions of Laws (13,14) for the case when the dynamic frictional stress τ_d equals the static yielding stress (τ_s) (experimentally $\tau_d \approx 0.85 \tau_s$) and is significantly less than the estimates of Allen (14) and Aveston (2) which do not include stress concentration effects at crack surfaces. The efficiencies are well above the estimate of Aveston (2) for the case when no crumbling occurs at crack surfaces and stress concentration effects are largest.

The agreement between theory and experiment must be regarded, however, with some caution because of the limitations of the strand pull-out test. It is not clear at

Figure 4   Theoretical estimates of efficiency of reinforcement.

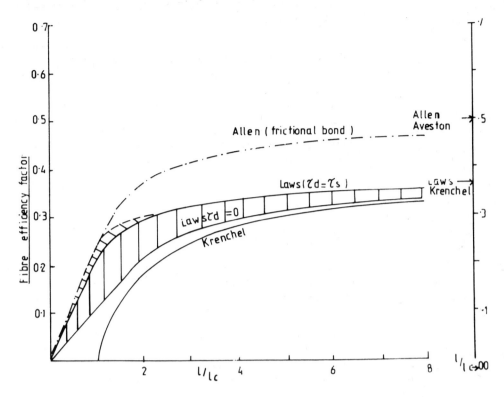

present exactly what is the effect of neighbouring fibres on pull-out length and whether multiple cracking has a marked effect on pull-out and hence efficiency at failure. There is some evidence that the critical length estimate is reasonable from experiments in which the fibre length has been varied (JA Lee, private communication) where the ultimate strength changed with fibre length much as predicted by Laws' upper curve. Further work is planned to confirm this behaviour.

At this stage the limited data on strength of strands in a cement environment have meant that direct values of the strength efficiency factor K_4 have only been obtained at one month. By assuming that this factor remains constant with age, as does the stiffness factor K_3, it is possible to estimate how the actual strength of strands in the composite changes with age. Alternatively, the observation that, in cement paste and low sand content mortars, the stiffness and strength efficiency factors at early ages are identical (ie, $K_3 = K_4$) enables equations (11) and (12) to be combined to give

$$\sigma_f = \frac{E_f}{E_{1V}} \sigma_c \qquad (13)$$

Thus an estimate of fibre strength in the strands may be made directly from measurements of post-cracking stiffness and composite strengths even if fibre orientation, content and geometry factors are unknown.

Apart from the two boards having the highest sand contents, where the efficiency factors were not equal, both methods indicate very similar, and small, changes in utilised fibre strength with age of material. Results are given in Figure 3c and show a very similar variation to that observed by Majumdar (17) for strengths of single filaments extracted from composites stored in laboratory conditions.

CONCLUSIONS

The effectiveness of glass strand reinforcement in controlling the stress at which multiple cracking begins in the matrix (bend-over point) is markedly dependent on the orientation of the reinforcement. It is little affected by sand content up to sand/cement ratios of about 0.7. On the other hand the effectiveness of strands in producing and controlling multiple cracking is unaffected by orientation over the range of anisotropy covered in these experiments.

Strand-matrix bond strengths in the composites decrease somewhat ($\sim 30\%$) during laboratory storage over one year as evidenced by changes in crack spacing and bend-over point. A drop in bond strength is observed at the highest sand contents: this drop appears to be controlled by the structure of the matrix rather than the sand content itself.

In the final post-cracking region the reinforcement is equally effective in providing both strength and stiffness up to sand/cement ratios of about 1. The strength efficiency then falls off indicating increasing fibre pull-out. Effectiveness is again orientation dependent. Values of strength efficiency factors at the lower sand contents agree quite closely with the theoretical estimates of Laws (16,17) for the assumption of equal dynamic and static frictional bonds.

ACKNOWLEDGEMENT

The authors wish to thank the Directors of Pilkington Brothers Limited and Dr DS, Oliver, Director of Group Research and Development, for permission to publish this paper.

REFERENCES

1. Aveston, J, Cooper, GA and Kelly, A, "Single and multiple fracture", The Properties of Fibre Composites, IPC Science and Technology Press, 1971, pp 15–26.
2. Aveston, J, Mercer, RA and Sillwood, JM, "Fibre reinforced cements — scientific foundations for specifications", Composites — Standards, Testing and Design, IPC Science and Technology Press, 1974, pp 93–102.
3. Proctor, BA, Oakley, DR and Wiechers, W, "Tensile stress/strain characteristics of glass fibre reinforced cement", Composites — Standards, Testing and Design, IPC Science and Technology Press, 1974, pp 106–107.
4. Majumdar, AJ and Nurse, RW, "Glass fibre reinforced cement", Materials Science and Engineering, Vol 15, 1974, pp 107–127.
5. Steel, BR, "Glass fibre reinforced cement", Proc. Int. Building Exhib. Conf. "Prospects for fibre reinforced construction materials", Building Research Establishment, 1972, pp 29–39.
6. Bartos, P, "Glass reinforced cement", CIRIA Report, 1970.
7. Majumdar, AJ, "The role of the interface in glass fibre reinforced cement", Cement and Concrete Research, Vol 4, No 2, March 1974, pp 247–268.
8. Lawrence, P, "Some theoretical considerations of fibre pull-out from an elastic matrix", Journal of Materials Science, Vol 7, 1972, pp 1–6.
9. deVekey, RC and Majumdar, AJ, "Interfacial bond strength of glass fibre reinforced cement composites", J. Materials Science, Vol 5, 1970, pp 183–184.
10. Pell, FR, "The effects of pozzolanas and autoclaving on the mechanical properties of glass fibre reinforced Portland cement", MSc Thesis, University of Bristol, 1969.
11. Aveston, J and Kelly, A, "Theory of multiple fracture of fibrous composites", J. Materials Science, Vol 8, 1973, pp 352–362.
12. Brown, JH and Pomeroy, CD, "Fracture toughness of cement pastes and mortars", Cement and Concrete Research, Vol 3, 1973, pp 475–480.
13. Hibbert, A, "A method for assessing the quality of glass fibre in an opaque matrix", J. Materials Science, Vol 9, 1974, pp 512–514.
14. Allen, HG, "The strength of thin composites of finite width with brittle matrices and random discontinuous reinforcing fibres", J. Phys. D: Appl. Phys., Vol 5, 1972, pp 331–343.
15. Krenchel, H, "Fibre Reinforcement", Akademisk Forlag, Copenhagen, 1964, pp 11–38.

16 Laws, V, "The efficiency of fibrous reinforcement of brittle matrices", J. Phys. D: Appl. Phys., Vol, 4, 1971, pp 1737–1746.
17 Laws, V, Lawrence, P and Nurse, RW, "Reinforcement of brittle matrices by glass fibres", J. Phys. D: Appl. Phys., Vol 6, 1973, pp 523–537.

7.6 Structural properties of carbon fibre reinforced cement

S Sarkar and M B Bailey
*Department of Civil Engineering,
The Hatfield Polytechnic*

Summary *This paper reports the results of an experimental and theoretical investigation on the structural properties of fibre reinforced cement (CFRC). In all 54 tensile and flexural CFRC specimens were tested. The specimens were reinforced with 60 to 150 mm wide continuous carbon fibre tapes produced by Courtaulds Limited. The range of the volume fraction of carbon fibres used was between 2 and 10%.*

Résumé *Cette communication présente un rapport sur les résultats d'une étude expérimentale et théorique sur les caractéristiques structurales du ciment renforcé de fibres de verre (CFRC). On a mis à l'essai 54 éprouvettes de CFRC pour essais de tension et de flexion. Ces éprouvettes sont renforcés de rubans de carbone continus de 60 à 150 mm fabriqués par Courtauld. La gamme des teneurs en fibres de carbone était de 2 à 10%.*

MATERIALS

In this investigation rigid control was maintained on the properties of the constituent materials. Brief descriptions of the cement matrix and carbon fibres and the properties relevant to this study are given below.

Cement matrix

Finely ground Portland cement, commercially known as Swiftcrete, with an average fineness of 800 m²/kg was used in all the specimens. The water/cement ratio by weight (w/c) was kept at 0.5 without any additives. It was estimated that the maximum practical fibre content (1) would be about 13% by volume (V_f). In practice, a V_f of 10% for the tensile specimens, where the fibres were dispersed uniformly in the matrix, was obtained with care whereas for the flexural specimens, the practical limit was a V_f of 8% when it was intended that the fibres were to be concentrated on the tensile face as far as possible.

Carbon fibre

The carbon fibre used was Courtaulds GRAFIL A produced in twist free tows containing 10000 and 20000 filaments in the form of tapes 75 mm and 150 mm wide respectively. The filaments were of a mean diameter of 9.2 μm with a standard deviation of 0.38 μm. Some relevant properties of the fibres (extracted from the GRAFIL data sheets) are given below:

Average density = 1750 kg/m³
Weight/unit length of tow = 1000 mg/m
Ultimate tensile strength (f_{ft}) = 1900–2600 N/mm²
Elastic modulus (E_f) = 180–220 GN/m².

The tapes were in the form of layers of single filaments fixed in a water soluble medium.

FABRICATION OF SPECIMENS

The tensile specimens were of rectangular cross-section of uniform thickness with enlarged widths at the ends, whereas the flexural specimens were of uniform cross-section over the whole length. The tensile and the flexural specimens were 250 mm and 300 mm long respectively. The nominal dimensions of the test zone for the tensile specimens were 30 mm x 6.4 mm and those for the flexural specimens were 60 mm x 6.4 mm. The widths were decided by the width of the tapes which was 60 mm. The thickness was decided from practical considerations of the ease of fabrication and the probable thickness in actual use of carbon fibre reinforced cement (CFRC). The V_f for the tensile specimens as 4 to 10% and that for the flexural specimens was 2 to 8% (Table 1).

The dimensional tolerance in fabrication was ± 0.25 mm. The moulds were fabricated from Perspex sheets.

Method of fabrication

The specimens were fabricated by laying a thin layer of the cement paste using a brush and a layer of CF tape alternately, the paste being worked well into the fibre to achieve complete wetting. Vibration of very low amplitude was tried but was found to have no significant effect on the density of the specimens and therefore was not used. Finally a finishing layer of the paste was applied and smoothed to the correct thickness.

Curing

The specimens were left in the moulds for 24 hours in a humid atmosphere of about 95% relative humidity, after which the specimens were demoulded and stored in water at 21 °C until the time of testing.

TEST PROCEDURE

A DARTEC M 2502 universal testing machine was used to carry out both the tensile and the flexural tests.

Tensile test series

Linear voltage displacement transducers (LVDT) were used in conjunction with an automatic plotter for recording the load-extension characteristics of the tensile specimens. These tests were conducted under displacement control at 0.02 mm/sec.

Each of the specimens was loaded beyond cracking, unloaded and reloaded to a higher point than the previous one and this cycle was repeated on the average four times until the failure was reached. This sequence was adopted to obtain moduli of the composites (E_{t1}, E_{t2}, E_{t3}) as defined later.

Control Specimens: Tensile control specimens with holes at enlarged ends, which were of the same cement paste as the composite ones and were cast in similar moulds, were tested using dead weights.

Flexural test series   Flexural specimens were tested over a span of 256 mm giving span/depth ratio of 40. Third point bending was adopted to give a zone of constant bending moment. LVDT's were used to obtain continuous load/deflection (mid-span) plots and the loading was cycled to progressively higher values, as in the tensile series, to obtain E_{b1}, E_{b2}, and E_{b3} moduli.

Table 1 (a)   **Results**

Tensile Specimens

Specimen number	V_f (%)	Age at test (days)	Density (kg/m³)	Ultimate tensile load (kN)	Ultimate tensile stress (N/mm²)	E_{t1} (N/mm²)	E_{t2} (N/mm²)	E_{t3} (N/mm²)
T10.1	10	14	1 880	19.61	104.0	30 200	14 800	21 300
.2	10	55	1 870	10.37	54.0	34 400	20 800	26 000
.3	10	55	1 900	19.11	100.0	31 200	13 300	26 000
.4	10	28	1 900	19.28	100.0	32 200	14 750	24 700
.5	10	14	1 860	18.92	100.0	31 000	16 200	—
.6	10	55	1 960	20.65	108.0	30 200	15 000	19 600
.7	10	14	1 920	15.71	82.5	27 000	15 400	—
.8	10	28	1 800	16.18	84.5	29 200	16 180	24 600
T8.1	8	50	1 780	16.51	86.0	30 200	12 500	18 700
.2	8	28	1 815	16.10	84.0	30 600	14 200	—
.3	8	50	1 870	20.25	105.5	—	12 800	—
.4	8	14	1 710	12.10	64.0	32 300	13 700	—
.5	8	28	1 760	13.97	73.0	24 500	12 600	—
.6	8	14	1 790	15.08	80.0	26 000	12 600	17 200
.7	8	14	1 790	16.30	86.0	27 000	11 400	—
.8	8	28	1 810	17.77	92.5	28 500	14 200	—
T6.1	6	14	1 885	10.67	55.7	24 500	10 700	20 200
.2	6	14	1 840	12.11	63.2	27 000	7 300	16 600
.3	6	28	1 900	10.45	54.5	32 500	11 450	18 400
.4	6	28	1 885	14.30	74.5	—	8 900	15 200
.5	6	7	1 885	15.10	78.5	21 000	8 750	—
.6	6	7	1 885	17.04	89.0	19 500	9 900	15 200
T4.1	4	14	1 840	11.97	62.3	18 800	6 250	9 100
.2	4	14	1 810	9.07	47.4	16 200	6 350	15 200
.3	4	28	1 835	9.70	50.5	25 000	6 550	10 400
.4	4	7	1 920	5.50	28.7	27 500	7 350	—
.5	4	7	1 840	7.30	38.0	23 400	6 550	13 450
.6	4	28	1 900	9.86	51.5	17 200	6 650	13 600
.7	4	28	1 910	10.82	56.3	—	6 250	—
.8	4	28	1 915	10.68	55.6	—	6 750	11 700

Table 1 (b)

Flexural Specimens				Modulus of rupture (N/mm²)	E_{b1} (N/mm²)	E_{b2} (N/mm²)	E_{b3} (N/mm²)
F8.1	8	14	1 870	113	28 700	15 100	23 200
.2	8	14	1 850	87	16 000	10 500	14 300
.3	8	28	1 900	130	22 000	13 300	18 300
.4	8	28	1 900	130	–	–	21 000
.5	8	7	1 840	106	40 000	16 500	18 700
.6	8	7	1 780	114	31 800	16 500	20 400
F6.1	6	28	1 850	127	–	14 300	19 300
.2	6	14	1 830	93	23 500	14 300	17 000
.3	6	14	1 810	114	27 500	12 700	16 500
.4	6	28	1 840	94	–	11 600	15 400
.5	6	40	1 860	115	–	15 700	20 600
.6	6	40	1 860	128	–	14 300	18 800
F4.1	4	40	1 840	132	22 500	13 300	16 000
.2	4	40	1 840	110	19 300	12 100	15 200
.3	4	28	1 850	89	16 700	9 100	12 700
.4	4	28	1 870	94	25 500	12 100	15 600
.5	4	14	1 870	115	–	12 100	–
.6	4	14	1 820	103	–	15 200	–
F2.1	2	14	1 860	99	18 700	9 400	12 100
.2	2	28	1 880	115	–	13 100	15 500
.3	2	14	1 860	80	23 600	8 700	13 200
.4	2	28	1 900	99	18 900	12 100	14 300
.5	2	7	1 850	73	17 000	10 900	16 000
.6	2	7	1 800	79	–	–	15 500

Age at testing

The ages of the specimens when tested were in the range of 7 to 55 days as shown in Table 1.

RESULTS

A summary of the results is given in Table 1. The three tensile moduli E_{t1}, E_{t2} and E_{t3} refer to the gradients of the initial, cracked matrix and reloading portions of the curves as shown in Fig 1(f).

Similarly E_{b1}, E_{b2} and E_{b3} refer to the three moduli obtained from the flexural tests. f_t and f_b refer to the ultimate tensile strength and the modulus of rupture of the composites.

Strength properties

Fig 1 (a) shows the relationship between f_b and V_f. The full curve is obtained by ignoring the results of the flexural tests for $V_f = 2\%$ at 7 days whereas the dotted line includes these results. Fig 1 (b) shows the relationship between f_t and V_f.

Fig 1 (d) shows the relationship between f_b/f_t and V_f shown as a full line, whereas the dotted line represents f_b/f_t versus V_f but with f_b values for F4, 6 and 8 series taken to correspond with f_t values for T6, 8 and 10 series respectively.

Stiffness properties

The three moduli E_1, E_2 and E_3 for both flexural and tensile tests are shown in Fig 1 (c). The ratios E_{b2}/E_{b1}, E_{b3}/E_{b1}, E_{b2}/E_{b3} and the corresponding tensile modular ratios are shown in Fig 1 (e).

Fig 1 (f) is the schematic representation of a typical load-deformation curve showing the location of E_1, E_2 and E_3. An enlarged detail at X in this diagram shows the actual

Figure 1 Variation of properties of the composite with V_f.

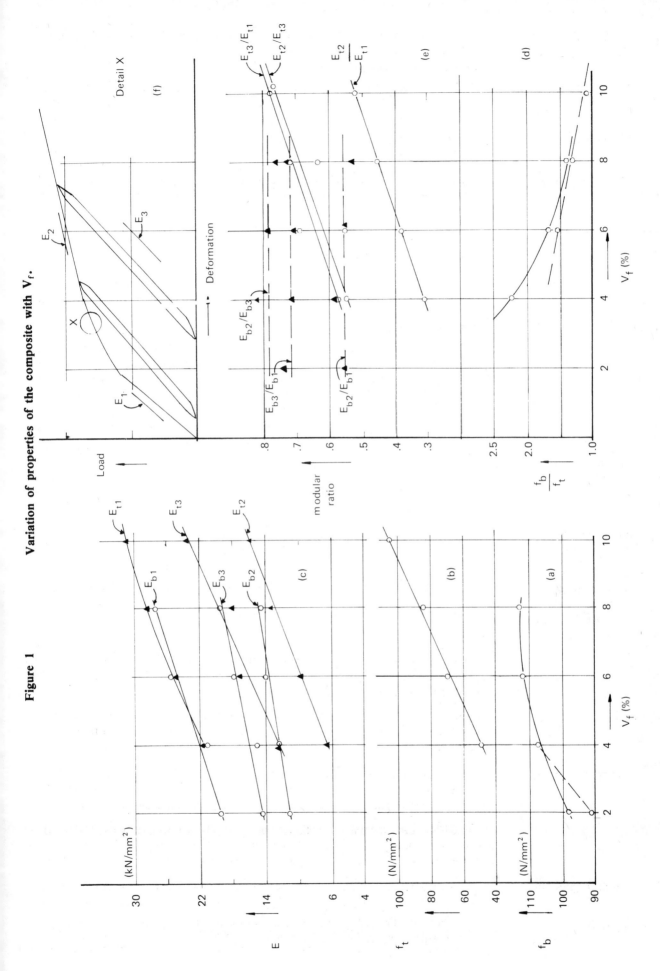

nature of the curve in the cracked matrix zone. Further relationships obtained between tensile and flexural moduli E_{b1}/E_{t1}, E_{b2}/E_{t2} and E_{b3}/E_{t3} were 0.999, 1.47 and 1.13 respectively (not shown in Fig 1).

Cement matrix

The properties of the unreinforced cement matrix obtained from control specimens were found to be E_m = elastic modulus = 15 KN/mm², f_{ct} = tensile strength = 4 N/mm².

Modular ratio

The modular ratio, m = E_f/E_m, is taken to be 200/15=13.3 for all computations in this paper.

Photo-micrograph

Figure 2 gives a typical photo-micrograph of a failure surface (T10.4) at a magnification of 525 where some cracks in the matrix are visible. It was estimated that the crack spacing was about 0.105 mm for this specimen. Several other micrographs gave an approximate estimate of the crack spacings for different V_f

THEORETICAL CONSIDERATIONS

In this section an attempt is made to appraise theoretically the experimental findings presented in the previous sections. The theoretical basis is the principle of mixtures applied to composite materials as detailed in the works published by Aveston, Cooper, Kelly, Sillwood and Kollek (2, 3, 4, 5) and Holister and Thomas (6). The assumption is that the composite consists of continuous aligned fibres bonded ideally to the matrix, the overall strain in the composite being equal to the strain in the fibre and in the matrix. Ideally, the phases of a composite should exhibit strain compatibility. However, for the CFRC the matrix cracks, exposing the fibres to stress concentrations. When $V_f > V_{fc}$ here, V_{fc} is the critical volume fraction of fibres, the failure of the composite is normally gradual as can be associated with multiple fractures. The derivation of various theoretical parameters is given in the Appendix.

Critical volume faction V_{fc}

Using equation A.5, V_{fc} is given by:-

$V_{fc} = 0.0021$ for $f_{ft} = 1900$ N/mm²

and $V_{fc} = 0.0015$ for $f_{ft} = 2600$ N/mm².

Interfacial shear stress

The crack spacing obtained from photo-micrographs of the T10 series was found to be about 0.105 mm. Using equation A.7 the interfacial shear stress is found to be equal to 0.77 N/mm². Using this value of τ the crack spacings x^1 for fibre concentrations of 0.08, 0.06 and 0.04 are found to be 0.134, 0.185 and 0.283 mm respectively.

$\epsilon_{mu}, \epsilon_{mc}, \epsilon_{cu}$

Using $f_{ct} = 4$ N/mm² and $E_m = 15$ KN/mm²

$$\epsilon_{mu} = 0.000266.$$

Using equation A.10 ϵ_{mc} is found to lie between 355 x 10⁻⁶ and 626 x 10⁻⁶ for the tensile series as detailed below:

V_f	$\epsilon_{mu}(1 + \alpha/2) < \epsilon_{mc} < \epsilon_{mu}(1 + \frac{3}{4}\alpha)$	
0.1	355 x 10⁻⁶	401 x 10⁻⁶
0.8	380	436
0.06	421	498
0.04	505	626

The ultimate strain in the composite is obtained using equation A.11 and is given by:

$0.01091 < \epsilon_{cu} < 0.010955$ which may safely be taken equal to $\epsilon_{fu} = 0.011$

$$\text{where} \quad \epsilon_{fu} = \frac{f_{ft}}{E_f} = 0.011.$$

E_{t1}, E_{t2}, E_{t3}

The three moduli E_{t1}, E_{t2}, E_{t3} obtained for the tensile tests using equations A.1, A12 and A.13 respectively are shown in Table 2.

Table 2   Comparison between experiment and theory

Specimen number	Experimental results				Theoretical results			
	E_{t1} (N/mm²)	E_{t2} (N/mm²)	E_{t3} (N/mm²)	Ultimate tensile load (kN)	E_{t1} (N/mm²)	E_{t2} (N/mm²)	E_{t3} (N/mm²)	Ultimate tensile load (kN)
T10	30670	15800	23700	19.9	30500	18000	22900	38.1
T8	28440	13000	17900	16.2	29800	14400	20800	30.4
T6	24900	9500	17100	13.2	26100	10800	16500	21.0
T4	21300	6670	12240	9.6	22400	7200	11800	15.2

Ultimate tensile strength

The failure stress of the composite is given by $f_{cu} = E_{t2} \cdot \epsilon_{cu}$ and the failure load is then equal to $A \cdot f_{cu}$, where f_{cu} = the ultimate tensile strength of the composite and A = area of cross-section of the specimen. These computed values are shown in Table 2.

DISCUSSION OF RESULTS

This discussion relates to four main aspects of the investigation:

(i) General behaviour of the specimens
(ii) Strength characteristics
(iii) Stiffness characteristics
(iv) Fracture characteristics.

General behaviour of the specimens

The load-deformation characteristics of both the tensile and the flexural specimens generally followed the theoretical model given by Aveston et al (2). The multiple fracture characteristics were evident as shown in detail, Fig 1 (f).

Strength characteristics

The ultimate tensile strengths obtained from the experiments are on the average about 0.56 of the theoretical values. This may be attributed to the fact that the ultimate failure is a combination of bond failure associated with progressive fibre failure but as the fibres do not attain the ultimate strain simultaneously the degree to which each contributes is not known. An interesting phenomenon emerges from comparison of the f_b and f_t values plotted in Fig 1 (a, b, d). It appears that f_b reaches a maximum value at $V_f = 8\%$ whereas for the tensile specimens f_t shows a linear increase up to 10%. This may be because the fibres were placed as near the tensile face of the flexural specimens as possible. This also resulted in much higher strengths than those of the tensile specimens with the same V_f. The dotted line in Fig 1 (d) represents the relationship between f_b/f_t and V_f when V_f values of 4, 6, 8% are used to plot the actual f_b values for $V_f = 2, 4$ and 6% respectively. Even with this modification the ratio f_b/f_t ranges from 1.5 to 1.1 for V_f from 6% to 10%.

A further point of interest is that for the flexural specimens the optimum V_f appears to lie between 7 and 8% which when extrapolated for tensile specimens may give an optimum V_f of about 15% but such a high volume concentration may create fabrication difficulty. The effect of age on f_t or f_b is not clear from the results of these tests.

Stiffness characteristics

It is quite apparent that the correlation between the theoretical and the experimental moduli E_{t1}, E_{t2} and E_{t3} is good. This is of great practical significance as it would make accurate prediction of deformation possible.

Within the range of the experiments, the initial flexural and tensile stiffness moduli have excellent correlation. The increases of E_{t2}, E_{t3}, E_{b2} and E_{b3} are all linear with the increase of V_f. However, no definite relationship could be obtained between E_{t2} and E_{b2}, and E_{t3} and E_{b3}. The ratios $E_{b2}/E_{b1}, E_{b3}/E_{b1}$ and E_{b2}/E_{b3} shown in

Fig 1 (e) are 0.55, 0.71 and 0.78 respectively and are independent of V_f. The ratios E_{t2}/E_{t1}, E_{t3}/E_{t1} and E_{t2}/E_{t3} increase linearly with V_f. There has been a tendency for an increase in the stiffness with age but no clear conclusions could be drawn from the limited scope of these experiments.

Fracture characteristics

The multiple fracture characteristics were clearly shown in all the load-deformation plots as shown in detail X in Fig 1 (f). The final failure was always abrupt but predictable, the failure surface being rather jagged in nature.

Photo-micrographic study was limited to a few specimens at this stage and further study is in progress. Figure 2 shows that the fibre–matrix bond is intermittent and some longitudinal cracks along the fibres are also visible. The computation of the stiffness based on crack spacing has been based on a very limited number of observations and needs further investigation.

CONCLUSIONS

From the study it is evident that substantial gains in the strength and stiffness properties can be achieved with relatively low volume fractions of continuous carbon fibres. The behaviour of the composites during cracked matrix and reloading conditions have been linear and predictable. The handling and the fabrication creates some problems at volume fractions above 10% for the type of specimens used in this study.

Figure 2 Photo-micrograph of failure surface of T10.4 (x525).

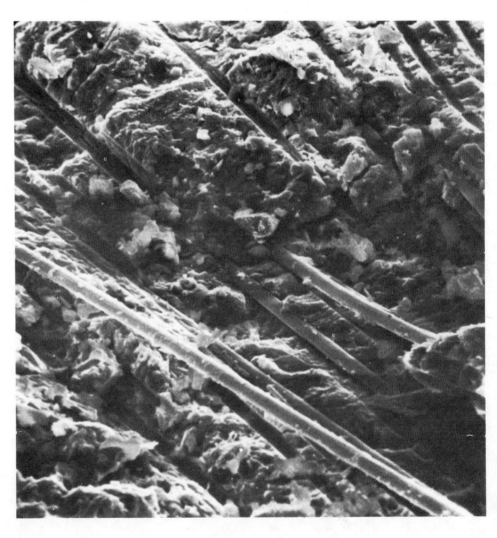

ACKNOWLEDGEMENTS

This study was conducted in the Department of Civil Engineering and Construction of the Hatfield Polytechnic.

The authors would like to record their thanks to Mr J Waller of Messrs Oscar Faber and Partners and Courtaulds Limited for making the carbon fibres available for the project.

REFERENCES

1. Waller, J, Carbon Fibre Cement Composites, Fibre Reinforced Concrete, ACI Special Publication SP 44, 1974, pp 143-161.
2. Aveston, J, Cooper, G A and Kelly, A, Single and Multiple Fracture, Proc. Conf. on the properties of fibre composites, National Physical Laboratory, 1971, pp 15-24.
3. Kelly, A, Microstructural Parameters of an Aligned Fibrous Composite, Proc. Conf. on the properties of fibre composites, National Physical Laboratory, 1971, pp 5-14, Research Report.
4. Cooper, G A and Sillwood, J M, Multiple Fracture in a Steel Reinforced Epoxy Resin Composite, Journal of Material Science, Vol 7, 1972, pp. 325-333.
5. Kollek, J J, Fibres in Cement-Based Materials—An Appreciation, Paper presented at the 'Advances in Concrete' Symposium, Birmingham, September 1971, London, the Concrete Society, 1972, p. 10.
6. Holister, G S and Thomas, C, Fibre Reinforced Materials, Elsevier Publishing Company Limited, London, 1966.

APPENDIX

Uncracked condition

The elastic modulus of the composite is given by:

$$E = E_f V_f + E_m (1 - V_f) \qquad \text{A.1}$$

where E is the elastic modulus of the composite.

From strain compatibility considerations:

$$\epsilon = \frac{f_m}{E_m} = \frac{f_f}{E_f} \qquad \text{A.2}$$

where, ϵ = strain in the composite.

Combining equations A.1 and A.2

$$\frac{f}{f_m} = 1 + V_f (m - 1) \qquad \text{A.3}$$

where. $m = E_f/E_m$. Equation A.3 represents the upper bound of composite strength for $V_f \geqslant V_{fc}$. However, a cracked matrix would reduce equation A.3 to the lower bound given by:

$$\frac{f}{f_m} = m V_f \qquad \text{A.4}$$

The critical volume fraction may be obtained by using the following relationship:

$$f_{ft} V_{fc} = f_{ct} (1 - V_{fc}). \qquad \text{A.5}$$

Multiple fracture condition

Multiple fractures will occur if

$$f_{ft} V_f > f_{ct} V_m + f_f' V_f \qquad \text{A.6}$$

where f_f' = stress in fibres at a strain equal to the rupture strain of the matrix, ϵ_{mu}.

Interfacial shear stress

If the failure strain of the fibres is sufficiently large the matrix will be fractured into lengths between x^1 and $2x^1$

where
$$x^1 = \frac{V_m \, r \, f_{ct}}{V_f \, 2 \, \tau} \qquad \text{A.7}$$

where
  r = radius of the fibres and
  τ = maximum shear stress that the interface can sustain.

Stress-strain relationship during multiple fracture of matrix

The load-deformation or stress-strain relationship shown in Fig 1 (f) has three distinct parts (i) uncracked, (ii) cracked matrix and (iii) cycled loading.

(i) *Uncracked Condition*

$$E_{t1} = \frac{E_f E_m}{V_f E_m + V_m E_f} \qquad \text{A.8}$$

(ii) *Cracked Matrix Condition*

The increase in stress in the fibres is assumed to be linear between zero at x^1 from the crack to a maximum of $f_{ct} V_m / V_f$ at the crack. The average increase in strain over $2x^1$ is then given by $\tfrac{1}{2}\alpha \epsilon_{mu}$ where,

$$\alpha = \frac{1 - V_f}{m \, V_f} \qquad \text{A.9}$$

With further increase of stress the distance between cracks is reduced to between $x\tfrac{1}{2}$ and x^1, the maximum stress increase being $f_{ct} V_m / V_f$ and the average strain being increased to $\tfrac{3}{4} \alpha \epsilon_{mu}$. The overall strain at the limit of multiple fracture, ϵ_{mc}, is given by

$$\epsilon_{mu} \left(1 + \frac{\alpha}{2}\right) < \epsilon_{mc} < \epsilon_{mu} \left(1 + \frac{3\alpha}{4}\right) \qquad \text{A.10}$$

Further increase in load will cause slipping of fibres through the blocks of matrix and the ultimate strain in the composite, ϵ_{cu} is given by

$$(\epsilon_{fu} - \tfrac{1}{2}\alpha \epsilon_{mu}) < \epsilon_{cu} < (\epsilon_{fu} - \tfrac{1}{4}\alpha \epsilon_{mu}) \qquad \text{A.11}$$

and the modulus of the specimen will become $E_f V_f$. \qquad A.12

(iii) *Cycled Loading*

During unloading of a specimen it is necessary that τ be overcome. The mean compressive strain in the matrix for crack spacings of $2x^1$ would be $\tfrac{1}{2}\epsilon_{mu}$, and the permanent set in the specimens is given by $\tfrac{1}{2}\alpha \epsilon_{mu}$. It can be shown that for a crack spacing of $2x^1$ the modulus will be

$$E_o = \frac{E_{t1}}{1 + \alpha/2} = E_{t3} . \qquad \text{A.13}$$

It can further be shown that if the crack spacing is between x^1 and $2x^1$ the modulus becomes

$$E_o = E_{t3} = \frac{E_{t1}}{\left(1 + \frac{\alpha}{2}\right)\left(2px^1 + \frac{1 - 2px^1}{1 + \alpha/2}\right)} \qquad \text{A.14}$$

where p is the number of cracks per unit length.

Section 8

Applications

8.1 Opening Paper: Inflation forming of steel fibre reinforced concrete domes

Gordon B Batson,
Department of Civil and Environmental Engineering, Clarkson College of Technology, Potsdam, New York, USA

Dan J Naus and Gilbert R Williamson
Construction Materials Branch, US Army Construction Engineering Research Laboratory, Champaign, Illinois, USA

Summary

The feasibility of inflation forming steel fibre concrete shell structures suitable for shelters and fortification is shown to be a viable construction method.

The technique of construction is based on the experience gained building several 2.75 m diameter domes of variable shell thickness. The technique developed is to be used for the construction of 8.5 m diameter domes.

The preliminary stress analysis for a 8.5 m diameter dome of variable thickness using a finite element method showed that the steel fibre concrete was not highly stressed for the loads, openings and boundary conditions investigated. The buckling analysis was not completed in time to be included in this report; however, preliminary values are presented.

The inflation forming technique has the possibility of reducing the time of construction, man-hours of labour and level of skill needed to construct shelters and fortifications.

Résumé

On a montré que l'emploi de formes gonflées pour la construction de coques en béton armé de fibres d'acier est une méthode de construction valide.

Cette technique est basée sur l'expérience obtenue lors de la construction de plusieurs dômes de 2.75 m de diamètre, avec une épaisseur de coque variable.

L'analyse préliminaire des contraintes pour un dôme de 8.5m de diamètre avec épaisseur variable, par la méthode des éléments finis, a montré que les contraintes dans le béton de fibres d'acier n'étaient pas élevées pour les charges, ouvertures de baies et conditions limites envisagées. L'analyse de flambement n'ayant pu être achevée à temps pour figurer au rapport on a dû se contenter de présenter des valeurs préliminaires.

Cette technique permet de réduire le temps nécessaire à la construction ainsi que la main d'oeuvre et le niveau de qualification requis pour la construction d'abris et de fortifications.

INTRODUCTION

The random dispersion of discrete metallic, glass and plastic fibres in plain concrete or mortar has improved many of the mechanical properties of plain concrete. Low carbon steel fibres have significantly improved more of the mechanical properties of plain concrete than other kinds of fibres for the least cost of fibres; as a result steel fibre reinforced concrete is the predominant type of fibre concrete being evaluated for military construction and maintenance. The US Army Corps of Engineers has been conducting research and development on fibre concrete since the mid 1960s.

There is a continual need by the military for construction methods that reduce the time, the level of skill, the man-hours, and volume and/or weight for materials that must pass through the supply system for military construction projects in the theatre|of operations. A potential solution for the needs of military construction of shelters and fortifications in the theatre of operations is the use of steel fibre concrete and the technique known as inflation forming.

The military construction needs are in general equally valid for civil construction where these techniques might be used for constructing emergency shelters for food supplies and equipment after a natural disaster or for the storage of grain crops.

The technique of inflation forming of structures has been used for many years (1, 2), but the recent developments by Dante Bini (2) of Italy has attracted wide attention as a rapid economical means for the construction of large concrete dome or shell type structures. Bini's method consists of placing concrete and a special patented (3) expandable reinforcing mesh of soft coil springs between two fabric reinforced rubber membranes and anchoring the system to a rigid concrete ring beam. A low pressure, less than 3.5 kN/m^2, high volume air system applies pressure to the bottomside of the bottom membrane and lifts the encased mass of concrete. As the membranes are lifted, the expandable mesh of reinforcing drags and spreads the concrete as the dome continues to inflate. At the desired rise, about one third or less of the diameter, the system is stabilized and the concrete allowed to harden. Twenty hours later a hole is cut into the concrete dome and the bottom (now the inside) membrane is retrieved and the process can be repeated using the same membranes on another ring beam.

Construction Engineering Research Laboratory (CERL) has modified the Bini method by using steel fibre concrete to eliminate the expensive and time consuming effort of placing the expandable mesh of coil springs of the Bini method. From the viewpoint of the needs of the military the use of steel fibre concrete:

1. reduces the volume and weight of materials that must be transported in the military supply system,

2. reduces the time of construction, since approximately 50 percent of the construction time of the Bini method is consumed in placing the expandable mesh,

3. reduces the man power and level of skill.

The research and development programme at CERL was to determine the feasibility of the technique of inflation forming of steel fibre concrete domes for shelters and fortifications. The technique was evaluated by constructing a series of 2.75 m diameter domes and performing the analysis for the stresses and the buckling load by the finite element method for 8.5 m diameter domes.

The results of construction of the prototype domes and the theoretical analysis showed that the inflation forming technique of domes with steel fibre concrete is a viable method of construction which may be suitable for military applications.

BUILDINGS AMENABLE TO INFLATION FORMING IN THE THEATRE OF OPERATIONS

The applicability of fibre concrete to the construction of facilities in the theatre of operations was appraised by considering the functional characteristics of the structures used to support the Army that would be amenable to the inflation forming technique.

Since the inflation forming technique provides a shell or dome type structure free of interior supports, structures with functional characteristics of large open areas and a minimum of electrical and mechanical requirements would be ideal for inflation forming with fibre concrete. The functional requirements suggest structures classified as shelters and fortifications. Shelters provide protection from the environment for supplies, equipment and personnel; whereas fortifications provide protection from the environment and hardenability against ballistics. The requirements of a shelter can be met by a thin shell structure, 50 to 100 mm thick, whereas a fortification would require a thick shell structure 300 mm or more in thickness.

Since there already exist in the military supply system prefabricated structures with clear spans up to about 7.6 m, the smallest inflation formed structure should be greater than 7.6 m in diameter and be available in sizes up to 36.6 m in diameter. The rise of the dome normally decreases with increasing diameter; the maximum rise being one-third the diameter for the smallest diameter dome.

To take advantage of the rapid rate of construction and re-use of the inflation forming equipment, a rear area casting yard should be established for the construction of the smaller domes that can be transported by helicopters to forward areas. It is probable that a steel fibre concrete dome about 7 m in diameter can be kept to a minimum weight of 6800 kg. Large shelter type structures and fortifications would have to be constructed at the desired location, but the volume of equipment and materials to be moved would be less than for a similar structure made of timber or steel.

TECHNIQUES OF INFLATION FORMING OF STEEL FIBRE CONCRETE STRUCTURES

The objective of the research and development programme was to construct and test load a steel fibre reinforced concrete dome 8.5 m in diameter with a rise 3 m at the crown. To evaluate the feasibility of the inflation forming technique, a series of 2.75 m diameter domes with a rise of 0.9 m at the crown were constructed. The 2.75 m diameter domes were of a size that could be constructed using equipment, material and personnel available within CERL.

Ten 2.75 m diameter domes were attempted with varying degrees of success. However, each successive dome generated valuable experience which reflected improved procedures for each succeeding dome. The experience generated was used as input for the planning of the construction and procurement of materials and equipment for the 8.5 m domes.

2.75 m diameter domes

A sketch of the geometry of a 2.75 m diameter dome is shown in Figure 1. Note that the dome is a spherical sector of one base; not a hemisphere.

The construction base or foundation for the inflation forming technique required a heavy concrete ring beam. The weight of the base (ring beam) must be greater than the maximum force applied by the air pressure during inflation, otherwise the base will lift and air will leak out under the edge of the base. Assuming the maximum air pressure to be 3.5 kN/m^2, the dead weight of the ring beam for a 2.75 m diameter dome is 2074 kg. The ring beam had a dead weight of about 1223 kg and was tied to plywood sheet by metal straps, the inside of the ring beam being filled with sand to ensure sufficient dead weight as shown in Figure 2.

A water manometer tube was used to monitor the air inflation pressure. Depending on thickness and type of membrane material, the inflation pressure varied from 150 to 330 mm of water for the 2.75 m diameter domes.

Laboratory air supply was applied either directly through a needle valve or was diverted through a pressure regulator valve. Usually there was too much leakage of the air and the necessary volume of air could only be maintained through the direct line. The exhaust line was capped and the manometer tube attached to it. A surge in the air pressure, 12.7 to 25 mm, was noticed when the laboratory compressor was turned on

Figure 1 Geometry of 2.75 m (9 ft.) dome of variable thickness with a rectangular opening.

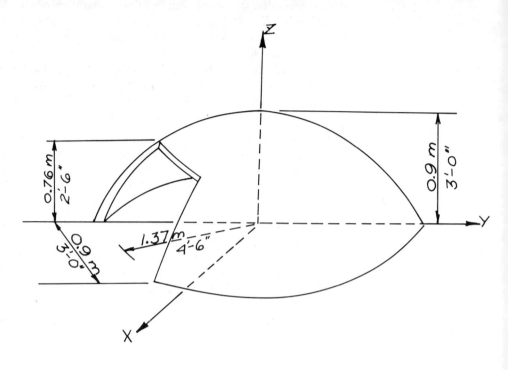

Figure 2 Detail of ring beam and membrane anchorage method.

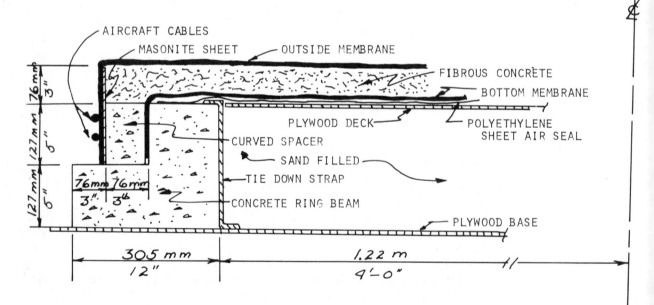

by the low pressure control switch.

The effect of a change in the air pressure on the dome depended on whether the fibre concrete was in a plastic or hardened state. If the fibre concrete was in a plastic state, pressure changes of 2.5 mm of water would produce deformation of 12 to 25 mm at the crown. Also there was a time lag after decreasing or increasing air flow through the needle valve before the change was noticed on the manometer. The large volume of enclosed air easily deformed the plastic fibre concrete and the system acted like an accumulator damping out pressure changes by the large elastic deformations. If the fibre concrete was in a hardened state, no time lag of manometer reading was observed in the shape of the dome. There was little or no damping of air pressure changes in the hardened state.

Inflation membranes, both inside (bottom) and outside, were initially made from neoprene rubber sheeting bonded together with adhesive. These membranes were made by cutting triangular segments and bonding them together with lap joints to form a circle. The concept for this shape was to keep the membrane symmetrical so it would inflate symmetrically. Later membranes were made by bonding sheets straight with a lap joint into a large square sheet and cutting out a circular shape. This method did not appear to cause noticeable unsymmetrical deformation during inflation.

Initially 1.6 and 0.8 mm thick neoprene was used for the bottom and outside membrane respectively; but later the inside membrane was increased to 3.2 mm thick material. The thicker material required a greater inflation pressure, 100 to 127 mm, and provided a stiffer shape for surface vibrating the fibre concrete after inflation. The outside membranes for later domes were made of polyurethane, a transparent plastic sheeting 0.2 mm thick. The particular polyurethane had a trade name, "Tuftane", and was manufactured by B F Goodrich. It had a greater resistance to tearing than neoprene and was easily bonded with a lap joint using tetrahydrofuran (THF).

Several schemes were tried for anchoring of the membranes to the ring beam. The method shown in Figure 2 was found to be convenient for the 2.75 m diameter domes. The inside membrane was bonded by adhesive to the ring beam. The outside membrane was anchored by aircraft cables wrapped around the outside of the curved spacer of the ring beam and tightened with turn-buckles. A different anchorage method was used for the 8.5 m domes.

The slump of the fibre concrete is a critical parameter for constructing good domes. Too large a slump allowed the concrete to flow and slide down the sides of the membranes during inflation; too low a slump and the concrete would "tear" or separate excessively during inflation. In general, a slump of between 50 and 75 mm was necessary for fibre contents used. The fibre concrete was surface vibrated after inflation to heal "tears" that developed during inflation. A Vibco surface vibrator operating at a frequency of 150 Hz and developing a force of 2000 N could heal most of the tears completely through the thickness of the fibre concrete.

Construction sequence The bottom membrane was attached to the ring beam by an adhesive and curved concrete spacing segments were positioned. A masonite dam to provide a uniform 55 mm thickness of fibre concrete was placed against curved spacers and the outside membrane was secured by 6 mm aircraft cable tightened by a turn-buckle (see Figure 2).

The quantity of materials for the fibre concrete mix was weighed out separately; cement, fibres, sand, pea-stone aggregate, water and water reducer. The sand, pea-stone aggregate, cement and steel fibres were mixed dry in a 0.4 m^3 turbine mixer. Sufficient water and a water reducer agent were added to yield the desired slump concrete. The ratio of sand to pea-stone was approximately 3. Fibre contents of approximately 1.5 per cent by volume of concrete, 90 kg per cubic yard, were used. The fibres were the flat type 0.25 x 0.50 mm in cross-section by 20 mm long.

The fibre concrete was placed directly on the bottom membrane and was screeded off using a vibrator attached to wood timbers riding on the edge of the masonite dam. The outside polyurethane sheeting was secured to the outside of masonite by another 6 mm aircraft cable and turn-buckle. The air pressure was applied and the dome inflated in about 20 minutes. At maximum rise the pressure was adjusted to stabilize the system and surface vibration applied to heal the "tears" in the fibre concrete that occurred during inflation. Approximately 8 hours after inflation the fibre concrete had hardened sufficiently so that air pressure could be decreased to atmospheric. The inside membrane collapsed when the air pressure was reduced to atmospheric and the drilling of the holes for the lifting points did not puncture the inside membrane. Twenty-four hours after inflation the outside membrane was removed and the dome lifted from its

Figure 3 Lifting FR concrete dome at age of 48 hours.   Figure 4 Stored FR concrete domes.

base by a three-point lift from an overhead crane.

The completed domes were stored out of doors; some of the domes stacked or "nested" and others rested on four concrete blocks. Also rectangular openings were sawed in the domes using a demolition or concrete saw.

STRESSES AND BUCKLING LOAD OF FIBRE CONCRETE DOMES

The particular procedure used for inflation forming provides the shell with a variable thickness. The fresh fibre concrete is placed in a uniform thickness. During inflation the membrane stretches with the greatest stretch occurring at the crown and reducing nonlinearly to zero at the edge where the membrane is restrained. Since there is a fixed volume of fibre concrete, the concrete thickness is least at the crown and increases nonlinearly down to the edge. The thickened edge of the variable thickness shell eliminates the need of a heavy ring beam for stability of the dome.

The thickness at the crown of the 2.75 m domes reduced from 75 mm to about 20 mm. The measured thickness at the crown averaged about 25 mm and increased 75 mm near the edge along longitudinal lines at right angles to each other. A slight bulge at the edge on the inside surface of the dome was caused by surface vibration forcing some of the fibre concrete down to the edge.

The variable thickness shell is better suited to military needs than a uniform (constant) thickness shell. The thickened edge provides stability so that the dome can be lifted off the casting base ring beam, and if the dome is less than 6800 kg in weight it can be transported by a helicopter.

Finite element analysis

An 8.5 m dome with a rise of 3 m and a variable thickness was assumed. The finite element programme, "EASE" was used to analyze the dome for static loads equal to one dead load, three dead loads, and 28 °C temperature rise. The boundary conditions were considered to be the ideal simple support type at the edge of the shell. None of the computed stresses were large enough to be of concern because steel fibre reinforced concrete can resist tensile stresses up to 6900 kN/m^2.

Besides a static load analysis, stresses created by a 28 °C temperature change were computed for a simple edge and a fixed edge support. The stresses in the shell for a fixed edge support with a 28 °C temperature change were too large to be resisted by the fibre concrete.

Table 1 **Static and buckling stresses**

a) Maximum stress for one dead load

Closed dome	207 kN/m² at edge
Rectangular opening	4140 kN/m² at point "a"
Vertical opening	6900 KN/m² at point "b"

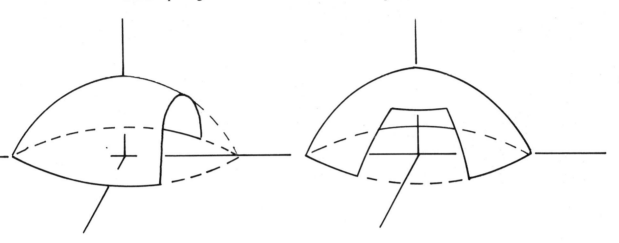

Vertical opening   Rectangular opening

b) Buckling stresses (preliminary)

Closed dome	241 kN/m²
Vertical opening	21 kN/m²
Rectangular opening	62 kN/m²

From static analyses it appears important to insure that the edge of the shell is free to deform under applied and thermal loads. In the field, the shell should be supported as nearly as possible to meet the ideal conditions of a simple support; the edge should not be rigidly restrained.

The "EASE" programme analysis considered only a closed symmetrical shell without openings. The stresses due to openings, the buckling load for closed shells and shells with openings were analyzed using the "NASTRAN" finite element programme. A necessary prerequisite of the buckling analysis is a static analysis. The results of the "NASTRAN" and "EASE" programmes for the static analysis agreed very well using different numbers and shapes of finite elements.

The preliminary results of the buckling load analysis for the closed shells are given in Table 1 and the buckling analysis for shells with openings has not been completed. The theoretical buckling load is generally larger than the actual buckling load (4). Several design codes recommend that the theoretical buckling stress be multiplied by a constant ranging from 0.1 to 0.6 for concrete shells which implies that the critical buckling stress for the closed shell may be as low as 240 kN/m². Preliminary calculations for shells with openings indicate that the critical buckling stress may be reduced by an order of magnitude to about 143 kN/m².

CONCLUSIONS

The feasibility of the inflation forming technique has been demonstrated to be a viable method for the construction of shell type structures using steel fibre concrete.

The experience gained from the construction of a series of 2.75 m diameter domes with a rise of 0.9 m served as input for the procedure to construct 8.5 m diameter domes with a rise of 3 m.

The inflation forming technique has the possibility of reducing the time of construction, man-hours of labour, and level of skill needed to construct shelters and fortifications.

ACKNOWLEDGEMENTS

Acknowledgement is made to the US Army Construction Engineering Research Laboratory and Clarkson College for sabbatical arrangements that made it possible for G Batson to participate in the fibre concrete R&D programme of the US Army. During this investigation Col R W Reisacher was Director and Dr L R Shaffer was Deputy Director. E A Lotz was Chief of the Materials Division.

Dr Ronald Merritt, Civil Engineer, CERL, is acknowledged for his assistance in using the NASTRAN finite element programme for the buckling analysis.

REFERENCES

1 Frei, Otto, Tensile Structures, The MIT Press, Cambridge, Massachusetts, Vol I (1967), Vol II (1969).
2 Dent, R N, Principles of Pneumatic Architecture, Halsted Press Division, John Wiley and Sons, Inc, New York, 1972.
3 US Patents 3 462 521, (August 19, 1969) and 3 686 817, (August 29, 1972).
4 Ramaswamy, R G, Design and Construction of Concrete Shell Roofs, McGraw Hill, 1969.

8.2 Full-scale trials of a wire-fibre-reinforced concrete overlay on a motorway

John M Gregory, John W Galloway and Kenneth D Raithby
*Highways Department, Transport and Road Research Laboratory,
Department of the Environment*

Summary   *To evaluate the potential of wire-fibre-reinforced concrete for thin overlays on heavily trafficked concrete roads in Great Britain, a full-scale experiment was carried out on a stretch of reinforced concrete carriageway on the M10 motorway, which had been in service for 15 years and had suffered some cracking. The overlay was constructed in two thicknesses, 60 mm and 80 mm, and three fibre contents were used, 2.7, 2.2 and 1.3 per cent by weight of 0.5 mm x 38 mm duoform wires. 16 test sections were laid, some of which were bonded to the original pavement slab and some only partially bonded. The total length was 200 metres.*

Mixing was performed at a nearby "ready-mix" concrete plant and a slip-form paver was used for laying. Expansion joints were formed in the wet concrete over the expansion joints in the original pavement. Some additional contraction joints were also formed in the overlay.

Samples of material were taken from the paver to make up beam specimens for flexural strength and fatigue tests, which were carried out after curing for 3 months.

The best performance, in terms of surface cracking, after 6 months heavy traffic appears to be given by the thinner bonded sections.

Résumé   *Dans le but d'évaluer l'utilisation possible du béton armé par des fibres métalliques dans la construction, en Grande-Bretagne, de couches de recouvrement minces pour les routes en béton supportant une circulation intense, on a mis en oeuvre une section expérimentale de 200 m de long au-dessus d'un tronçon de chaussée en béton armé de l'autoroute M 10. Cette autoroute, mise en service il y a quinze ans, montrait des signes de fissuration. La couche de recouvrement a été construite en deux épaisseurs, 60 mm et 80 mm et on a utilisé trois teneurs en fibre, 2.7, 2.2 et 1.3 pour cent du poids de fils "duoform" de 0.5 mm x 38 mm. On a mis en oeuvre 16 sections expérimentales, parmi lesquelles quelques unes adhéraient complètement à la dalle originale de la chaussée tandis que d'autres n'y adhéraient que partiellement.*

On a utilisé une installation de production de béton prêt à l'emploi pour le mélange du béton et une machine à coffrage glissant pour la mise en oeuvre. On a confectionné les joints de dilatation dans le béton humide au-dessus des joints de dilatation de la chaussée originale. On a aussi confectionné des joints de retrait supplémentaires dans la couche de recouvrement.

On a prélevé des éprouvettes de matérial dans le paveur pour confectionner des poutres d'essai, dans le but d'effectuer des essais de résistance à la flexion et des essais à la fatigue. Les essais ont été effectués après soumission du béton à une cure de trois mois.

La meilleure performance obtenue en ce qui concerne la fissuration superficielle, après six mois de circulation intense, semble être offerte par les sections adhérentes les plus minces.

INTRODUCTION

Many existing roads are becoming inadequate for present-day traffic because of structural and surface deterioration. The usual method of extending the life of such roads has been the addition of a bituminous overlay; concrete overlays have rarely been used. Escalating costs of bitumen have given impetus to the consideration of other materials which could be used to overlay concrete pavements. One such material is steel-fibre concrete and this paper describes the design, construction and early performance of an experimental overlay in this material constructed in May 1974 on the M10, a dual two-lane motorway carrying heavy commercial traffic.

DETAILS OF TEST SITE

Selection of site

A heavily trafficked road was considered desirable for an overlay experiment so that meaningful results could be obtained in a relatively short time. The site on the M10 was chosen after several other sites had been rejected because of inadequate traffic intensities.

The total length of the experimental overlay (just over 200 m) was limited by the funds available and its location was influenced by the need to minimise traffic diversion during construction of the overlay. The site selected was on the two-lane northbound carriageway close to the roundabout at the beginning of the motorway, on a stretch of road where the concrete pavement had developed fairly extensive cracking. The estimated traffic flow at this point is 2200 commercial vehicles a day in one direction, equivalent to about one million standard axles (1) in a year. As the experimental site is on an uphill gradient and is close to the entry roundabout the proportion of the commercial traffic using the offside lane is larger than usual. Fig 1 shows typical traffic on the test section.

Constructional details of existing pavement

The M10 was constructed as a reinforced concrete pavement in 1959 and opened to traffic in November of that year. The slab thickness was a nominal 275 mm laid on a polyethylene sliplayer; the sub-base was 175 mm of hoggin on which there was a 25 to 40 mm sand regulating layer. The full two-lane width of 7.9 m was laid in one operation and the slabs were reinforced with steel mesh weighing 5.4 kg per square metre, placed a nominal 60 mm below the surface. All the transverse joints were of the expansion type and at the experimental site these joints were at 36.6 m intervals; load

Figure 1 Traffic on test section.

transfer was provided by 32 mm diameter steel dowel-bars 685 mm long at 300 mm centres. Longitudinal and transverse joints were formed by sawing, the longitudinal joint being 3.7 m from the nearside edge. Delay in sawing the longitudinal joint groove led to a considerable amount of longitudinal cracking, roughly along the centre of the carriageway, before the joint was formed. The concrete had an aggregate-cement ratio of 7.9 and was not air-entrained; the average 28-day compressive strength was 38 N/mm^2. The omission of entrained air led to surface spalling at joints and cracks because of frost action: this was made worse by the use of de-icing salt.

Condition of pavement before overlaying

The original pavement had suffered a considerable amount of cracking, especially in the nearside lane, in the 15 years that it had been carrying traffic. Longitudinal and transverse cracking appeared in the slabs shortly after their construction and spalling occurred, particularly after the severe winter of 1962-63, necessitating a considerable number of surface repairs. These repairs took the form of thin bonded concrete patches, many of the cracks being sealed with a reinforced neoprene sealing strip. Fig 2 shows typical cracked and patched areas. In the length to be overlaid the average spacing of transverse cracks more than 2 m long was about 1.2 m in the nearside lane and about 5.6 m in the offside lane. About one third of the cracks had been repaired and in many cases cracks had developed in the repair patches. There was a more or less continuous longitudinal crack running the whole length of the experiment within about 300 mm of the centre line of the carriageway.

Crack patterns in two of the test sections before overlaying are illustrated in Fig 3, which also shows how cracking developed in the overlay during the first 3 months.

DESIGN OF THE OVERLAY EXPERIMENT

In an experimental length of only 200 m it was not possible to incorporate many variables. The factors investigated were overlay thickness, fibre content and interface condition. The layout of the experiment is given in Fig 4.

Figure 2 **Typical length of existing carriageway showing concrete repairs and longitudinal cracking (nearside lane is on right of figure).**

Figure 3   **Examples of crack development—nearside lane.**

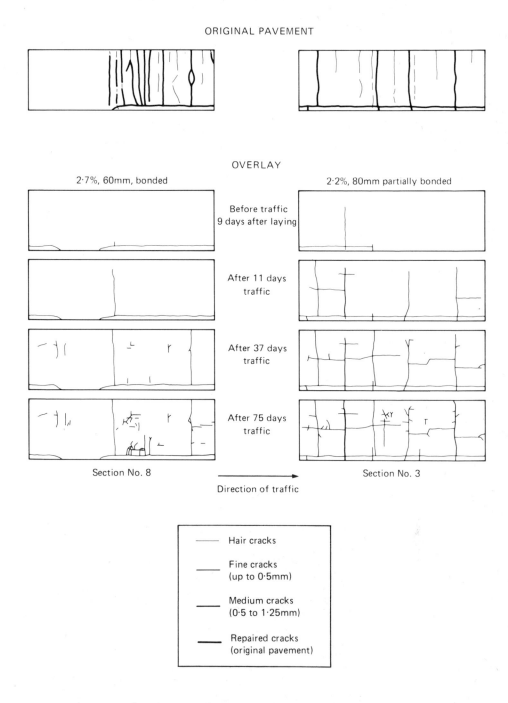

Thickness

There are no established design procedures at present for steel-fibre concrete overlays but there are proven methods for the design of overlays in conventional concrete, for which thicknesses of 125 and 150 mm have given good performance. Evidence from American experiments with steel-fibre concrete in pavement applications (2) has indicated that slab thicknesses can be reduced to about one-half of those required with normal concrete. On this basis it was decided to use two overlay thicknesses in the experiment, 60 and 80 mm.

Fibre content

Again because of the restricted length of the experimental section, only two fibre contents were used in each lane; the upper limit of fibre content was fixed at 3 per cent by weight, because higher concentrations would be likely to cause difficulties in mixing. It was originally intended that fibre contents of 3 and 2 per cent should be used in the nearside lane and 2 and 1 per cent in the less heavily trafficked offside lane, each fibre

Figure 4 Layout of experimental steel-fibre concrete overlay.

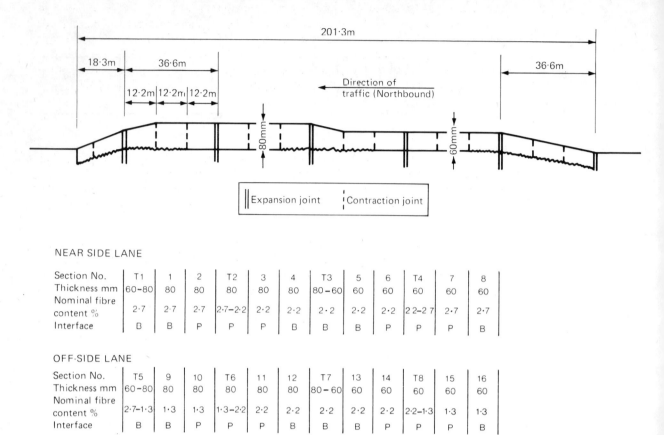

content being used for half the total length of the experiment. The ramp sections at the ends of the experiment were to have 3 per cent of fibres. In practice, for ease of batching and to avoid undue "balling" problems, the nominal fibre contents were changed to 2.7, 2.2 and 1.3 per cent.

Interface conditions   A completely unbonded overlay has been recommended by the US Army Corps of Engineers for applications where the base pavement is in poor structural condition (2). This ought to reduce the tendency for reflection cracking by avoiding the build-up of high local strains in the overlay. Partially bonded overlays are recommended if the base pavement is in fair condition. Although the M10 contained many cracks, its structural condition was considered good for a reinforced concrete pavement.

It was decided to investigate partially bonded and fully bonded interface conditions. Inclusion of a completely unbonded overlay was not considered practicable. For half the length of the experiment, the overlay was placed directly onto the cleaned, but otherwise untreated surface. Over the other half the surface was lightly scabbled, damped, and coated with cement slurry immediately prior to paving.

Joints   Knowledge relating to suitable slab lengths for steel-fibre concrete overlays is sparse. For the present application expansion joints were provided in the overlay at intervals of 36.6 m to coincide with those in the existing pavement. Transverse contraction joints were then formed at 12.2 m intervals, chosen to give test slabs of a reasonable length enabling all the variables to be covered. Transitional lengths were introduced where changes of either thickness or fibre content between experimental sections occurred. No dowel-bars or tie-bars were used in any of the overlay joints.

Concrete specification The limited time available for arranging the construction of the experiment precluded the preparation of a detailed specification. The requirements for the concrete were that the fibre content should be as stated and that the workability should be suitable for placing by slip-form paver. The water-cement ratio was not to exceed 0.55 and the specified minimum cube strength at 7 days was 28 N/mm^2. Mix design was based on preliminary tests at TRRL (see below) but the size and type of fibre and the method of mixing were left to the contractor's choice; they were governed by the availability of materials and plant.

CONSTRUCTION

An over-riding factor in the construction was the limit of 18 days during which the carriageway could be closed to traffic. The contractor planned to complete all preparatory work and concrete laying in 11 days, which meant that the last concrete to be laid (on the offside lane) would be only 7 days old when it was subjected to traffic: this programme was met.

Preparatory work After removal of "cats-eyes" and edge-markers, the surface was roughened, where bond was required, by multi-tool scabbling machines, the sealant in the transverse joints was removed and the joint grooves cleaned. The end ramp sections required the removal of concrete down to reinforcement level in order to be able to lay overlays to the correct thickness. It was found that the scabbling machines were too slow for this purpose because of the flint aggregates used in the original concrete. To accelerate removal a drop-hammer concrete breaker was employed on the long ramp; however the drop on the machine was set too high with the result that the slab beneath the reinforcement was extensively shattered.

Mixing The mix adopted after preliminary trials by the contractor had a flint gravel aggregate of 10 mm maximum size, an aggregate-cement ratio of 4.3 and a water-cement ratio of 0.55. The materials were batched at a "ready-mix" plant about 3 km from the site. The steel fibres were hand-fed onto the dry materials as they passed on a conveyor from the weighing apparatus to the truck-mixer and water was added at the plant. In the first two mixes with a nominal 3 per cent fibre content there was considerable "balling", as a result of which the maximum fibre content was reduced to 2.7 per cent.

Laying A conforming-plate type of slip-form paver, the CPP 60, controlled by guide wires, was used to lay the steel-fibre concrete in two equal widths of 3.95 m. The joint in the overlay was not therefore directly over the sawn longitudinal joint of the original carriageway.

The more heavily trafficked nearside lane was laid first to allow the concrete in this lane the maximum time to cure before reopening the road to traffic. Paving commenced at the northern end and the lane was completed in about eight hours. As many fibre "balls" as possible were removed during the discharge from the truck-mixer but some passed through the paver; many of these were then removed by hand and the slab made good. The offside lane was paved four days after the nearside lane with one track of the paver running in the centre reserve and the other on the nearside overlay, which was protected by plywood sheets.

Texturing of the overlay was by transverse wire brushing but the texture achieved was less than that now required for concrete road surfaces. Efforts to obtain a deeper texture produced tearing of the surface. Curing was effected by spraying with a standard aluminised curing compound.

Contraction joints were formed by cutting a slot in the concrete with a trowel, placing a preformed plastic insert into the slot and reinstating the concrete by hand around the insert, using a bull float, and hand-trowelling to finish. After the concrete had hardened the top portion of the plastic insert, which was a push fit on the bottom part, was removed and the joints sealed with a hot-poured sealing compound.

Immediately before the paver reached the position of an expansion joint a timber filler was inserted into the top of the existing groove and held vertical until the paver had passed over it. Concrete above the filler board was removed by trowel and a capping strip of timber secured to the filler: finishing of expansion joints was similar to that for contraction joints. After the concrete had hardened the capping strip was removed and the joint sealed with hot-poured joint sealant. Sealing of the longitudinal joint between the two lanes was by a neoprene strip fixed with an adhesive to the edge of the nearside slab.

LABORATORY STRENGTH TESTS

Mix design tests

Previous work had indicated that a suitable mix for wire-fibre-reinforced concrete would be 1 part by weight of cement, 2.55 of 10 mm coarse aggregate and 2.05 of sand, with a water-cement ratio of 0.5.

In preliminary tests at TRRL on such a mix the effects of fibre content and fibre geometry on strength and workability were determined for a range of wire diameters and lengths.

These indicated that for a reasonable workability the maximum flexural strength was likely to be obtained with 0.25 mm diameter wires having a length to diameter ratio of 100 but a good compromise between strength, cost, workability and freedom from "balling" could be achieved with 0.5 mm diameter wires. With 3 per cent by weight of such wires increases in flexural strength of about 30 per cent might be expected. A very limited amount of flexural fatigue testing had indicated that the addition of wires was likely to improve the fatigue performance proportionally more than the flexural strength.

The basic concrete mix finally chosen by the contractor was slightly richer than that used in the TRRL preliminary trials, being a 1:2.15:2.15 mix with a water-cement ratio of 0.55. Fibres used were 0.5 mm x 38 mm duoform wires supplied by the National Standard Company. The mix proportions were chosen on the basis of 4-day and 8-day cube strengths determined from trial mixes made up at the plant.

Tests on material used in overlay experiment

Several samples of the mixed concrete were taken as the concrete was discharged from the mixer trucks. This material was then made up into standard cubes and beams for compressive and flexural strength tests, all specimens being compacted with a vibrating hammer and cured under water. Actual fibre contents of some samples were determined by washing and extracting the fibres with a magnet. Mean values were found to be 1.15, 1.85 and 2.33 per cent, compared with nominal values of 1.3, 2.2 and 2.7 per cent.

Compressive strength tests were carried out at 3, 7, 28 and 91 days and modulus of rupture and flexural fatigue tests at 91 days. It was not possible to make a simple comparison between the strength of the fibre-reinforced material and an equivalent plain concrete, because by removing the wires from the mix the workability was increased to such an extent that the mix would not have been acceptable for use in a slip-form paver. An alternative approach of keeping the same slump in both concretes by reducing the water-cement ratio from 0.55 to 0.4 in the conventional material resulted in a concrete having a basic strength considerably increased because of the lower water content.

The results of the laboratory tests are given in Table 1. Neither compressive strength nor modulus of rupture was significantly affected by variation in the quantity of fibres in the mix, and the strengths achieved were rather less than had been expected on the basis of the preliminary tests on mixes prepared in the laboratory. However, examination of the specimens after test indicated a considerable non-uniformity of dispersion of the fibres and there was little apparent difference in overall fibre concentration on the fracture faces between the different nominal fibre contents. In several cases there were few fibres present at the point where failure is likely to have initiated.

The fatigue test results indicate that the fibre-reinforced mixes all have a considerably better performance than would be expected from plain concrete of

Table 1 Summary of laboratory test results

Nominal wire content % by wt	Compressive strength Mean value, N/mm², at age, days				Flexural strength Age of concrete, days 91				Fatigue performance Age of concrete, days 91			
	3	7	28	91+	Number of results	Mean value N/mm²	Coefficient of variation %	Maximum stress N/mm²	Number of results	Geometric mean life cycles	Coefficient of variation of log. life %
1.3		36.91	52.03	61.76	8	4.85	2.5	3.72	4	>421,990	12.3
2.2	24.30	34.75	50.15	58.77	7	4.97	8.6	3.91	5	>148,280	10.3
								3.91	6	>126,560	26.0
2.7	24.40	32.60	53.58	59.72	4	4.84	4.1	3.91	4	>158,120	11.8

> Before "life" signifies that group includes one or two specimens that did not fail

+ "Equivalent cube" strength of broken beams.

equivalent strength. At about 80 per cent of the modulus of rupture the mean fatigue life of each of the fibre mixes was some five times greater than the probable life for plain concrete (3).

PERFORMANCE OF OVERLAY IN SERVICE

Condition of overlay before trafficking

When the overlay was approximately one week old and before the road was reopened to traffic, a considerable amount of hair cracking had developed in many of the test sections in the nearside lane. No cracks were observed in the offside lane. Such cracks must have been caused by shrinkage and thermal movements.

Many of the sections had developed a longitudinal crack near the longitudinal joint between the nearside and offside lanes. This crack closely followed a similar crack in the original pavement and appeared to be independent of whether the overlay was bonded or partially bonded.

All the other cracks were transverse and, of five which exceeded 2 m in length, four could be identified as reflection cracks from the underlying pavement. The total length of cracking in the nearside lane, before traffic, was equal to 23 per cent of that observed after 11 weeks of trafficking.

Crack development under traffic

The road was reopened to traffic 11 days after completing the overlay on the nearside lane and 7 days after that on the offside lane.

After 11 days of traffic the number of transverse cracks exceeding 2 m had increased to 15 in the nearside lane; with one exception these were all reflection cracks. A pattern of longitudinal cracking had begun to develop in some sections, mostly in the wheel paths but sometimes in the mid-slab position (see Fig 3). In the offside lane 11 transverse hair cracks greater than 2 m in length had appeared, almost all being reflection cracks. As the cumulative traffic has increased a network of cracks has developed in most of the test sections but, with a few exceptions, these have remained as hair cracks.

After 11 weeks of traffic the nearside lane had 32 transverse cracks exceeding 2 m in length, of which 31 were reflection cracks. A few of these cracks had increased in width to the "fine" category and one in Section 3 had become a "medium" crack. Half of the total length of fine cracks occurred in Section 3 (2.2 per cent fibre, 80 mm, partially bonded). The total length of longitudinal cracking in the overlay exceeded that of the original pavement and longitudinal cracks tended to be more pronounced in the partially bonded sections of the nearside lane, particularly in Section 3, which finally split up into 12 rectangular sections (see Fig 3) each having a width approximately equal to half the lane width.

Between 11 weeks and 6 months after opening to traffic there was little further change in the crack pattern. Up to November 1974 there had been no evidence of spalling or other surface damage around cracks in any of the test sections except for the "medium" crack in Section 3 which had suffered slight local spalling.

CONCLUSIONS

The performance to date shows no very clear relationship between fibre content and resistance to cracking but there does seem to be a tendency for the thinner bonded sections to perform better than the others. On the whole the bonded overlays in the nearside lane had a lower total crack length than the partially bonded sections, possibly because the overlay would have been restrained against separate warping and would therefore be less likely to crack under local bending stresses imposed by traffic loads.

Most of the cracks are still in the "hair" category and can be seen only by close inspection. Some of the cracks may be expected to open up during the winter period. The future performance will depend on how successfully the fibres spanning the cracks hold the individual pieces of concrete together. Particular attention will be paid to the penetration of water and salt through the cracks and the extent to which the wires may corrode and become ineffective in bridging the gap formed by the crack.

Table 2 Summary of nearside lane crack lengths

Total crack length, m, (all crack categories)

Section number	Overlay thickness mm	B = Bonded P = Partially bonded	Nominal fibre content %	Original pavement T	Original pavement L	Before traffic T	Before traffic L	Overlay After 11 days' traffic T	Overlay After 11 days' traffic L	Overlay After 5 weeks' traffic T	Overlay After 5 weeks' traffic L	Overlay After 11 weeks' traffic T	Overlay After 11 weeks' traffic L
8	60	B	2.7	43.1	6.9	0.3	9.6	3.0	9.8	11.0	11.3	20.3	15.2
7	60	P	2.7	57.9	11.7	0	4.7	3.8	6.1	10.7	14.3	33.5	34.7
T4	60	P	2.7-2.2	49.4	14.3	5.0	6.1	7.0	11.1	10.4	20.1	20.3	34.4
6	60	P	2.2	28.5	0	0	12.2	2.0	13.4	7.8	14.5	16.6	20.6
5	60	B	2.2	35.5	0	0	12.2	5.3	12.2	17.2	15.4	34.6	16.6
T3	60-80	B	2.2	27.6	0	0	11.6	3.8	12.2	16.6	13.1	26.5	15.2
4	80	B	2.2	27.6	12.2	0	12.2	0	12.2	10.8	15.2	17.2	21.9
3	80	P	2.2	31.5	12.3	3.2	4.9	14.6	17.4	20.1	25.5	27.1	28.8
T2	80	P	2.2-2.7	8.2	12.0	3.2	8.5	3.5	12.6	5.5	16.9	14.9	20.9
2	80	P	2.7	40.1	9.4	5.0	9.1	5.9	11.4	20.1	13.4	52.9	19.2
1	80	B	2.7	54.3	12.0	0.3	12.0	9.4	12.5	13.6	17.1	29.7	25.0
T1	80-60	B	2.7	48.2	12.3	0	12.3	1.4	12.3	3.7	12.3	7.5	12.3

Notes: T = Transverse   L = Longitudinal

Such a limited trial can give only some partial answers to the question of whether wire-reinforced concrete can be used as a thin overlay for concrete roads in the British Isles. Results after six months traffic are promising. If little further deterioration takes place in the coming winter it would be worthwhile mounting a more extensive trial to improve both mix design and construction techniques for situations where a thin overlay is desirable and to obtain more realistic cost data.

ACKNOWLEDGEMENTS

The work described forms part of the research programme of the Special Pavement Research Branch (Dr D Croney) and the Pavement Design Division (Mr N W Lister) of the Transport and Road Research Laboratory. The paper is published by permission of the Director. Thanks are due to Mr H M Harding for his assistance in the experimental work.

The co-operation of the staff of the County Engineer, Hertfordshire County Council, and of Robert McGregor & Co in setting up and running the experiment is gratefully acknowledged.

REFERENCES

1. Road Research Laboratory, A guide to the Structural Design of Pavements for New Roads, Department of the Environment, Road Note 29, London, 1970, (HM Stationery Office) 3rd Edition.
2. Rice, J L, Pavement Design Considerations. Fibrous Concrete, Construction Material for the Seventies, US Army Construction Engineering Research Laboratory, Conference Proceedings M.28, December 1972, pp 161-176.
3. Raithby, K D and Galloway, J W, Effects of Moisture Condition, Age and Rate of Loading on Fatigue of Plain Concrete, Abeles Symposium on Fatigue of Concrete, American Concrete Institute, SP 41, 1974, pp 15-34.

Crown Copyright 1974. Any views expressed in this paper are not necessarily those of the Department of the Environment. Extracts from the text may be reproduced, except for commercial purposes, providing the source is acknowledged. Reproduced by permission of Her Majesty's Stationery Office.

8.3 The use of fibre-reinforced concrete in hydraulic structures and marine environments

George C Hoff
Concrete Laboratory,
US Army Engineer Waterways Experiment Station,
Vicksburg, Mississippi, USA

Summary   *Four applications of fibre-reinforced concrete in hydraulic structures are reviewed. Included are repairs to sluiceways, spillways, and spillway aprons associated with dams and also the construction of breakwater armour units. The use of fibre-reinforced concrete in these structures appears to provide improved cavitation and erosion resistance and improved impact strength.*

Résumé   *L'auteur révise quatre applications de béton avec renforcement de fibres dans des ouvrages hydrauliques. Il inclut les réparations des galeries de fond de décharge, déversoirs et tapis de protection du sol, associés avec les barrages, et aussi la construction d'éléments de protection des brise-lames. Il paraît que l'emploi du béton avec renforcement de fibres dans ces ouvrages améliore la résistance à la cavitation et à l'érosion et également augmente la résistance au choc.*

INTRODUCTION

The concrete used in the discharge elements of locks and dams is subjected to waterborne forces of cavitation, erosion, and impact not normally experienced by structural grade concrete. These elements include the ogee crest of a spillway, the spillway itself, the spillway apron, sluiceways or outlet works, and the guide walls and lock walls. High-velocity water passages over spillways and through sluiceways or outlets can cause cavitation and erosion of the concrete. In areas of lesser-velocity water movement, such as the spillway apron, erosion and impact can be a problem. Impact from waterborne objects, such as debris and ships, can damage exposed walls.

Cavitation occurs when vapour bubbles that have formed in the water flow enter an area of higher pressure. These bubbles then collapse with a great impact, and for a short interval of time, the concrete adjacent to the bubble is subjected to tremendous compressive and tensile stresses. The concrete will then fail locally, with small pieces breaking out of the concrete and being carried off by the flowing water. The irregular surface that results aggravates the problem, and more cavitation then results at an increasing rate. Erosion is the actual wearing of the concrete surface due to the abrasive action of the water and the particles that it is carrying. The particle size varies from sands upward to boulders in excess of 91 kg. Pieces of steel and other man-made debris also contribute to the problem. Impact loading can result from larger particles of submerged debris bouncing along the bottom of a submerged element, from floating debris in fast-moving water, and from large ships moving at any speed through locks. Impact loading also occurs from severe wave action on such breakwater armour units as tetrapods or dolosse.

Fibre-reinforced concrete has been reported as having improved impact and abrasion resistance compared with plain concrete (1,2,3). Evaluations as to its cavitation resistance have not been reported. The fibre-reinforced concrete itself cannot be expected to resist severe cavitation loading as even steel cannot do that, but its use may raise the lower limit of water velocity at which cavitation begins in plain concrete. Also, with local surface failures, randomly dispersed fibres tend to hold the pieces of concrete together even though it has cracked, and thus may reduce the rate at which damage may occur by not allowing an irregular surface to form.

Scope

This paper reviews the use of fibre-reinforced concrete at three dam locations and in the construction of some breakwater armour units called dolosse. In most instances, it was difficult to establish whether a singular phenomenon was the cause of the damaged concrete in these structures. Cavitation damage may have precipitated erosion of the concrete or vice versa. Impact damage may have precipitated either erosion or cavitation. The processes involved are complex, but the end result is damaged concrete.

SLUICEWAYS

Libby Dam, located on the Kootenai River in northwest Montana, USA, is a major multipurpose project constructed as an integral unit of the water development plan for the Columbia River Basin in the United States and Canada. Water is passed through and over the structure by means of the intakes for an electrical power generating system, a two-gated spillway, and three sluiceways. The problem of cavitation and erosion discussed in this section occurred in a sluice (Fig 1) which had operated approximately 452 days at the time the damage was noted. The other two sluices had been in operation for an average of only 312 days at that time.

The sluice is of rectangular cross-section 3 m wide by 6.7 m high with its invert on a parabolic curve. The original concrete in the invert and walls of the sluice was of excellent quality. It contained 38 mm maximum size aggregate and 374 kg/m³ cement, and was made at water-cement ratios in the range 0.34 to 0.42 by weight. It was also air-entrained (4–7 percent). The thickness of the concrete varied from 0.8 m minimum at the invert to approximately 3 m in the walls. The concrete in the walls was placed against absorptive form lining. The concrete in the invert was hand-finished. Average compressive strength of the test cylinders at 90 days age was 43.0 N/mm².

Figure 1   **Section through Libby Dam.**

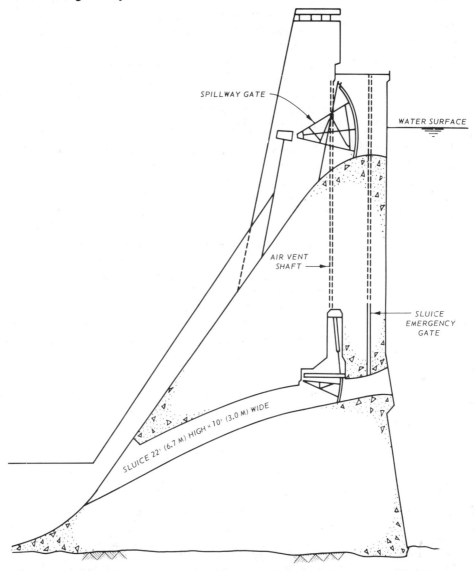

Indications that a problem existed were given by the surging behaviour of the water passing through the sluice. The sluice gate was closed and the concrete surface inspected. Two significant damaged areas were found. In the invert or floor, the damaged area began at a point approximately 41 m downstream from the start of the sluice and extended approximately 15 m further downstream. The area had a maximum width of approximately 2 m and a maximum depth of damage of approximately 660 mm. The damage in the right wall began at a point approximately 46 m downstream from the start of the sluice and extended approximately 12 m further downstream. The damage zone reached a height of 2 m on the wall and a maximum depth of 790 mm. The reinforcing steel in these areas of concrete had also been removed by the action of the water. Cores obtained from both the damaged and undamaged areas of the sluice indicated that all concrete around the sluice had attained a minimum compressive strength of 34.5 N/mm² at 28 days and considerably more strength by the time the cavitation and erosion problem developed. Based on core test results, it was concluded that the damage which occurred was not the direct result of poor strength of the concrete.

A review of several options for repair led to a decision to replace the damaged section totally with fibre-reinforced concrete. The decision was influenced by both favourable economics and the good performance of fibre-reinforced concrete in another cavitation and erosion prone structure, Lower Monumental Dam, discussed in following paragraphs. A section through the repair area is shown in Fig 2. All damaged

Figure 2   Libby Dam repair section and reinforcement.

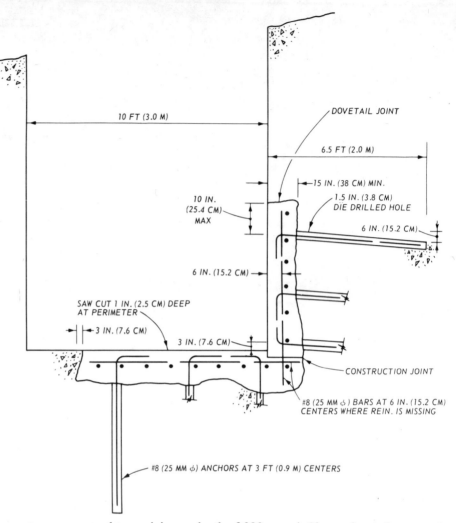

concrete was removed to a minimum depth of 380 mm. A 50 mm dovetail was used to key the repair concrete into the original concrete. The original wall and invert reinforcement consisted of 19 mm ϕ (No 6) horizontal and 25 mm ϕ (No. 8) vertical bars on 300 mm centres and 36 mm ϕ (No 11) normal and 25 mm ϕ (No 8) vertical bars on 300 mm centres, respectively. The missing steel was replaced by 25 mm ϕ (No 8) bars at 150 mm centres. Hook bars 25 mm ϕ (No 8) on 910 mm centres were embedded to a depth of 1980 mm as a final means of tie-in.

The basic fibre concrete proportions selected were as follows:

Material	Batch quantity kg/m^3
Cement	432
Sand	884
Coarse aggregate (19 mm MSA)	817
Water	173
Steel fibres (0.25 x 0.56 x 25 mm cross-section)	80
Air-entraining admixture	0.213 kg
Water-reducing admixture	1.021 kg

The fibre content was 1 percent by volume. The concrete was batched at a central mixing plant and transported to the top of the dam in transit-mix trucks. Each truck carried 2.3 to 3.0 m³. Agitation was continued both during transit and while discharging. The concrete was discharged into a 0.34 m³ bucket which was then lowered through the air vent shaft (see Fig 1) to the top of the sluice. Here it was discharged through the bottom gates of the bucket into wheeled buggies. Vibration was necessary to move the fibre concrete from the bucket. Without vibration, the concrete stopped flowing and bridged across the gate openings. The loaded buggies were connected by cable to an air-operated winch which was operated in a manner which allowed the manually guided buggies to slowly roll down the sluiceway to the placement area. Vibration was again needed to remove the concrete from the tipped buggy during discharge. The concrete was also vibrated once in place.

The invert placement was completed first. Grades and elevations were extremely critical on the curved invert and great care was exercised in setting forms. The finishers used standard magnesium floats and had excellent results. The wall placement was done one week after the invert and required a form finish. The concrete was placed through openings in the form and was vibrated internally. No honeycomb or other surface deficiencies were observed.

The results of the evaluation of each batch of concrete are as follows:

Location	Batch quantity m³	Air content %	Water-cement ratio	Workability (visual)	Compressive strength, N/mm²	
					7 days	28 days
Invert	2.3	8.0	0.409	Too wet	—	—
”	3.1	6.0	0.370	Fair	—	—
”	3.1	9.0	0.375	Too wet	—	—
”	2.3	—	0.365	Good	—	—
”	2.3	3.5	0.357	Fair	35.2*	50.0*
”					41.0	55.8
”	2.3	5.5	0.363	Good	—	—
”	2.3	5.0	0.360	Good	40.2	49.8
”	3.1	5.0	0.362	Good	—	—
Wall	2.3	3.5	0.363	Fair	—	—
”	2.3	3.5	0.390	Good	37.4	50.1
”	2.3	5.0	0.400	Excellent	—	—
”	2.3	5.5	0.398	Excellent	32.7	47.0
”	2.3	5.5	0.400	Excellent	—	—
”	0.8	5.5	0.403	Excellent	—	—

* Field-cured.

The two batches for the invert that had the very high air contents (8.0 and 9.0 percent) were discarded. In general, it was found that the batches with a water-cement ratio of about 0.400 with entrained air contents of approximately 5 percent produced the optimum placeability when assisted by vibration.

After one year elapsed time after completing the repair, with the sluice operating intermittently for approximately six of those twelve months, an inspection of the walls and invert revealed that little or no wear had occurred on the fibre concrete surfaces. This success has led to the planned usage of fibre-reinforced concrete for a repair of a similar problem area at Dworshak Dam in the western portion of the State of Idaho, USA.

SPILLWAYS

The Lower Monumental Lock and Dam, located on the lower Snake River in the State of Washington, USA, contains eight spillway bays. Four of these bays have been modified by the addition of prototype spillway deflectors (Fig 3) called "flip lips". This section reports on the construction of the first of these flip lips (4). The function of the flip lips is to change direction of the flowing water, thus causing turbulence and energy dissipation. In performing this function, the potential for cavitation and erosion damage is quite high on the surface of the flip lip. Fig 4 shows the general configuration of the flip lip. As the existing structure was operational at the time of the flip lip addition, the area of the proposed construction was underwater. A caisson or bulkhead was then constructed and dewatered, and all concrete removal, forming, steel placement, and concrete placement were completed within the bulkhead.

Figure 3   **Cross-Section – Lower Monumental spillway.**

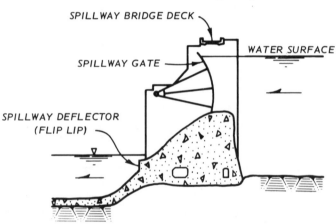

Figure 4   **Cross-Section – flip lip.**

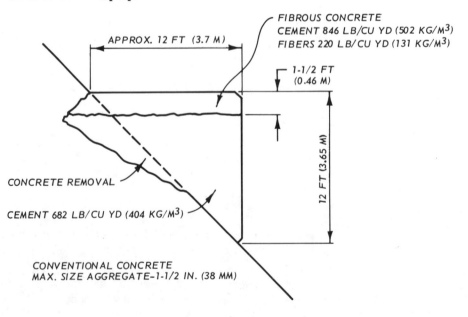

As the potential damage zone was only on the top surface of the flip lip, fibre-reinforced concrete was used only as a topping over conventional concrete. Approximately 183 m³ of conventional 38 mm maximum size aggregate concrete was used to fill the majority of the section with approximately 16 m³ of fibre-reinforced concrete providing a 460 mm topping. The concrete was batched at a central plant approximately 45 kilometres away and mixed at the plant and during transit in transit-mix trucks.

The delivery of the concrete was to a roadway on top of the dam, at which point it was discharged from the trucks into the hopper of a double-acting positive-displacement pump. It was then pumped through a 130 mm line for a 9 m horizontal run which then turned down for a vertical distance of 37 m. The discharge from the line was directly into the formwork.

The fibre concrete was delivered from the truck to the pump at slumps ranging from 38 to 76 mm. Some balling of the fibres occurred in the concrete, but the fibre balls were screened out prior to the concrete entering the pump. The basic mixture used was as follows:

Material	Batch quantity kg/m³
Cement	384
Sand	842
Coarse aggregate (9.5 mm MSA)	366
Water	155
Steel fibres (0.41 ϕ x 19 mm cross-section)	100
Water-reducing admixture	1.42 kg
Air-entraining admixture	0.26 kg
Pumping aid	0.04 kg

Quality control specimens were cast from each truckload of fibre concrete batched. The specimens were prisms 100 x 100 x 360 mm and were evaluated in flexure at 28 days age. These results and some batch data are as follows:

Batch no.	Batch quantity m³	Water-cement ratio	Slump, mm Batch plant	Slump, mm At the pump	28-day flexural strength N/mm²
1	4.2	0.40	110	80	4.0
2	3.5	0.39	80	50	5.9
3	3.5	0.38	50	40	5.9
4	3.5	0.39	80	50	4.8

This construction showed that fibre-reinforced concrete can be formulated for use in pumping and that conventional pumping equipment can be used without modification.

After two seasons of major spillage associated with the spring season runoff, the flip lips appear to be functioning satisfactorily. As the top surface of the flip lip remains approximately 2 m underwater, inspection was made by divers who reported that no visible erosion was present. Based on this promising behaviour, three additional flip lips containing fibre-reinforced concrete were also constructed at Lower Monumental Dam at a later date. Their results are not included in this paper.

In another spillway bay of the Lower Monumental Dam, three rows of steel dentates, in the form of T-beams, had been embedded in the spillway concrete. These dentates protruded into the flowing water and were supposed to perform the same function as the flip lip. The cavitation and erosion forces associated with the discharging water had, in some cases, completely destroyed the steel protrusions and, in other cases, had severely eroded the concrete under each beam. The areas of eroded and damaged concrete were repaired by cleaning up the eroded areas and removing the damaged concrete. Fibre-reinforced concrete was used as the repair material. The repair area was primed with a bonding agent prior to hand-trowelling the fibre concrete in place. The fibre concrete in this case was delivered to the repair area by concrete bucket instead of by pumping. The mixture was the same as for the flip lip construction except no pumping aid was included and the water-cement ratio was reduced to 0.35 in order to aid in placing on the sloping spillway.

One 1.5 m^3 batch was used for this repair. It had a 25 mm slump at the batch plant. The average flexural strength of the 100 x 100 x 360 mm prisms at 28 days age was 7.1 N/mm^2. After two major spillage seasons over this repair, it was still serviceable but had experienced some damage. Its performance was excellent, however, compared with the plain concrete or with two other areas under dentates which had been repaired with polymer materials. These other materials had completely eroded away after the first season spillage.

SPILLWAY APRONS

The spillway apron begins at the toe of the spillway and extends downstream to the stilling basin. The water moving over the apron is generally not at velocities which would cause cavitation damage. Large boulders and cobbles along with man-made debris can collect, however, in eddy currents over the apron and, through continual rolling and bouncing along the bottom, erode the concrete apron away. Such was the problem at Kinzua Dam. Kinzua Dam is part of a multipurpose project located on the Allegheny River in the northwest portion of the State of Pennsylvania, USA. Water can be passed over the four-gated spillway, through six lower and two upper sluices, or through the power generation intakes. The problem of eddy currents develops when sluices on only one side of the dam are discharging.

The spillway apron was originally constructed as a 1.5 m concrete slab. Near the toe of the spillway, nine large reinforced concrete baffles had been constructed on top of the apron. These baffles act as energy dissipators for the high-velocity water coming off the spillway. The major damage to the baffles was generally limited to the upstream face where a considerable amount of concrete had been eroded. A lesser degree of damage occurred on the sides and tops. Fibre-reinforced concrete was used to repair the upstream face of each baffle. Forms were offset from the baffle and the concrete placed between the form and baffle. This fibre concrete facing had a minimum thickness of 300 mm from the original face of the baffle. The original face was eroded to a depth of 100 to 130 mm in some locations. The fibre concrete was tied into the original concrete by 25 mm ϕ (No 8) bars 0.9 m on-centre. An epoxy bonding compound was also used on the surface between the old and new concrete. The sides and top of each baffle were repaired with an epoxy material.

The damage to the apron itself was quite severe in some locations. The maximum depth of erosion or scour was 1270 mm, which was approximately 83 percent of the original thickness of the apron. Across the entire centre of the apron, there was a general reduction of thickness. In one location, where a persistent eddy occurred, a doughnut-shaped trench of 1070 mm maximum depth had occurred in the concrete.

The repair work to the apron consisted of a fibre-reinforced concrete overlay. A cofferdam was built around one-half of the apron, the interior dewatered, and the surface prepared to receive the overlay. The surface was prepared by the use of air jackhammers and wet sand blasting to sound concrete. The surface was then washed clean of all rubble and sand. After drying, the surface was coated with an epoxy bonding compound with fibre concrete immediately being placed on the epoxy. The overlay was anchored to the prepared surface by the use of inverted U-shaped 25 mm ϕ (No 8) reinforcing bars embedded on 0.9 m centres in both directions. The fibre concrete was placed in alternating 6.4 by 9.1 m sections until the entire surface was covered.

The concrete was batched at a central batch plant and mixed and transported in transit-mix trucks. At the dam site, the trucks discharged into concrete buckets which were lowered to the placement area. After discharging from the concrete buckets, the concrete was consolidated by standard probe-type vibrators and brought to a finish grade by a vibratory screed. A final magnesium float finish was then given the surface. Curing was by wet burlap. The minimum overlay thickness achieved was 300 mm with the average thickness being 460 mm. In the areas of deep scour, a fill concrete was used to bring the concrete level up to that of the surrounding area. Approximately 60 percent of the apron was repaired in an initial effort. The cofferdam was then relocated and the remainder of the apron exposed and repaired in the same manner.

The basic fibre-reinforced concrete proportions used were as follows:

Material	*Batch quantity* kg/m^3
Cement	446
Sand	933
Coarse aggregate (9.5 mm MSA)	648
Water	178
Steel fibres	119
Air-entraining admixture	0.33
Water-reducing admixture	0.84

The fibre content was 2 percent. Slumps varied from 57 to 89 mm. Air contents varied from 5 to 6 percent. The total amount of fibre concrete placed was 1070 m³. The compressive strengths of 150 x 300 mm cylinders at 28 days age from the entire job varied from 27.6 to 41.4 N/mm². Flexural strengths of 100 x 100 x 400 mm prisms at the same age varied from 5.5 to 7.6 N/mm².

Plans have been made to inspect the fibre-concrete surface a number of times each year using divers. At the time this paper was prepared, the first planned inspection had not yet been made.

BREAKWATER ARMOUR UNITS

In the selection of an ideal material for breakwater armour units, the physical properties that should rank high are density, strength, toughness, resistance to impact and abrasion, and resistance to deterioration in a marine environment (5). Fibre-reinforced concrete appears to possess these attributes. At Eureka, California, USA, two jetties which protect the entrance to Humboldt Bay had approximately 5500 dolosse placed on them for protection from the 12 m waves which are prevalent during the winter season. A typical dolos resembles a double-ended tee whose cross members, called flukes, are oriented at right angles to each other and are connected by an intermediate stem or trunk. The dolos configuration and dimensions are shown in Fig

Figure 5 Typical dolos armour unit.

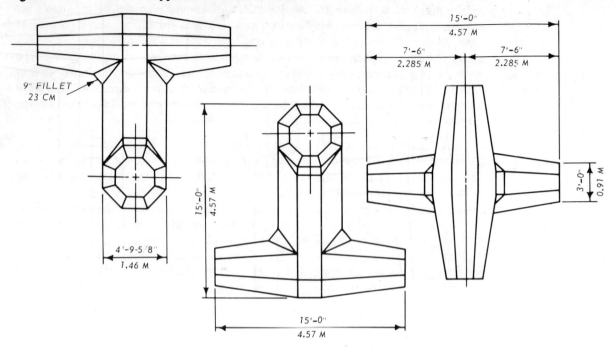

5. Each unit contains 15 m³ and weighs 38 or 39 metric tons, depending on the mixture proportions.

The impact forces of the wave action on an unreinforced dolos can cause the concrete in the fluke to crack and with repeated loadings cause the crack to propagate until the fluke falls off, thus reducing the effectiveness of the dolos. Even with conventional reinforcement, the crack can extend to the reinforcement where the seawater and air can then attack the steel. In time, the steel will be corroded to a point where it will separate under the impact loading, and the fluke can again fall off. Fibre-reinforced concrete can improve the first-crack strength and impact resistance of the concrete and, thus, should prolong the life of the dolosse compared with unreinforced and conventionally reinforced sections.

In order to compare the effectiveness of fibre-reinforced concrete for this type application, 21 fibre-reinforced dolosse were cast using the following two mixture proportions:

Material	*Batch quantity*	
	Mixture 1 kg/m^3	*Mixture 2* kg/m^3
Number of dolosse cast	10	11
Cement	323	390
Sand	748	842
Coarse aggregate:		
19.0 mm	807	504
38.1 mm	439	480
Water	132	159
Steel fibres: (0.25 x 0.056 x 25 mm cross-section)	47	119

All batching and mixing was done in a central batch plant with the concrete being transported to the formwork for the dolosse in concrete buckets. The buckets were lifted by crane so they could discharge directly into the top of the formwork. The concrete was consolidated by using internal vibration. At one day age, the forms were removed, the concrete sprayed with curing compound, and the dolosse allowed to remain where cast until placed on the jetty. In all, 338 m³ of fibre-reinforced concrete were used to construct the 21 dolosse. Eighteen of these dolosse were placed in exposed positions on the north jetty. Two of the other three dolosse were tested to evaluate prototype performance, and one dolos, containing 36 kg of fibre, was broken in handling.

The concrete was not air-entrained. Air contents of 2 to 3 percent were observed, however. Slumps varied between 0 and 25 mm. The 28-day strengths of laboratory batches of the two mixtures were as follows:

	Mixture 1* N/mm^2	*Mixture 2*** N/mm^2
Compressive strength	49.0	53.1
Flexural strength:		
First crack	4.9	5.8
Ultimate	5.3	6.7

*   Fibre content = 47 kg/m³
** Fibre content = 119 kg/m³

The prototype testing involved both static loading and impact testing. The static testing was achieved by yoking two full-size dolosse together in two steel tension frames and then jacking them apart by means of a 454 metric ton manually operated hydraulic jack. One dolos contained the 47 kg/m³ fibre content concrete and the other the 119 kg/m³ fibre content concrete. The dolosse were placed opposite each other with their trunks in line in a horizontal position. Opposite horizontal flukes were placed next to each other in parallel and were held together by the tension frames at their outer edges. The jack was centred in a horizontal position between the horizontal flukes. The force applied by jacking then caused balance reactions to occur at the tension frames. Increasing the jacking force resulted in increasing concrete stress, which led to cracking at the root of the flukes and ultimately to complete failure of the member. After one test, the dolosse were rotated to test the remaining flukes on each dolos. Mechanical strain gauges and dial gauges were used to collect the data on each leg of each fluke.

As expected, the concrete with the 47 kg/m³ fibre content failed at lower loads than the dolos with the higher fibre content. By using the total load, lever arm reaction distance, and section modulus at the point of cracking, the ultimate flexural stress in the flukes was determined to be 4.6 and 4.2 N/mm² for the 47 kg/m³ fibre content concrete and 4.8 and 6.4 N/mm² for the 119 kg/m³ fibre content concrete. A comparison of total loads at the fluke ends for nonreinforced, conventionally reinforced, and fibre-reinforced dolosse is as follows:

Description	*Ultimate load* kg
Nonreinforced plain concrete unit	62,100
Nonreinforced plain concrete unit	67,600

Conventionally reinforced unit

Reinf 43.6 kg/m³ *	67,600
Reinf 47.5 kg/m³ (computed)	77,100
Reinf 119 kg/m³ (computed)	188,000

Fibre-reinforced units

Reinf 47.5 kg/m³	91,600
Reinf 47.5 kg/m³	83,500
Reinf 119 kg/m³	95,700
Reinf 119 kg/m³	127,000

* Sustained increased loading to 92,986 kg at which point the initial crack had widened to 16 mm.

Specimens cored and cut from uncracked portions of each fluke had the following strengths at 90 days age:

150 x 300 mm cylinders

Fibre content kg/m³	Compressive strength N/mm²	Splitting tensile strength N/mm²	Flexural strength * N/mm²
47.5	53.8	4.2	5.7
	47.2	3.3	4.0
119.0	53.2	4.2	6.0
	56.3	3.8	5.5

* 150 x 150 x 500 mm beams

The impact testing was achieved by up-ending a 119 kg/m³ fibre content concrete dolos with the trunk in a vertical position on a concrete test pad. The unit was then toppled, causing the upper fluke to impact on the concrete test pad in the manner of a giant hammer blow. This was repeated for each fluke. Except for minor localized spalling, the impact produced no visible damage to the specimen.

Although the amount of testing on the fibre-reinforced dolosse was very limited, it did indicate that the use of fibre-concrete in this type of elements did have considerable potential. A visual inspection of the dolosse after two winter seasons revealed no damage to the fibre-reinforced dolosse. Some of the unreinforced dolosse on the jetties had failed after the first season. Once the dolosse had nested tightly together after the first season, no further damage to the unreinforced dolosse was observed.'

CONCLUSIONS

Although each of the jobs described above was not of exceptional magnitude with regard to the amount of concrete used, each has shown that there are distinct benefits associated with the use of fibre-reinforced concrete in areas which may be subjected to waterborne cavitation, erosion, and impact. Consideration should be given to using fibre-reinforced concrete as a tough-skin facing material in all new construction where these problems are anticipated.

ACKNOWLEDGMENTS

The information contained in this paper was developed through the efforts of the Walla Walla District, Pittsburgh District, and San Francisco District of the US Army Corps of Engineers. Special appreciation is extended to:

Ernest K Schrader,

Anthony V Munch,

Richard A Kaden,

John C Gribar,

G Coletti,

Dean Hanson,

Saul Barab.

Permission was granted by the Chief of Engineers to publish this information.

REFERENCES

1 Te'eni M et al, "Fibre-Reinforced Cement Composites", The Concrete Society, Technical Report 51 067, July 1973, London, 77 pages.
2 ACI Committee 544, "State-of-the-Art Report on Fiber-Reinforced Concrete," Journal of the American Concrete Institute, Vol 70, No 11, November 1973, Detroit, pp 729-744.
3 Johnston, C D, "Steel Fiber Reinforced Mortar and Concrete: A Review of Mechanical Properties", Fiber-Reinforced Concrete, ACI SP-44, 1974, Detroit, pp 127-142.
4 Kaden, R A, "Pumping Fibrous Concrete for Spillway Test", Fiber-Reinforced Concrete, ACI SP-44, 1974, Detroit, pp 497-510.
5 Barab, S and Hanson, D, "Investigation of Fiber-Reinforced Breakwater Armor Units," Fiber-Reinforced Concrete, ACI SP-44, 1974, Detroit, pp 415-434.

8.4 Steel fibre reinforced concrete pavement—second interim performance report

C D Johnston
Department of Civil Engineering, University of Calgary, Calgary, Alberta, Canada

Summary

The paper compares the performance of steel fibre reinforced concrete and plain concrete paving slabs laid on a typical granular base. The 55 m test section is located on a portion of a city bus route passing through the university campus, and is subjected to about 1000 bus passes per week. It consists of fifteen 3.65 x 3.35 m slabs ranging from 76 to 178 mm in thickness and containing 0, 40 or 79 kg/m³ of 19 x 0.25 mm diameter brass-coated steel fibre. The report discusses performance over the period of 12 months following construction. Comparison of performance over this period with predictions for plain concrete, based on long-term relationships developed from the AASHO Road Test in America and the Alconbury Hill test programme in England, indicates that fibre concrete behaves substantially better than would be expected on the basis of its flexural strength. If the trend continues in the long term it has considerable promise as a paving material.

Résumé

Une comparaison est faite entre les comportements des dalles de pavé faites en béton avec ou sans armature de fibre d'acier. La section d'essais de 55 m de longueur est sur la route d'un autobus qui travers l'enceinte de l'Université de Calgary et passe 1000 fois par semaine sur cette section. Celle-ci est composée de 15 dalles de 3.65 x 3.35 m, d'épaisseurs qui varient de 76 à 178 mm, et contienment soit 0.40 ou 79 kg/m³ de fibre d'acier enrobée de laiton, leur diamètre est de 19 x 0.25 mm.

Ce rapport étudie le comportement de ces dalles pendant les 12 mois qui suivirent la construction. La comparaison de leur comportement durant cette période avec le comportement de longue durée d'un pavé en béton non-armé a été faite d'après les résultats de "AASHO Road Test" des états unis et d'après les essais de l'Alconbury Hill de l'Angleterre. Le béton armé de fibres a un comportement nettement supérieur à ce qu'on peut prévoir en se basant sur sa résistance à la flexion. Si cette supériorité est continuelle, ce matériel promet un succès considérable pour la construction de pavés.

INTRODUCTION

Steel fibre reinforced concrete has been employed in a number of paving experiments in recent years, primarily as overlay to existing concrete pavement or bridge decks in need of repair. Notable examples are the two 53 x 7.6 m sections (1) laid at Tampa airport, Florida, the thirty-three 122 x 6.7 m sections (2) laid on a rural highway in Green County, Iowa, and the 29 x 9 m bridge deck overlay (3) constructed at Winona, Minnesota. Slab-on-grade experiments have been much less numerous, the only example known to the writer being the 30 x 7.6 m section (4,5) built by the US Army at Vicksburg, Mississippi, and tested under wheel loads simulative of the landing gear of the C5A military transport and the Boeing 747 jetliner.

The emphasis on restoration of existing pavements and bridge decks is understandable, particularly in Europe and N. America where much of the existing concrete highway networks were built prior to 1940, because the original structures have in many cases reached the end of their useful life, and require repair or restoration to cope with the demands of present-day traffic. However, the recent momentous increase in the price of oil products has tended to alter the relative economic merits of cement-bound and asphalt-bound materials for new construction in favour of the former. In view of this, the objective of the study is to establish comparative performance-thickness relationships for plain and fibre concrete in order to determine whether the performance-cost characteristics of concrete for new pavement are further improved by the inclusion of steel fibres.

EFFECT OF INCLUSION OF FIBRES

Steel fibres improve a number of the mechanical properties of concrete, the increase in flexural strength (6) being most important from the point of view of pavement performance in the field, as expressed by relationships developed for plain concrete in the AASHO Road Test (7) and the Alconbury Hill Test (8,9). Pavement performance, defined in the former (7) as the number of equivalent 18 000 lb. (8 160 kg) axle loads to give an AASHO Present Serviceability of 2.5, $N_{2.5}$, and in the latter (8,9) as the number of loads to failure, N, is related to flexural strength, f, and thickness, h, by the equations $N_{2.5} \alpha f^4 h^5$ and $N \alpha f^{3.2} h^6$ respectively. In terms of performance and cost, increased flexural strength can be exploited in two ways, either pavement life can be increased for a given slab thickness, or slab thickness can be decreased without reducing pavement life. The former implies an increase in initial cost, but reduced maintenance costs in the long term. The latter offers the possibility of reduced initial capital cost without loss of performance if the reduction in material volume and labour costs more than offsets extra costs associated with the fibres. Furthermore, in situations where there is a supporting structure, as for a bridge deck, the reduction in pavement thickness and the consequent decrease in dead load on the supporting structure may result in an additional capital cost saving.

Properties other than flexural strength which are substantially improved by the inclusion of fibres are flexure toughness, flexural impact resistance, flexural fatigue strength and uniaxial tensile strength. Details are given in previous papers by the writer (6,10) and others (11). However, the data relate to specific laboratory test conditions, and, although improvements in any or all of these characteristics can only be beneficial in pavement, their combined quantitive effect in the field, where the loading involves simple flexure, fatigue, tension and possibly impact, is not yet known. The increase in tensile strength reported (10), although much smaller than for flexure, can be expected to improve resistance to the effects of thermal change thereby allowing the normal joint spacing for plain concrete to be enlarged. The higher fatigue strength reported by Batson et al (11) for regular repeated loading must also be beneficial under the random repeated loading of traffic. The higher impact resistance reported (6) can be expected to be advantageous where bumping occurs due to the slight level differential that often exists between adjoining slabs after construction or develops in service.

Economic analysis

A rational economic comparison of the unit costs of fibre and plain concrete must take into account not only the additional costs associated directly with the fibres but also changes in thickness and joint spacing. As costs can vary considerably from place to place and from country to country, a basis for comparison is proposed which is quite general and allows numbers to be substituted as appropriate for a particular job. Using the following terms,

C—cost per m³ of plain concrete delivered to the site
F—cost per kg of fibre delivered to the batch plant
W—fibre concentration kg/m³
H—fibre handling cost per kg in the batch plant
L—labour, placement and finishing cost per m³ of concrete (plain or fibre)
J_c —cost of joints per m for plain concrete
J_f —cost of joints per m for fibre concrete
h_c—slab thickness required in plain concrete
h_f—slab thickness required in fibre concrete
K—constant relating thickness to area (1000 for millimetres and square metres),

the unit area costs of finished pavement are:

$$\text{Plain concrete:} \quad \frac{h_c (C+L)}{K} + J_c \text{ per m}^2$$

$$\text{Fibre concrete:} \quad \frac{h_f (C+L+WF+WH)}{K} + J_f \text{ per m}^2.$$

The difference between the two depends on the relative magnitudes of four independent factors (i) costs directly associated with the fibre, W(f+H), (ii) material and labour costs for placing concrete (plain or fibre), (C+L), (iii) the reduction in pavement thickness made possible by inclusion of fibres ($h_c - h_f$), (30-50% on the basis of this and other work (5)), (iv) the reduction in joint costs made possible by inclusion of fibres ($J_c - J_f$), (not yet clearly established). The first two are essentially fixed by local prices and wages and geographical proximity to a source of fibres, these being relatively heavy and bulky to transport. The last two are to some extent a function of fibre characteristics, mainly aspect ratio insofar as tensile and flexural strength are relevant (6,10), the maximum permissible aspect ratio being governed by the ability to achieve adequate workability without fibre balling in the fresh state.

TEST PROGRAMME

The test section occupies a 55 m length of one lane of an existing two-lane 7.3 m asphalt concrete road which forms a city bus route through the University of Calgary campus, as shown in Fig 1. The arrangement of slabs selected in this, the first phase of a continuing test programme, represents a compromise between the number of variables to be incorporated, the minimum size of slab practicable, and the total cost of the project. The selection was based on the premise that little would be learned if none of the slabs failed in the short term. It consists of fifteen 3.65 x 3.35 m slabs ranging in thickness from 76 to 178 mm and containing 0, 40 and 79 kg/m³ of 19 x 0.25 mm diameter brass-coated steel fibres. The minimum of 76 mm in plain concrete corresponds to the greatest degree of underdesign, while the maximum of 178 mm is slightly underdesigned when compared with the 203 mm pavement thickness required by local highway specifications. A cross-sectional view of the layout is given in Fig 2, two of the thickest slabs being positioned in the transfer region between asphalt and concrete. All slabs are keyed together in the transverse direction to promote continuity and minimize rocking.

SITE CONDITIONS

Site conditions are described in detail in an earlier paper (12), and only a summary is provided here.

Figure 1 General view of test section.

Figure 2 Cross-section of test slabs and adjacent strata.

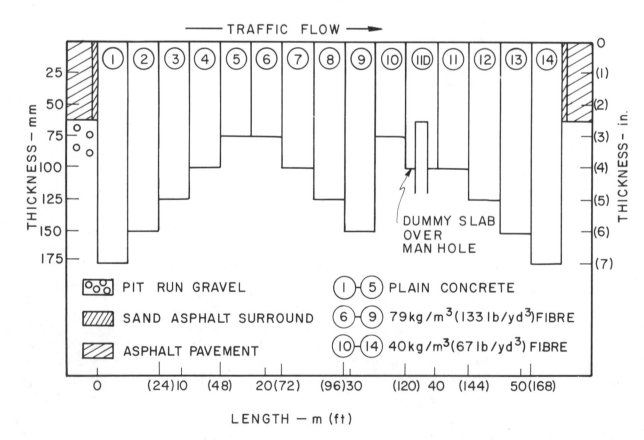

Substrata The subgrade is a well-drained sandy silt. The base is pit-run gravel topped with a 50 mm layer of crushed gravel. The mean CBR value measured on the crushed gravel prior to placement of concrete was 34.7% at a moisture content of 3.2%, conditions nearly identical to those in the AASHO programme (7). Procter density at optimum moisture content, 6.2%, was 83.1 kg/m^3. Actual density, measured by nuclear densometer, was 73.4 kg/m^3, 88.3% of the Proctor value.

Traffic Heavy vehicle traffic is exclusively buses, and the vehicle count is easily computed from transit system schedules. Each bus weighs from 9 280 kg unladen to 12 700 kg laden, approximately 71% being carried on the single rear axle, giving a rear axle load equivalence factor (13) of from 0.41 (unladen) to 1.09 (laden).

Exposure conditions Temperatures range from 32 °C in summer to –40 °C in winter. Frost penetration is in the range 0.9–1.5 m, and the number of freeze-thaw cycles is relatively high, with only about 100 frost-free nights, on average, per year. Mild spells sufficient to thaw surface snow are frequent throughout the winter months, and road grit and de-icing salts are employed to a considerable extent. Overall, with respect to rapid thermal changes and freeze-thaw cycles, the environment is a severe one for concrete.

CONCRETE MIXES

The plain concrete references mix was designed to comply with the local City of Calgary specification for paving concrete which requires a minimum cement content of 326 kg/m^3. The fibre concretes, subsequently referred to as Type F$\frac{1}{2}$ and Type F1, contained $\frac{1}{2}$% or 1% by volume, 40 kg/m^3 or 79 kg/m^3, of fibre respectively, and were designed using 297 kg/m^3 of cement and 119 kg/m^3 of fly ash. Aggregate maximum size was 13 mm. Air-entraining and water-reducing admixtures were included. Complete mix proportions, a description of construction procedures, the results of tests

Table 1 **Comparison of performance of predictions (7,8) with observed behaviour**

Material	Flexural strength N/mm^2	Relative strength	Relative life for constant pavement thickness		
			Predicted (7)	Predicted (8)	Observed (text*)
Plain	5.43	1.0	1.0	1.0	1.0
Type F½	6.68	1.23	2.29	1.94	4.0 (ii)
Type F1	7.63	1.41	3.90	2.96	>12.0 (v)
Type F½	6.68	1.0	1.0	1.0	1.0
Type F1	7.63	1.14	1.70	1.53	4.0 (iv)

Material	Flexural strength N/mm^2	Relative strength	Relative thickness for constant pavement life		
			Predicted (7)	Predicted (8)	Observed (text*)
Plain	5.43	1.0	1.0	1.0	1.0
Type F½	6.68	1.23	0.85	0.90	0.75 - 0.80 (vi), (vii)
Type F1	7.63	1.41	0.76	0.84	0.50 - 0.67 (ix), (x)
Type F½	6.68	1.0	1.0	1.0	1.0
Type F1	7.63	1.14	0.90	0.93	0.60 (viii)

* Relevant point number in text section entitled Comparative Performance Summary.

for slump, Vebe time and air content of the fresh concretes, and values of compressive and flexural strength at ages of 7, 14, 28 and 91 days are detailed in an earlier paper (12). Terminal flexural strength, probably the most important material property measurable in the laboratory, is approximated by the 91-day values given in Table 1.

INTERIM PERFORMANCE REPORT

In this investigation, performance is qualitatively illustrated by tracings of the crack patterns at various ages built up from photographs. Damage is quantified in terms of total crack length for each slab and a damage index defined as the ratio of total crack length to the root mean slab dimension ($\sqrt{ab}$ for a rectangular slab of length a and width b). Unlike terms such as crack length per unit length of lane or crack length per unit area of finished pavement, which differ for the imperial and SI unit systems, the damage index is nondimensional and therefore the same in any system of units. Moreover, its physical meaning is very obvious. For example, a value of 1.0 corresponds to a total crack length equal to the length or width of the slab if it is approximately square, as is normally the case in paving.

Crack development

The data given in this report relate to pavement performance over a period of service of 12 months during which the slabs were subjected to just under 1000 bus passes per week for a total of 50 000 passes. Crack patterns after 3 months and 12 months in service are shown in Fig 3. As expected, major cracking (damage index > 1) developed at an early stage in the thinner plain concrete slabs, and was well advanced after 3 months, as shown in Fig 3 and 4. At this time, mid-October, the 76 mm slab was clearly unserviceable (damage index > 3), and cracking in the 102 mm and 127 mm slabs was severe enough to give rise to doubt regarding their ability to last through the upcoming six months of winter, during which replacement would be impossible, without creating an intolerably rough ride and possibly causing closure of the section. In view of this, the 76, 102 and 127 mm plain concrete slabs were replaced after $3\frac{1}{2}$ months in service. In contrast to their counterparts in plain concrete, the majority of slabs in the 76-127 mm range for both types of fibre concrete exhibited little or no damage at this stage, only the 76 mm Type F$\frac{1}{2}$ slab showing moderate cracking (damage index $\simeq 1$), as illustrated in Fig 3 and 4.

Following the initial four months of summer and fall exposure during which the subgrade was unfrozen and relatively dry, the test section was subjected to the effects of the winter period of subgrade freezing, surface exposure to freeze-thaw cycles and de-icing salts, and subgrade thawing prior to spring. In the latter part of this period, the most severe test condition of the year from the point of view of subgrade weakness, cracking progressed significantly in the 76-127 mm Type F$\frac{1}{2}$ fibre concretes, the 76 mm slab approaching an unserviceable condition from the point of view of rideability and the 102 and 127 mm slabs exhibiting major cracking (damage index > 1), as illustrated in Fig 3 and 4. In contrast, the 76 mm Type F1 slab exhibited only minor cracking (damage index < 1) and the 102 and 127 mm slabs were undamaged.

Cracking in the 178 mm plain concrete slab, No 1 in Fig 2, has not been mentioned in the above discussion, and is considered to be unrepresentative of the general behavioural trend because it developed at an early stage and was associated with depression of the sand asphalt surround in the wheel paths following opening to traffic. Damage is thought to be due primarily to the abnormally severe loading resulting from its function as a transfer slab.

Comparative performance summary

Comparing slabs of equal thickness, the significant differences in performance apparent from the data in Fig 4 are as follows:

(i) After 3 months, the 76, 102 and 127 mm Type F$\frac{1}{2}$ slabs exhibit respectively 32%, 8% and 0% of the damage incurred by their counterparts in plain concrete. Corresponding percentages for the Type F1 slabs are 5, 0 and 0.

(ii) The 76, 102 and 127 mm Type F$\frac{1}{2}$ slabs after 12 months are in a condition very similar to their counterparts in plain concrete after 3 months.

Figure 3 Crack development after 3 and 12 months in service.

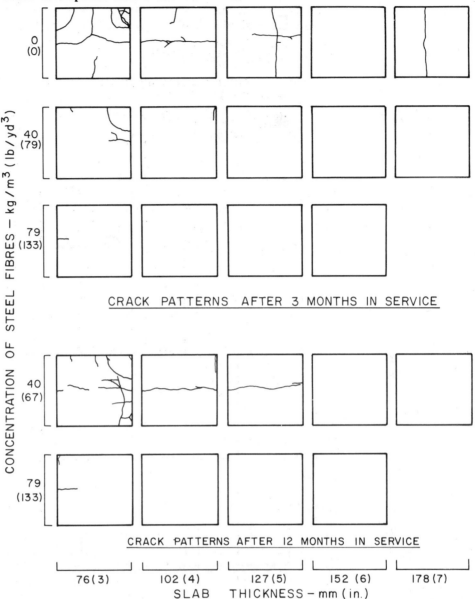

(iii) After 12 months, the 76, 102 and 127 mm Type F1 slabs exhibit respectively 10, 0 and 0% of the damage incurred by their Type F$\frac{1}{2}$ counterparts. (Relative to plain concrete the 10% value would be even smaller than for Type F$\frac{1}{2}$.)

(iv) The 76, 102 and 127 mm Type F1 slabs after 12 months are in similar or better condition than their Type F$\frac{1}{2}$ counterparts after 3 months.

(v) The 76, 102 and 127 mm Type F1 slabs after 12 months are in significantly better condition than their counterparts in plain concrete after 1 month.

Comparing slabs of differing thickness from the point of view of establishing the thickness reductions made possible by inclusion of fibres, the important results are as follows:

(vi) Up to 6 months (the time scale over which the data can be compared), 76 mm of Type F$\frac{1}{2}$ is as good as, or better than, 102 mm of plain concrete.

(vii) Up to 12 months, 102 mm of Type F$\frac{1}{2}$ is as good as, or better than, 127 mm of plain concrete.

(viii) Up to 12 months, 76 mm of Type F1 is as good as, or better than, 127 mm of Type F$\frac{1}{2}$.

Figure 4   Quantitative crack data after 1, 3, 6 and 12 months in service.

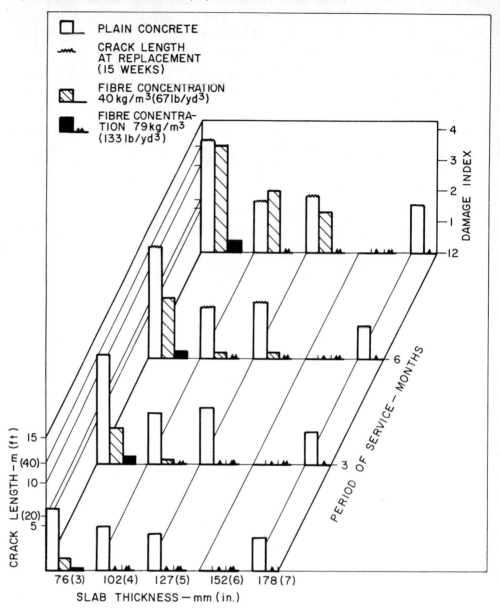

(ix) Up to 12 months, 102 mm of Type F1 is as good as 152 mm of plain concrete, both being undamaged to date.

(x) Up to 12 months, 76 mm of Type F1 is almost as good as 152 mm of plain concrete.

Comparison with predictions based on flexural strength

When the plain, Type F$\frac{1}{2}$ and Type F1 concretes are analysed in terms of the relationships established for plain concrete in the AASHO programme (7) ($N_{2.5} \propto f^4 h^5$) and the Alconbury Hill programme (8) ($N \propto f^{3.2} h^6$) using the 91-day flexural strengths as values of F, relative life predictions based on constant pavement thickness and relative thickness predictions based on constant pavement life can be derived as shown in Table 1. Comparing the predictions with the observations noted in the previous section, it is clear that fibre concrete performs significantly better than expected on the basis of flexural strength alone. Therefore, the improvement in other properties, flexure toughness, tensile strength, impact resistance and flexural fatigue strength, referred to earlier, must contribute significantly to its performance in pavement. Moreover, the standard flexural strength test does not on its own appear appropriate for evaluation of the potential performance of fibrous concrete in pavement, as predictions based on it appear to be overly conservative.

CONCLUSIONS

For concretes with 13 mm aggregate, cement contents of the order of 297-326 kg/m^3 and steel fibres of aspect ratio 75 in concentrations of $\frac{1}{2}$ and 1% by volume, 40 and 79 kg/m^3, the following behavioural trends have been observed for paving slabs on grade.

(i) For constant pavement thickness, Type F1 fibre concrete has a life about four times that of Type F$\frac{1}{2}$ which in turn has a life about four times that of plain concrete. Also, Type F1 has a life more than twelve times that of plain concrete.

(ii) For equivalent performance in terms of cracking, and by implication equivalent pavement life, Type F1 fibre concrete can be used in less than 60%, and possibly as little as 50%, of the thickness needed for plain concrete, and Type F$\frac{1}{2}$ in 75–80% of the thickness needed for plain concrete.

(iii) Predictions of the performance of fibre concrete from relationships established for plain concrete (7,8) using the standard flexural strength value as the measure of concrete quality are overly conservative. In this programme, fibre concrete has performed significantly better than expected on the basis of flexural strength alone, and this suggests that alternative or supplementary laboratory tests are needed to evaluate its potential performance in pavement.

ACKNOWLEDGEMENT

This work is part of a continuing study of fibre reinforced concrete funded mainly by grants from the National Research Council and the Defence Research Board of Canada. Materials and services donated by the Streets and Construction Division of the City of Calgary and Consolidated Concrete Ltd are gratefully acknowledged, as is the co-operation of the contractor, Standard-General Construction Ltd.

REFERENCES

1. Parker, F, "Construction of Fibrous Concrete Overlay—Tampa International Airport", Conference Proceedings M-28, Fibrous Concrete—Construction Material for the Seventies, US Army Construction Engineering Research Laboratory, Champaign, Illinois, December 1972, pp 177-197.
2. Editorial, "Iowa Fibrous Concrete Test is Largest to Date", Concrete Products, Vol 77, No 1, January 1974, pp 72-74.
3. Editorial, "A New Dimension in Bridge Deck Construction", Concrete Construction, Vol 18, No 7, July 1973, pp 321-324.
4. Editorial, "Steel Fibres in Airport Runways", Concrete, Vol 6, No 8, August 1972, pp 34-35.
5. Gray, R H and Rice, J L, "Pavement Performance Investigation", Conference Proceedings M-28, Fibrous Concrete—Construction Material for the Seventies, US Army Construction Engineering Research Laboratory, Champaign, Illinois, December 1972, pp 147-157.
6. Johnston, C D, "Steel Fiber Reinforced Mortar and Concrete—A Review of Mechanical Properties", ACI Publication, SP-44, Fiber Reinforced Concrete, July 1974, pp 127-142.
7. Vesic, S A and Saxena, S K, "Analysis of Structural Behavior of AASHO Road Test Rigid Pavements", National Cooperative Highway Research Program, Report No 97, Highway Research Board, 1970.
8. Gregory, J M, "The Effect of Strength on Performance and Economics of Concrete Pavements", UK Dept of the Environment, Transport and Road Research Laboratory, Report LR 423, 1971.
9. Nowak, J R, "The Full Scale Pavement Design Experiment at Alconbury Hill—Performance During the First Ten Years", UK Dept of the Environment, Transport and Road Research Laboratory, Report LR 193, 1968.
10. Johnston, C D and Coleman, R A, "Strength and Deformation of Steel Fiber Reinforced Concrete in Uniaxal Tension", ACI Publication, SP-44, Fiber Reinforced Concrete, July 1974, pp 177-193.

11 Batson, G, Ball, C, Bailey, L, Landers, E and Hooks, J, "Flexural Fatigue Strength of Steel Fiber Reinforced Concrete", ACI Journal, Proceedings, Vol 69, No 11, November 1972, pp 673-677.

12 Johnston, C D, "Steel Fiber Reinforced Concrete Pavement—Construction and Interim Performance Report", Proceedings of a Symposium on Paving, American Concrete Institute Convention, San Francisco, April 1974 (to be published in an ACI Special Publication, SP series, 1975).

13 Highway Research Board, "The AASHO Road Test: Report No 5—Pavement Research", Highway Research Board Special Report 61E, 1962.

8.5 Investigations of fibre-reinforced materials in the USSR

B A Krylov and
V P Trambovetsky

Summary *This paper reviews recent developments in research on fibre reinforcement of cement and concrete in the USSR.*

Résumé *Cette communication donne une revue de synthése sur le béton et le cement renforcés en URSS.*

INTRODUCTION

Concretes, mortars and other multi-component materials dependent on cement binding are relatively cheap, possess a sufficient durability and a good atmosphere-resistance and at the same time they can have high strength indices. Already today, for example, there have been achieved the strength of 100 N/mm² and higher and by the year 2000, the American experts predict (1), compressive strengths of concrete will have achieved approximately 400 N/mm², while a strength of 140 N/mm² will be common.

However cement concretes have a rather low tensile strength and at present it constitutes at best about an eighth of the compressive strength.

In accordance with the above-mentioned prediction, tensile strength in the year 2000 could be considerably increased and in some necessary cases could achieve half the compressive strength value.

The quest for such a considerable increase in tensile strength of concrete has been caused by the wish to make this material more economical due to a reduction in the consumption of reinforcing steel and more generally with the aim to widen the field of its application.

Overcoming an insufficient tensile strength of concrete is achieved by means of normal and prestressed reinforcement as well as by design of structures in which tensile stresses are either excluded or in which engineers allow the presence of cracks of limited width. A considerable expansion of fields and outlets for applications of concrete materials and among other things the development of road, runway slabs and other special kinds of constructions, in combination with an intensification of their service requirements, increased the need for rational ways to increase crack-resistance, impact strength, abrasion resistance, frost-resistance and other properties. One of the possible means for the solution of such problems is the use of dispersed reinforcement of cement stone (hardened cement paste) by fibres from different materials. At present the investigations in this field are very active in many countries.

The idea of fibre reinforcement application is not new. As early as 1910 H Porter (2) made some attempts to apply short lengths of steel wire to increase the concrete tensile strength. Somewhat later attempts to realize the idea were carried out in our country by V P Nekrasov (3). Since that time some experts have periodically returned to this idea and the study of concrete with dispersed reinforcement has lately been stirred to activity. The investigations were carried out in different directions but the main are the following: concrete reinforcement by unidirectional fibres; concrete reinforcement by randomly-oriented short lengths of wire (up to 40 mm) or by a combination of mineral and organic fibres; cement stone reinforcement by short lengths of mineral fibres (up to 1 mm). The available test results show that dispersely reinforced concrete and cement (fibro-concrete) possess higher strength properties than some normal concretes and will obviously find a proper application in construction.

K L Biryukovich and colleagues (4) have developed a new building material, glass-cement, containing high-strength glass-fibre and cement. Having investigated a wide range of problems concerning the physico-mechanical characteristics of this material, they created a production technology for glass-cement and have manufactured and studied a series of structures, where its properties are used most effectively. Glass-cement is based on the principle of randomly-distributed reinforcement, ie glass-fibres are distributed uniformly throughout the full section of the material and are mainly directed parallel to the tensile forces. As a result an anisotropic material is formed, in which very thin layers of cement stone are closely reinforced by fibres. In the correct conditions, certain combinations of glass-fibres and cement stone provide a new material with the ability to undergo considerable deformation without crack-formation, as well as to work under large stresses in an elastic state. It is characteristic that for the authors' data the ratio between ultimate tensile and compressive strengths for some types of glass-cements amounts to 0.4.

Later investigations both in the USSR and abroad have shown that the alkali medium of cement stone has a detrimental effect on the properties of glass-fibre, resulting in a considerable strength reduction with time. This circumstance was the reason for searching on the one hand for other kinds of fibre and on the other hand—for developing alkali-resistant glass-fibres.

As it is known from contributions to the Conference there has been created in England an alkali-resistant type of glass-fibre that has already found practical application.

In the USSR the works of V V Timashov, Yu M Butt and their colleagues resulted in the use of monocrystalline filaments of MgO, SiC, Si_3N_4 60 μm thick and 1 mm long and glass-fibres of a special composition 9–11μm in diameter and 1–3 mm in length that are considerably less subjected to alkali attack and corrosion. Rupture strength for monocrystalline filaments is as great as 5 kN/mm^2, while that for glass-fibres reaches 2 kN/mm^2. Introduction of 1–3% monocrystals in cement stone composition increased the rupture strength of cement stone 1.5–2 times, while the reinforcement by glass-fibres (2–10%) increased the strength of the matrix by 30–80% (10,11).

The variety of non-metallic fibre reinforcement most widely applied in practice is asbestos fibres. The properties of these fibres have been thoroughly studied and applications for asbestos cements and peculiarities of their behaviour have been described in detail in literature, therefore this kind of fibre reinforcement is not considered in the report.

The principal types of fibre reinforcement being investigated in the USSR at present are short steel wires with different lengths and diameters. In the studies much attention is being given to theory with the aim of determining the main factors affecting the behaviour of fibre-reinforced materials and revealing the interrelation between their separate properties and the components of the composite.

At a meeting, held in Moscow in September 1973, on dispersely reinforced materials (5) some reports were devoted solely to this problem. G G Stepanova made an attempt to construct a theory, enabling prediction of the mechanical properties of fibre-reinforced concrete with different concentrations (V_f) of steel wires of various moduli (the ratio of length to diameter of the wire — ℓ:d). Using mathematical probability theory she has proposed a precise definition for inter-fibre distance, being similar to that derived earlier by Romualdi (6)

$$S = 1.25\, d\, \sqrt{\frac{\ell}{V_f}}$$

with an orientation factor, (ie ratios of forces along any axis, perceiving by unidirectional fibres), being equal to 0.25.

A V Nosarev in his report "Some theoretical problems of materials strengthening by dispersed reinforcement" has shown a theoretical possibility for increasing the ultimate tensile strength of dispersely reinforced materials. Proceeding from well-known theories of cracks and making definite assumptions the author suggested some formulae that allowed calculation of expected increases in strength of materials, including concrete with dispersed reinforcement. There is no doubt that the above approach is interesting, but it calls for a well-grounded experimental affiirmation.

In the majority of works as a rule the influence of separate factors on strength properties of fibre-reinforced materials has been investigated but A A Kalnais and F Ts Yankelovich have made an attempt to carry out a multifactor analysis. Using the corresponding mathematical techniques the authors express an increase in strength of fibre-reinforced concrete as a function of its components. In the general form the increase of concrete strength is a function of the properties of the reinforcement elements (V_f, d, ℓ), as well as the strength of concrete matrix R, ie:

$$k = f\,(V_f, d, \ell, R).$$

From definite assumptions compatible with experimental results the following expression has been obtained for the flexural strength increase in fibre-reinforced concrete:

$$K = 0.13 \frac{V_f}{d^2} - 18 \frac{V_f}{d\ell R} + 0.87,$$

where V_f/d^2 is the distance between fibre centres.

The formula can be used for predicting the properties of fibre-reinforced concrete and for designing its optimum composition.

Yu N Khromets and others (7) studied the behaviour of three-layer structures, consisting of external (up to 10 mm thick) gypsum layers, reinforced by glass-fibre and of an internal foamed plastic layer. As a result of their work the relationship between principal strength and strain parameters (compressive and tensile strength, elastic modulus, resilience, etc) and the strength of gypsum matrix and a reinforcing factor has been determined.

In general it must be said that the theoretical problems concerning fibre-reinforced materials studied in the USSR and abroad have not yet been conducted in sufficient volume; there are some vague and even contradictory conclusions and therefore works in this direction must be continued.

From experimental investigations one should notice the work by D S Abolinsh and others (8), devoted to investigation of the behaviour of fine-grained concretes, reinforced by wire fibres. In this concrete the authors applied Portland cement and sand with fineness modulus 2.1–2.2 in a proportion 1:3 and with w/c ratios from 0.45 to 0.50. Wire 0.3–0.4 mm in diameter and with ℓ =10–40 mm was used for reinforcement at percentages from 1 to 4%. The work showed that the maximum effect of dispersed reinforcement is obtained in bending (strength increases five times) and in tension (three times) compared with plain concrete. The compressive strength increases negligibly, only by 10–20%. Deformability and elastic modulus for dispersely reinforced fine-grained concrete in tension are higher than for plain concrete. These conclusions support as a whole the well-known facts.

A N Kulikov (5) in his investigations used deformed wires 0.25 mm in diameter and 25 mm in length (ℓ:d = 100) with percentages of reinforcement from 1 to 3%. It had been ascertained that with well designed mixes the crack-resistance increases essentially due to inhibition of crack-formation process. So, for example, with V_f=1.5–2.0% crack opening within 0.1–0.2 mm is achieved in fibre-reinforced concrete under tensile stresses about 2.5–3 times greater than those for plain and reinforced concrete. The fracture toughness for dispersely reinforced concrete increases even more (7–10 times), while the tensile strength rises about 3 times. Introduction of coarse aggregate in mix composition reduces the effect of the dispersed reinforcement.

L G Kurbatov and V P Vylegzhanin (9) have studied the behaviour of fibre-reinforced cement-sand mortar in thin-walled structures, a series of spherical shells 15–20 mm thick, 7 m in diameter with 4 m radius of curvature having been used for mobile houses of lens-shaped form. Reinforcing was by wires 0.25–0.35 mm in diameter and 25–40 mm in length. The composition of the mortar was C:S=1:2 at w/c = 0.40 and V_f = 2–3%. As a result of the experimental investigations it has been ascertained that at a constant relative content of fibres in concrete the strength increases with decrease of w/c, reduction of fibre diameter and the use of longer fibres.

Tensile crack-resistance depends in many respects on the quality of bond between the fibres and cement stone and increases up to 4 times in comparison with reinforced concrete and up to 2–2.5 times compared with ferro-cement have been observed. Flexural strengths increase 5–10 times compared with plain concrete, and toughness rises 1.5–2 times when compared with reinforced concrete and up to 5–8 times compared with plain concrete.

Yu V Izbash, V P Pustovoitov and I L Sergeeva (5) studied concrete, reinforced by steel fibres 0.3 mm in diameter and 30 mm long at a total content of 10, 20 and 30 kg per m^3 of concrete. When preparing the samples the sand and the crushed stone was mixed first and then wire fibres and water were introduced. As a result of mechanical tests it was shown that the concrete compressive strength did not increase practically; tensile strength was increased by 15–30% while the flexural strength increased by 80–210% compared with plain concrete samples.

Certain technological difficulties in preparing fibre-reinforced materials can affect results, so that the reinforcing effects are rather weak or do not show at all. For this reason N N Uskov's and M K Babinet's data (5) for reinforcing with steel fibres showed an increase in flexural strength of only 25%, while cube and prism strengths showed no practical variation. Similar data have been obtained by some other investigators.

Assessing as a whole the results of investigations being carried out at present, it should be noted that dispersed reinforcing manifests itself more effectively on cement paste and fine-grained concrete (mortar properties). To a lesser degree this phenomenon is observed in normal concretes or in reinforced concrete structures. This fact will define to a large extent the fields of application. However it should be emphasized that it is a complicated matter to achieve maximum benefit from fibre reinforcement and some technological problems, including rational proportioning of mix, mixing, compacting, etc require solving. All these problems are being studied now and the results will allow the application of fibre-cement and fibro-concrete to be extended considerably.

Experience of applications of fibre-reinforced concretes in practice in the USSR is not great up to now. They have been applied in manufacturing the spherical shells for mobile houses; tank walls were made from them; experimental pavement slabs and some for public roads were prepared, as well as some other structures. A wider distribution of fibre-reinforced materials in the USSR is impeded to a certain degree by the lack of appropriate methods for designing the optimum composition for such materials and because of the lack of mastered technology for preparation of materials with fibre reinforcement and of special equipment for these purposes. Up to now the cost of wire fibres is rather high (from 180 to 400 roubles per ton, depending on the diameter), which makes the application of fibre-reinforced concrete economically advantageous only in those cases when improvement of its performance can compensate the considerable initial expenditures.

REFERENCES

1. Concrete—Year 2000, ACI Journal, N 8, 1971, pp 581-589.
2. Porter, H F, Preparation of concrete from selection of materials to final disposition, ACI Proceedings, Vol 6, 1910, p 296.
3. Nekrasov, V P, Metod kosvennogo vooruzhenia betona, "Novij Zhelezobeton", Ch I NKPS, Moskva, 1925.
4. Biryukovich, K L, i dr. Steklocement, "Budivelnik", Kiev, 1964.
5. Trambovetsky, V P, Beton, armirovanij dispersnoj armaturoj, "Beton i Zhelezobeton", N 12, 1974, pp 40-42.
6. Romualdi, J R, Mandel, J A, Tensile strength of concrete affected by uniformly distributed and closely spaced lengths of wire reinforcement, ACI Journal, Vol 61, 1964, pp 657-670.
7. Khrometz, Yu N, i dr. Mechanicheskie svoistva gipsovich izdelij armirovannych steklovoloknom, "Stroitelnie materiali", N2, 1973, pp 21-22.
8. Abolin'sh, D S, i dr. Melkozernistij beton, armirovanij obrezkami provoloki, "Beton i Zhelezobeton", N 5, 1973, pp 9-11.
9. Kurbatov, L G, i dr. Mnogosloijnie sfericheskie obolochki dlia peredvizhnych domov, "Beton i Zhelezobeton", N 7, 1973, pp 19-20.
10. Butt, Yu M, i dr. Issledovanie cementnogo kamnia, armirovannogo steklovoloknom, Trudi, MCHTI im. Mendeleeva, Moskva, 1973, vup 72, p 148.

11 Timashev, V V, i dr. Issledovanie cementnogo kamnia, armiro-vannogo voloknistymi monokristallami, Sb. trudov "Legkie betoni na isskustvennysh i estestvennych poristych zapolniteliach Dal'nego Vostoka", Vladivostok, 1972.

8.6 Effective applications of steel fibre reinforced concrete

Kunio Nishioka, Noboru Kakimi, Sumio Yamakawa
and Kiyoshi Shirakawa
*Central Research Laboratories,
Sumitomo Metal Industries Ltd,
Amagasaki, Japan*

Summary

Steel fibre reinforced concrete and mortar have been used in a pavement and in beams to investigate the reinforcing effects of steel fibres. In addition, their effects on castable linings in high temperature applications were also studied.

Pavement: The slabs of steel fibre concrete (1.7 vol %) paved the yard of Kashima Works of Sumitomo where, as is general in steel mills, heavy wheel loadings travel on the pavements. The tested pavement was 15 m long and 150 mm thick (50 mm thinner than the conventional concrete pavement) without any meshes or joints.

The fibre was mixed as ready mixed concrete, transported by mixer truck and placed in the same way as conventional concrete. The loading tests and crack pattern disclosed that the pavement of steel fibre concrete was stronger than conventional concrete.

Beams: The investigation was to evaluate the effect of steel fibre percentage on the flexural characteristics of reinforced concrete beams under static and repeated loadings into their plastic range.
(1) Cracking, yielding and ultimate strength increased and the crack width was reduced as the volumetric percentage of the reinforcing fibre increased.
(2) The increase of fibre content remarkably raised the beams' strength against repeated loadings, showing that these beams were applicable in seismic design.

Castable lining: Because the door of a plate mill furnace is exposed to an extremely high temperature (1270°C) and is subjected to impulsive shock loadings, the castable lining of the door is apt to break and fall off before its expected life. Stainless steel fibre of ten percent by weight was mixed to the castable to increase its life. This test concluded that the fibre lining was stronger than the lining with no fibre reinforcement.

Résumé

Une comparaison de performance établie entre le cas où s'applique le béton de fibre en acier au pavage, à la poutre en béton armé et au revêtement réfractaire et le cas d'absence du béton de fibre en acier.

Pavage: Il a été fait un essai de pavage en béton de fibre en cier du parc dans l'aciérie de Kashima où passent fréquemment des poids lourds. Par rapport au pavage classique, ce pavage (d'une épaisseur de 150 mm), dépourvu des mailles métalliques, est moins épais de 50 mm, l'intervalle entre les joints étant allongé.

De même que le béton classique, le béton de fibre en acier a été appliqué sous sa forme préalablement mélangé en bétonnière et transporté en camion-mélangeur.

Un essai de chargement au poids et une étude complémentaire des fendilles sur le pavage ont démontré que le béton de fibre en acier était supérieur en résistance au béton classique.

Poutre: En vue d'améliorer la poutre en résistance, en rigidité, en propriétés de fendilles et en propriétés de la fatigue, il a été fait un essai de charge à la flexion statique et un essai de fatigue à bas cycle sur la poutre en béton armé mélangé de fibre en acier à raison de 1.2% du taux de capacité.

Sont les résultats obtenus: à mesure de l'augmentation du fibre en acier, la force de fendillement et de fléchissement ainsi que la résistance extrême ont été augmentées, la largeur des fendilles diminuant et l'endurance à la flexion répétée s'améliorant beaucoup, en même temps.

Revêtement réfractaire: En vue de prévenir l'écaillement du revêtement réfractaire de la face intérieure de la porte du four à réchaffer la brame par suite d'une haute température et d'une charge de choc, une étude du revêtement a été fait en y mélangeant du fibre en acier à raison de 10% du poids spécifique. L'étude a démontré que la vie du revêtement s'améliorait beaucoup par l'application de ce procédé.

INTRODUCTION

There have been many investigations into concretes and mortars which are reinforced with randomly distributed smooth steel fibre, and several findings have been given of the mechanical properties, the mechanism of fibre reinforcement, the mix proportions and the placing technique (1).

Although adequate applications of steel fibre concrete have been studied and it has been used for various purposes, there are problems to be investigated for the proper design concepts and applications where economic considerations will render it competitive.

Utilizing superior characteristics of steel fibre concrete and mortar in their bending, tensile and impulsive strength, they were applied for three different purposes.

This paper reports the findings obtained in the course of these basic studies and applications.

APPLICATIONS

Pavement

As this steel fibre pavement was the first pavement in Japan, it was carefully formed to investigate the placing procedures such as mix proportions and the structural characteristics such as the strength of pavement.

Description of experiments: The details of the tested pavement are shown in Table 1.

The thicknesses of the steel fibre concrete and the conventional concrete were determined to show the same safety factor to the static flexural strength. The pavement of the steel fibre concrete has no meshes and no joints, being economically competitive. The steel fibre concrete slab paved a yard that had been repaired often because it had been subjected to heavy wheel loadings such as the fork lift truck with a gross weight of 52 000 tonnes. The yard was in Kashima Works of Sumitomo Metals.

Table 1 also shows the mix proportions. Referring to the previously reported results, the fibre volume ratio of the concrete's mortar was two percent and the maximum size of coarse aggregate was 10 mm.

Table 1   **Mix proportions of concrete and condition of pavement**

Type of concrete	Maximum size of coarse aggregate (mm)	Proportions, by weight (kg/m³)					Type of pavement	Size of pavement			Interval of joint (m)	Mesh
		Cement	Sand	Coarse aggregate	Water	Fibre*		Thickness (mm)	Width (m)	Length (m)		
Fibre reinforced concrete	10	223	465	1110	480	132	A1	150	4.15	15	5	—
		"	"	"	"	"	A2	150	4.15	15	15	—
Conventional concrete	25	160	290	815	110	0	B	200	4.15	15	5	exists

* fibre: 0.25 x 0.55 x 25 mm

Findings:

(1) Mixing methods at concrete plant

It was found that in order to obtain a homogeneous steel fibre concrete having no fibre balls, concrete should be mixed by a drum-type-mixer (capacity of 1.5 m³) and by the following mixing procedures.

Step 1.
The cement, fibre, coarse and fine aggregates in this order are charged into the mixer in the usual manner.

Step 2.
The mixer, which should be dry, is rotated at normal mixing speed for one minute.

Step 3.
The water is added to the rotating mixer and the drum is rotated for thirty more seconds.

(2) Strength of pavement

The loading tests were performed one month after the concrete was placed. The stresses in the slab were measured by molded wire strain gauges. The maximum bending stresses measured on the surface at the corner edge of the slab, where the slab is severely loaded, were 3.75 N/mm² in the steel fibre concrete and 3.31 N/mm² in the conventional concrete.

As the experimental values were smaller than the flexural strength by standard bending test, ie, 6.85 N/mm² in the steel fibre concrete and 4.47 N/mm² in the conventional concrete, it was proved that both the steel fib. concrete and the conventional concrete would not be broken under the static bending loads.

However, from the examination of cracking pattern in the slabs after one year it was found that in the conventional concrete there was a large crack which went through the thickness near the corner and ten rather small cracks of 2~5 m long, and that there was only one hairline crack in the steel fibre concrete. The hairline crack was at the centre of the pavement A2 which had no joint and the crack was considered to be a shrinkage crack originated by the long (15 m) joint interval.

Consequently, it was concluded that the joint interval of steel fibre concrete pavement could be longer than that of conventional concrete, and that the life of the pavement became remarkably increased in proportion to the volumetric percentage of reinforcing fibres.

Beams

The purpose of beam experiments was to find out the structural effectiveness of steel fibres in reinforced concrete structures. Strength, rigidity and cracking characteristics under static loadings, as well as structural behaviour under repeated loadings, were studied. In place of concrete, more homogeneous mortar was used to obtain more rigorous data.

Static bending test: Figure 1 shows the experimental procedure of the bending test. The smaller beam depth was used to make the effect of fibre more noticeable. The deflections at the centre of span, the strains in the reinforcing bar and the crack width in the constant moment area were measured at various loading steps until the failure of beams occurred.

Figure 1   **Experimental procedure of static bending test.**

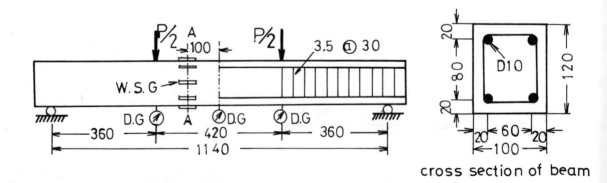

cross section of beam

Figure 2. Load-strain relationships.

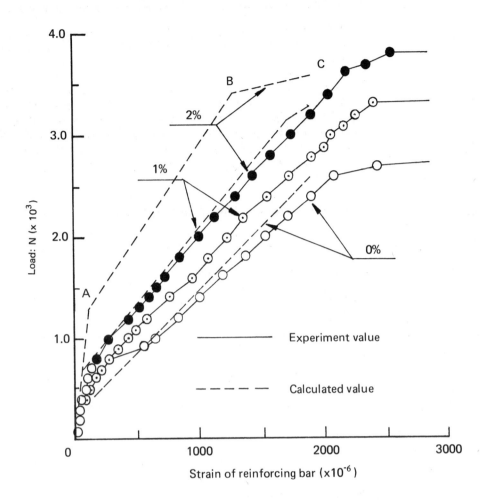

Table 2 Results of static bending test.

Volume of fibre (%)	Values at design load [1]						Loads					
	Stress of reinforcing bar		Maximum width of cracking		Deflection at centre		Cracking load		Yielding loads		Ultimate loads	
	(N/mm^2)	ratio	(mm)	ratio	(mm)	ratio	N	ratio	N	ratio	N	ratio
0	196	1.00	0.075	1.00	$(2.57)^{(2)}$ 1.90	(1.00) 1.00	(7550) 8830	(1.00) 1.00	(25700) 25500	(1.00) 1.00	31300	1.00
1	150	0.76	0.055	0.73	(1.86) 1.60	(0.72) 0.84	(9910) 11800	(1.31) 1.33	32400 32600	(1.27) 1.28	34400	1.10
2	112	0.57	0.035	0.47	(1.56) 0.90	(0.61) 0.47	(12200) 18600	(1.61) 2.11	35400 35300	(1.38) 1.38	37400	1.19

Notes: (1) design load: stress of reinforcing bar reaches to 196 N/mm^2

(2) calculated values in brackets.

Figure 2 and Table 2 show the experimental results, from which the following were concluded.

(1) Cracking, yielding and ultimate strength increased as the volumetric percentage of the reinforcing fibre increased. This fact indicated that the effect of steel fibres is applicable not only for plain concrete but also reinforced concrete structures.

(2) The stresses in the reinforcing bars, the widths of cracking and deflection of the beams at design loads decreased as the volumetric percentage of the reinforcing fibre increased.

These characteristics can be explained with the stress-strain relationship of the steel fibre mortar in tension side as shown in Fig 3 (1). When mortar is not reinforced by steel fibres, it fails at the moment of crack occurrence. However, steel fibre mortar does not fail at that moment, but the cracking strength is maintained until fibres are pulled out or broken. Furthermore, the cracking strength of the steel fibre mortar increases as the volumetric percentage of the reinforcing fibre increases.

Figure 3   **Stress-strain relationships of steel fibre mortar.**

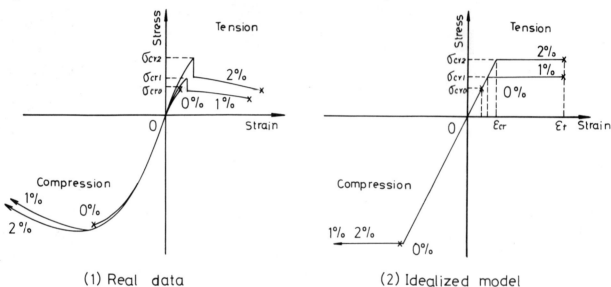

(1) Real data   (2) Idealized model

Some more discussions can be carried out as follows. It is assumed that the stress-strain relationships of the steel fibre mortar are as shown in Figure 3 (2) which indicates the perfectly elasto-plastic relationship and that the ultimate tensile strain is independent of the volumetric percentage of steel fibres. On these assumptions, and numerically solving equilibrium equations of stresses on the cross-section for various points of interest the load-strain relationships can be established as shown by broken lines in Figure 2. The point A of the steel fibre mortar shows the cracking load, the point B shows the load when the strain on the tension side of beam reaches the ultimate tensile strain (ϵ_t) and the point C shows the yielding load when the strains of reinforcing bars arrive at their tensile yield point. It can be seen in Figure 2 that the experimental results are explained approximately well by this calculating concept, although some differences exist quantitatively.

Therefore, it is concluded that the effect of steel fibre can be expected on the strength of reinforced concrete structures and of plain concrete structures.

Repeated loading test: The repeated loading tests were performed under conditions where the alternating deflection of the cantilever beams having the same section as simple beams shown in Figure 1 should reach the controlled deflection which is about four times greater than the yielding deflection. The strain on the compression side of beams extended beyond the ultimate tensile strain (ϵ_t).

Figure 4 shows the crack pattern at the fixed end after sixteen cycles. In the conventional mortar beam, the mortar near the fixed end has fallen out and the

Figure 4 Crack pattern in fixed end at sixteenth cycles.

Figure 5 Load amplitute vs repeated cycles.

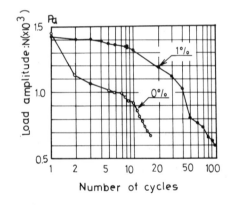
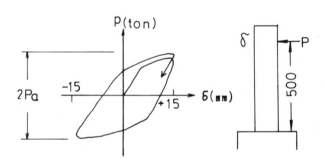

reinforcing bars are exposed. On the other hand, in the steel fibre mortar beam, although cracks have occurred, the steel fibre mortar has not fallen out.

Figure 5 shows the decrease of the maximum loads with the number of cycles. This figure illustrates that the strength of reinforced mortar beams which are subjected to large repeated deflections increases remarkably by adding the reinforcing fibres. The fact that the strength of the steel fibre mortar beam increased remarkably is due to the stress-strain relationship of the steel fibre mortar on the compression side (see two (0% & 2%) distinctive compression characteristics in Figure 3 (1)). The compression failure of plain mortar occurs suddenly having no plastic deformations when the plane of the maximum shear stress slips, while the compression failure of steel fibre mortar occurs slowly having a large plastic deformation, which is caused by the resistance of steel fibres on the maximum shear stress plane. It can be stated that the superior strength of the steel fibre mortar beam subjected to repeated loads is attributed to the difference of the stress-strain relationships of the steel fibre and plain mortars after the slips start on the maximum shear stress plane. It can be expected that this superior strength against repeated plastic loadings could be utilised in preventing the failure of reinforced concrete structures subjected to severe seismic loads, indicating the possibility of promising applications where the compressive properties of the steel fibre concrete are effectively utilized.

Castable lining

The castable lining conventionally used and exposed to high temperature on the door of a plate mill furnace is apt to flake off due to impulsive loads. Aiming to prevent this flaking, the model tests in high temperature were performed to investigate the reinforcing effect of stainless steel fibres and, in addition, the prototype of the specimen has been placed on the real furnace door.

Mechanical properties of stainless steel fibre castable: Firstly, to examine the temperature limit for the stainless steel fibres to be experienced, two types of thermally treated fibres were subjected to tensile tests. Uncovered fibres and buried fibres in 5 mm thick castable were heated to 1000–1300 °C for three hours and slowly cooled. Thus treated fibres were tested by a SHOPPER testing machine.

Table 3 Tensile strengths of stainless steel fibre

Heated temperature	Tensile strengths of stainless steel fibre			
	Uncovered fibre		Buried fibres in castable	
	N/mm^2	ratio	N/mm^2	ratio
at room	787	1.00	787	1.00
1000 °C	630	0.80	598	0.76
1100 °C	513	0.65	579	0.74
1200 °C	456	0.58	565	0.72
1250 °C	451	0.57	529	0.67
1300 °C	297	0.38	303	0.39

Test results are shown in Table 3 from which the following can be seen:
(1) Stainless steel fibres covered with castable lining whose thickness is less than 5 mm can be expected to have the reinforcing effect provided that they are not heated over 1250 °C.
(2) At high temperature (1300 °C), stainless steel fibres within a 5 mm thick castable layer are oxidised severely and their cross-sectional area is reduced, resulting in a remarkable (60~70%) decrease in their strength. Therefore, the reinforcing effects of fibres in this layer can no longer be expected.

As the second phase of the examination, the bending, compressive and impact loading tests were performed on standard castable specimens (40 × 40 × 160 mm) containing stainless steel fibres (zero up to ten percent by weight). The castable specimens were exposed to high temperatures (1000 °C and 1200 °C) for three hours and were subjected to various tests at room temperature.

From the test results shown in Table 4, it was proved that the impact and the flexural strengths of the stainless steel fibre castable were remarkably improved by adding fibres of more than three and five percent by weight respectively. However, no effects of fibres could be seen on the compressive strength. This tendency was confirmed even when steel fibre castables were heated up to 1200 °C.

An application to the actual furnace: To demonstrate these mechanical properties, the stainless steel fibre castable lining has been applied to the actual door of a plate mill furnace. The tested door is 2.1 m wide and 2.8 m high, the castable's thickness being

Table 4 Strengths of stainless steel fibre castable

Type of castable	Heated temperature	Weight of fibre (%)	Flexural strength N/mm²	ratio	Compressive strength N/mm²	ratio	Absorbed energy N/mm²	ratio
A	1000 °C	0	1.74	1.00	—	—	0.0210	1.00
		3	2.00	1.14	—	—	0.0500	2.38
		5	2.93	1.69	—	—	0.0661	3.15
B	1200 °C	0	5.20	1.00	25.30	1.00	—	—
		3	5.43	1.05	25.10	0.99	—	—
		5	6.00	1.15	25.60	1.01	—	—
		7	7.81	1.50	23.20	0.92	—	—
		10	8.72	1.68	27.70	1.09	—	—

250 mm. It is exposed to the extremely high temperature (1270 °C) for a long time and is often subjected to impulsive loads at the time of opening and closing. Usually the life of conventional castables is three months.

The observation of castables after four months of use proved that, in the castable lining reinforced by the stainless steel fibres of ten percent, by weight, the area to be repaired on the door decreased to 30% of that on the conventional castable with no fibre. Furthermore, about half the thickness of the conventional castable had flaked off while only the superficial layer of stainless steel fibre castable suffered from flaking off.

Consequently, it is concluded that, as stainless steel fibres increase the strength of a castable lining at extremely high temperatures, there is a wide possibility of applications of steel fibres even in severe conditions of this sort.

CONCLUSIONS

Aiming to utilize its superior properties, steel fibre concrete was applied for three purposes: pavements, reinforced concrete beams and castable linings.

Findings in these three applications are described herein. In any of these three cases, it was found that the steel fibres played a remarkable role of increasing the strength of the mortar, the concrete and the castable lining.

ACKNOWLEDGEMENT

The authors are grateful to Professor Emeritus M Kokubu of the University of Tokyo for invaluable guidance. They are indebted to Dr T Ikeshima, the Manager of Central Research Laboratories for his advice and encouragement.

REFERENCES

1. ACI Committee 544, "State-of-the-Art Report on Fiber Reinforced Concrete", ACI Journal, Proceedings, Vol 70, No 11, November 1973, pp 729-744.
2. Shah, S P and Rangan, B V, "Fiber Reinforced Concrete Properties", ACI Journal, Proceedings, Vol 68, No 2, February 1971, pp 126-135.

8.7 Tailoring fibre–concretes to special requirements

C D Pomeroy and J H Brown
*Cement and Concrete Association,
London*

Summary

The greatest potential applications of fibre-reinforced cement-based products are in areas in which concrete is not traditionally used or in areas where conventional concrete has limitations. Exploitation of the desirable properties of fibre-concretes should follow a logical pattern. A particular problem or need should be identified and a performance specification drawn up that defines the properties required from the product. These may include strength, modulus of elasticity, creep and shrinkage parameters, the ability for the product to withstand a given impact, fire resistance, durability and cost. From this specification various alternative solutions, which include traditional reinforced concrete designs may be considered.

The logical development of materials with special properties is discussed in relation to a programme that was undertaken to produce a deformable, but load bearing replacement for soft and hard wood timber baulks used in the construction of cribs for strata support in coal mining. It is shown that for this application a combination of traditional steel hoop reinforcement, a polypropylene fibre-reinforced mortar and block shape provided the specified design requirements.

As another example high strength polymer concretes are considered. These are brittle and cracks propagate easily, a feature that limits their use. The addition of steel fibres makes these materials more ductile and hence improves their usefulness.

Mention is also made of the development at the Cement and Concrete Association of a steel-fibre-reinforced, pressed concrete manhole cover and surround, to replace the traditional cast-iron assembly.

These examples emphasise the need to identify and define an objective and to apply the novel properties of fibre-concretes when considering alternative solutions.

Résumé

Les ciments et les bétons renforcés de fibres trouvent leur application là où le béton n'est pas normalement employé, ou là où le béton ne convient pas. On devrait exploiter d'une façon logique les propriétés du béton renforcé de fibres. Pour l'application choisie, on devrait préparer une liste detaillée des qualités requises et chercher diverses alternatives, y compris celle utilisant le béton armé classique.

Le développement logique des bétons spéciaux ayant les propriétés sous charge des bois tendres et des bois durs utilisés pour soutenir les couches de terrain dans les mine de houille et pouvant se déformer comme eux est exposé.

Un mortier à fibres de polypropylène moulé en blocs rectangulaires et renforcé d'une armature de frettes d'acier, offre une solution satisfaisante.

Comme autre example les bétons polymères de grande résistance sont considérés. Ceux-ci sont fragiles et les fissures se propagent facilement, ce qui limite leur emploi. L'utilisation de fibres d'acier rend ces matériaux plus ductiles et par conséquent les rend plus utiles.

Le développement à C & CA d'un obturateur d'égout et de son encadrement faits de béton comprimé renforcé de fibres d'acier pour remplacer la classique plaque de fonte, est également exposé.

INTRODUCTION

It is tempting to think of fibre-concrete as ordinary concrete with improved properties which allow it to be used in situations where conventional concrete is used but in a more adventurous way, by the use of lighter sections or by making longer spans with a possible cost benefit. In fact the replacement of conventional reinforced concrete with fibre-concrete is seldom likely to be worthwhile and the real potential of fibre-concrete lies in areas where concrete is not traditionally used or where concrete is in some way inadequate.

The effective performance of fibre-concrete depends on many factors, such as fibre type and characteristics, the concrete mix and the size of the structural member. Knowledge of the interactive effects of these parameters may allow solutions to particular problems to be obtained but the range of properties that can be obtained from fibre-concretes is so great that a simple compilation of the material properties—expressed as a proportional improvement over unmodified concretes—can obscure rather than advertise the possible uses of fibre-concretes.

It can be more rewarding to consider a specific problem for which a full performance specification is available and to compare different material and structural solutions. Several satisfactory solutions to a problem may be found when the final choice will be based on cost and user preference.

Certainly some basic research on fibre-concretes must be done, for otherwise the attempts to solve specific problems could be misguided. It is also likely that difficulties in solving particular problems could pose questions that require more basic research to be undertaken before a solution is obtained.

In this paper three examples are considered in which fibres have been used to help solve various problems. In the first example it is shown how the undesirable brittleness of high strength polymer concrete can be compensated by the inclusion of steel fibres. In this example a particular application was not chosen, but the use of fibres to remedy a deficiency is demonstrated.

The major part of the paper shows how cement based materials were tailored to reproduce the performance of timber crib assemblies that are used to control the roof movement in coal mines. The problem was particularly interesting since the traditional timber construction provided a performance specification to which the concrete replacement could be matched.

As a final example brief mention is made of the replacement of cast-iron medium duty manhole covers by a steel-fibre-reinforced concrete. In this case a design criterion was available in the form of a British Standard which stipulated the loads that the covers must withstand.

STEEL-FIBRE, POLYMER MODIFIED CONCRETE

This example highlights some of the problems associated with the measurement and assessment of the material properties of fibre concrete. The modification of concrete by the inclusion of polymers has been shown to increase the strength, particularly the tensile strength, above that of the basic concrete, especially for concretes of high w/c ratio (1). With the increase in strength there is an increase in the brittleness of the material and cracks propagate very rapidly. This characteristic can severely limit the potential use of high strength, polymer modified concrete and it removes one of the features that makes conventional reinforced concrete such an accommodating structural material, namely its inherent ductility whereby distress on being overloaded is shown by local breakage that does not normally result in the catastrophic failure of the structural element.

The question that is posed, therefore, is whether the inclusion of fibres in the polymer concrete would restore some ductility without loss of any of the benefits bestowed by the polymer treatment. Polymer modified concretes are inevitably expensive so that it is probable that they will be used in thin sections and so the tensile or impact strengths

Figure 1 Flexure tests of polymer concrete beams containing steel fibres.

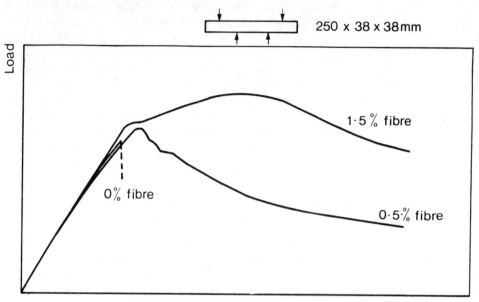

will be of greater importance than the compressive strength. For this reason in the laboratory investigation to be described the problem was simplified from a general study of the effect of fibre inclusions on the properties of polymer concrete to a much more restrictive objective that was directly linked to the probable use of the material, namely the performance of small beams in flexure. Full details of the polymer treatment are given elsewhere (1).

Flexural tests were made on a range of polymer (PMMA) modified mortar beams that were reinforced with up to 1.5 per cent volume of brass-coated steel fibres (0.13 mm dia x 15 mm long). The load deflection curves (Figure 1) show clearly how even 0.5 per cent of the fibres changes the failure from abrupt to docile. The stress at which first cracking was detected was almost independent of the fibre inclusion but it increased with polymer content. The "maximum effective strength" of the beams, calculated from the maximum load that can be carried, is very sensitive to fibre content and increases at an increasing rate as the quantity of fibres is raised. These observations are consistent with the generally held views that the flexural strength of concrete can be increased by the inclusion of polymers, but fibres in low volume concentrations have little effect until the matrix has cracked, when they become important load bearing elements as they are drawn from the material as the cracks open. In the experiments it was observed that the polymer treatment supplemented the gain in "maximum effective strength" provided by the fibres, possibly by increasing the bond between fibre and matrix.

The flexure tests thus showed that the fibres restored some ductility to the high strength polymer concrete, but it is difficult to relate these findings to a specific practical problem. The difference between the rising load characteristic of the beams containing 1.5 per cent of fibres (Figure 1) and the steady but continuous fall in load for the specimens containing 0.5 per cent could have practical significance, but the difference in failure pattern is due more *to the geometry* of the specimen (particularly beam depth) *than to the material properties*. This problem has been discussed elsewhere (2) and it is sufficient to note that behaviour of the beam containing 1.5 per cent of fibres will not necessarily be reproduced by beams of different size. It follows that values of modulus of rupture calculated from maximum loads will not apply to other thicknesses of beam and so are not material constants.

The resistance of the material to impact damage was assessed by the fracture energy, determined from the total work necessary to separate a specimen completely across a plane section in tension. This parameter is a more fundamental material property, although in practice multiple cracking in the matrix cannot be entirely avoided. However the values obtained for the fibre-reinforced, polymer modified mortar beams correlated closely with values of energy expended in bending beams some arbitrary distance (greater than that corresponding to the maximum load) but the shape of the flexural failure load curve could not be predicted. The energy expended in bending is affected by the multiplicity of cracks formed since the loads carried by fibres spanning the various cracks will vary widely. It is thus difficult to propose a simple acceptance criterion for fibre-reinforced concrete without making allowance for size. Many criteria can be used, such as the maximum failure load, the load at which cracking is first detected, the energy expended in deflecting the beam, etc, but the choice must depend upon the application of the material.

It is possible to predict a component behaviour from fundamental material properties, as has been done by Allen (3), but the manipulation of the data is complex and at the present time it is probably more satisfactory to relate the tests to specific applications.

The general conclusions from this work are that steel fibres can add ductility to a brittle polymer modified mortar, the presence of polymer apparently enhancing the bond between fibre and matrix. However it would not be possible to apply this information to a given problem without making proper provision for the effects of component size.

FIBRE-CONCRETE SUBSTITUTES FOR TIMBER CRIBS USED TO CONTROL ROOF MOVEMENT IN COAL MINES

Every year large quantities of timber are used sacrificially in the coal mines to control the progressive convergence that occurs between the roof and floor, formed when the coal seam is extracted. There has been a rapid increase in the cost and a serious shortage of certain types of timber so that a search has been made for a suitable substitute.

Conventionally timber cribs have been formed from either soft or hard woods by stacking rectangular baulks of timber as shown in Figure 2. The size of the baulks increases with the seam thickness, but typically they might be 450 x 100 x 100 mm, stacked to a height of about 1 m. As these crib assemblies satisfy the mining requirements it is possible to use their load/deformation performance to specify the performance required from the substitute. There are additional factors that must also be considered. The timber crib is built by one man in a confined and dangerous place. The alternative must be equally manageable. The system must be safe, not easily abused and unable to be assembled incorrectly. The components of the system must be portable and resistant to impact damage during transit. The cost must be competitive. It can be seen that gradually a requirement specification is being built up, both for the performance of the complete crib and of the individual components of the system.

The most challenging aspect of the specification is the rising load characteristic that is observed when a timber crib is crushed to about half its initial height (Figure 2). This means that in practice there is a progressive increase in the resistance provided by the crib as convergence occurs between the roof and floor. Various solutions to the problem were considered (4) but only those involving fibre-concrete are considered here. The requirement of considerable ductility seemed an ideal application for fibre-concrete, but the very large deformations that must be accommodated lifted the problem straight out of the fields for which data were immediately available.

In the first tests that were made a direct timber replacement was tried. It seemed logical to use a low density concrete or mortar that contained fibres to maintain some integrity as the material was crushed. A variety of Faircrete* was chosen as this class of material had low density, a strength in compression of the desired magnitude and

---

*Registered trade mark of John Laing Research & Development Ltd. Faircrete is a special form of architectural concrete containing a very small amount of polypropylene monofilaments.

Figure 2   **Failure characteristics of Faircrete "dog-bone" cribs.**

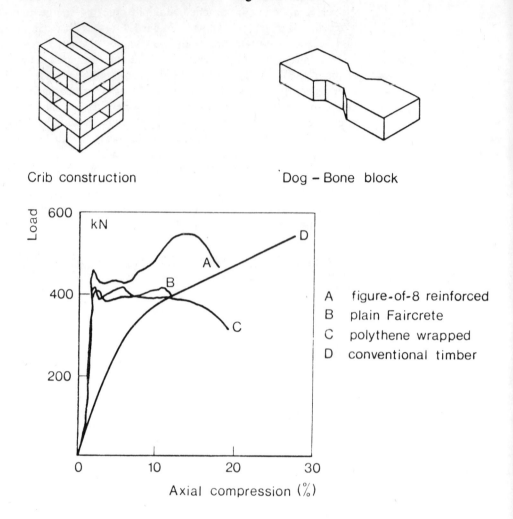

contained polypropylene fibres. When the timber crib is crushed the main restraint is provided by the four columns formed by the overlapping blocks close to the corners (Figure 2), the mid parts of the blocks effectively tying the corners together. With timber the long fibres maintain the stability of the system but when the Faircrete blocks, containing fibres only 15 mm long, were used cracks soon opened to an extent that fibre restraint was lost, the corner columns became unstable and buckled and the complete crib collapsed when it was compressed a mere five per cent.

A second crib made from Faircrete blocks that contained 38 mm long polypropylene fibres gave a better performance, and while the load fell steadily after the peak load had been reached the crib was still carrying more than half the peak load when it was compressed by over one third. Although at this stage the problem had not been solved the tests were considered to be very encouraging.

One of the obvious limitations of the crib construction is that only a small part of each block actually carries load. The blocks were therefore modified to the dog-bone shape shown in Figure 2. Three variations of the blocks were tried. They were all cast from a Faircrete with a porosity of about 30 per cent that contained 5 per cent by volume of 40 mm long polypropylene fibres. One variation consisted of the material as cast, the second of similar blocks which were encapsulated by shrink wrapping them in polythene sheets and the third by the incorporation of three figure-of-eight cords of 128 end-twisted rovings of polypropylene. These latter were used in attempts to reduce the spalling that occurs at the side of the blocks as crushing progresses. The load/deformation curves that were obtained are shown in Figure 2. Unfortunately the only rig that was available that enabled large compressive movements to occur did not apply the loads uniformly and in consequence the cribs suffered buckling instability. However, for all three variations there was further improvement in the crib performance and the loads remained sensibly constant as crushing reduced the crib height from 95 to 80 per cent of the original value. A more rigid loading system would almost certainly have given a more acceptable performance, but even if this is true it seemed unlikely that the blocks would be satisfactory without further development.

It seemed logical to pursue the idea of combining a low density mortar or concrete containing polypropylene fibres but some additional restraint at high deformations appeared to be essential. Further modifications that would reduce the spalling that occurs from the sides of the blocks were also considered.

Experiments were undertaken to see if hoop reinforcement could improve the performance of blocks at high deformation and to see if blocks with larger internal voids would collapse inwards rather than by loss of material from the outer surfaces, so that the general configuration of the crib would be maintained as the crushing

Figure 3 Failure characteristics of fibre-reinforced, polystyrene aggregate concrete cylinders. (A plain cylinder, B two half cylinders hollowed at mating ends, C two half cylinders hollowed at mating end and hoop reinforced with steel wire, D desired characteristic.)

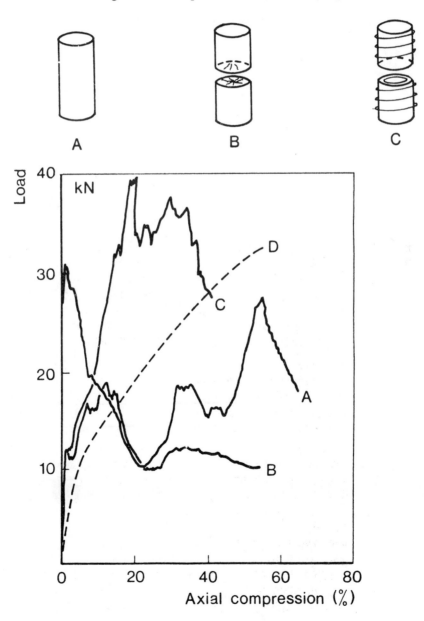

progressed. These ideas were examined by tests on different forms of cylinder in which the desired voids were obtained in different ways. In some, expanded polystyrene spheres were used as the aggregate and in others larger internal cavities were formed. Hoop reinforcement was provided by ductile steel, sisal or hemp. Some of the results are shown in Figure 3. It can be seen that these ideas, in combination, could be applied to the block designs, with the prospect of the manufacture of a crib that would deform substantially under a continuously rising load.

In the final part of the investigation cribs were constructed from rectangular Faircrete blocks that were reinforced with hoops of ductile steel and in which cylindrical cavities had been formed. Results of tests on two solid model cribs are shown in Figure 4. In test A the aspect ratio of the crib was about 2 and in test B it was increased to 3.7. It can be seen that the cribs deformed under a rising load as they were crushed to one half the initial height and there was no substantial fall off in load nor instability in the tests, even when very high aspect ratios were employed. At all times the load carried by the concrete crib exceeded that by the original timberform.

Tests were also made of a more open stack of these Faircrete blocks. Although, as shown in Figure 5, the load carried by this crib fell below that for a hard wood crib, it could be deformed by more than 60 per cent without a loss in load bearing capability and at greater compactions the load rose rapidly. This form of block thus appears

Figure 4 **Failure characteristic of a model crib of solid construction.**

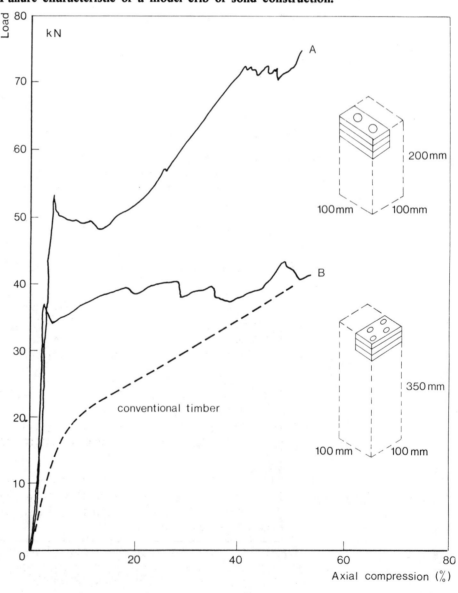

Figure 5 Failure characteristic of a model crib of open-centre construction.

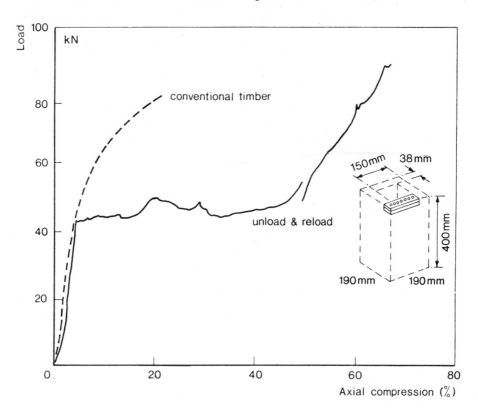

entirely satisfactory for the mining application, but small changes in block geometry could be made to improve the match between timber and substitute if necessary.

This example shows how certain known properties of fibre concrete can be exploited to solve a particularly demanding problem. The material, on its own, could not satisfy the requirements, but the additional knowledge that hoop reinforcement increases both compressive strength and strain capacity and the concept that the provision of internal voids would help to maintain the integrity of the blocks enabled a satisfactory solution to be reached.

WEXHAM MEDIUM-DUTY MANHOLE COVER

In the above examples two extreme cases have been considered. In one ductility was given to a very strong but brittle concrete. In the other, a concrete assembly which would support substantial loads after considerable deformation was developed. In the final example brief mention is made of the development by Hollington (of the Cement & Concrete Association) of a concrete replacement for medium-duty cast-iron manhole covers.

The attractions from the use of steel fibres are the increase in failure load of fibre-concretes in flexure, the control of cracking by the fibres and the significant improvement in impact strength. Further, a performance specification existed for the traditional cast-iron covers which could be used to judge the performance of the concrete replacement (5).

A 2:1 fine concrete mix with 7 per cent by weight of steel fibres was used in the prototype units, which had flexural strengths of about 20 N/mm^2. A viable replacement for cast-iron covers could be manufactured from this mix that complied with the existing standards for cast-iron covers.

Absorption and exposure tests on fibre-reinforced concrete specimens indicate that the units will have the high durability of pressed paving slabs. The use of the manhole cover has been approved by the Agrément Board (6).

DISCUSSION

Fibres of various kinds have the ability to limit the development of cracks in hardened cement pastes and concretes but it is debatable whether in small volume concentrations they significantly increase the stress at which crack initiation occurs. Crack inhibition implies that the fibres add ductility to the concrete and increase the resistance to impact. The magnitude of the benefits depends on the size of the structural element as well as on the fibre and concrete properties.

Plain concrete is seldom used in structures so that a comparison between plain and fibre-concrete is not very helpful and a comparison with reinforced concrete may show fibre-concrete in a less favourable light. It is suggested that the viability of fibre-concretes can only be judged when different solutions are sought to well defined problems. Sometimes the traditional solutions will remain superior but gradually the worthwhile applications of fibre-concrete will become established. To reach this situation, it is vital for full requirement specifications to be provided. The specification must be as thorough as possible and not only provide strength and stiffness requirements. Component weight, fire resistance, thermal and acoustic properties, durability and other factors may render unsatisfactory a material that possesses the strength and stiffness required.

In the example of the replacement for mining timber it is shown that more than one parameter may have to be considered before an acceptable solution is found. Here the use of fibre-concrete alone was insufficient, but if, in addition, some internal voids were formed in the blocks and ductile hoop reinforcement was used it was possible to match closely the very demanding specification.

CONCLUSIONS

In order to exploit the existing knowledge of the effects of fibres on the performance of concrete it is essential to specify the requirements for particular applications very stringently. Comparisons can then be made between alternative solutions and the fields in which fibre-concretes are viable can be established. Examples have been given of the ways in which fibres have been used to impart particular properties to concretes to satisfy different needs. Steel fibres were used to give ductility to a very strong and brittle polymer modified concrete. A highly deformable concrete was made from Faircrete, but additional hoop reinforcement and the provision of internal voids were essential to satisfy the mining requirements. A steel-fibre-reinforced concrete was also shown to satisfy the specification of a cast-iron manhole cover and to be economically viable.

There are thus many different potential applications for fibres, of various kinds, in concrete, but user requirements must be properly prescribed before the worthwhile applications of fibres in concrete technology are established.

REFERENCES

1. Brown, J H and Pomeroy, C D, Mechanical properties of some polymer (PMMA) modified concretes, Cement & Concrete Association (London) Technical Report (to be published).
2. Brown, J H, The failure of glass-fibre-reinforced notched beams in flexure, Magazine of Concrete Research, Vol 25, No 82, March 1973, pp 31-38.
3. Allen, H G, Stiffness and strength of two glass-fibre-reinforced cement laminates, Journal of Composite Materials, Vol 5, 1971, pp 194-207.
4. Pomeroy, C D, Taylor, H P J and Brown, J H, Concrete products for the replacement of timber for chock construction in coal mines. The development of ductile concrete assemblies, Cement & Concrete Association (London) Technical Report 42.491, May 1974.
5. BS 497 : 1967 (Cast manhole covers, road gully gratings and frames for drainage purposes).
6. Agrement Board Assessment Report No 120, Wexham Manhole Cover, 1974.

8.8 The design of glass fibre reinforced cement cladding panels

A J M Soane and J R Williams
Bingham Blades & Partners—Consulting Engineers

Summary

The practical application of glass fibre reinforced cement (GRC) technology to the building industry is in its infancy. The experience gained and problems encountered by the authors in their roles as structural designers of certain GRC products complements theoretical knowledge available on the subject. The composites described are made from ordinary Portland cement with 5% by weight of Cem-FIL glass fibre.

The paper discusses the care that is required when utilising material properties which have been obtained under laboratory conditions when designing mass-produced structural components. Various methods of using GRC are assessed with reference to cladding panel designs for buildings. There is as yet only partially complete information on the long-term behaviour of the material and the authors also stress the production difficulties that must be considered in the design of structural components.

Finally, fixing and waterproofing details and handling of glass reinforced cement units are discussed and comparisons made with conventional philosophies used when dealing with traditionally reinforced concrete. A number of examples are illustrated.

Résumé

L'utilisation pratique de la technologie du ciment renforcé par des fibres de verre (GRC) dans l'industrie de la construction ne fait que commencer. Les connaissances théoriques qui existent à ce sujet ont été complétées par l'expérience acquise et les problèmes rencontrés par les auteurs en tant qu'architectes de certains produits de GRC. Les produits renforcés décrits sont fabriqués à partir du ciment Portland ordinaire avec 5% en poids de fibres de verre Cem-FIL.

L'article traite des soins requis lors de la conception des pièces structurales lorsqu'on utilise des propriétés matérielles qui ont été déterminées sous des conditions de laboratoire. Il donne une évaluation des différentes méthodes d'utilisation du GRC dans la réalisation des divers modèles de panneaux pour façades. Jusqu'ici les données disponibles relatives au comportement à long terme du matériau ne sont qu'incomplètes et les auteurs soulignent les difficultés de production dont il faut tenir compte dans l'étude des éléments structuraux.

En dernier lieu on examine l'installation, l'imperméabilisation et la manutention des éléments de GRC et on compare ces particularités avec les philosophies classiques suivies dans l'utilisation du béton renforcé par les moyens traditionnels. Plusieurs examples illustrés sont présentés.

INTRODUCTION

Glass reinforced cement (GRC) is defined as a composite of ordinary Portland cement and Cem-FIL glass with possible additives in the form of fillers and binders. Cem-FIL is manufactured by Pilkington Brothers Limited of St. Helens and a number of manufacturers are licenced by them to produce composites. Bingham Blades & Partners have acted as consulting engineers on various projects involving GRC during the past four years and techniques utilising the material have been evolved.

The role of GRC in engineering is controlled by the license arrangements to produce a steady growth through non-structural and semi-structural components to eventual use of the material for major load-bearing elements. At present there is, quite rightly, an embargo on the use of any component whose failure could initiate a structural collapse. The reason for caution is simply the relative youth of the material and the lack of precise knowledge about long-term behaviour. The data from tests on samples which are several years old indicate satisfactory behaviour with a decline in performance that has a rate of change which decreases with age. All structural materials deteriorate to some degree in a time-dependent manner and the extreme care with which GRC is being monitored is a recognition of this sometimes neglected fact.

One of the semi-structural applications is for cladding panels and in addition to a number of designs for these, the authors have utilised GRC for air-ducts, air-intake louvres, fire resistant panels and permanent shuttering. The paper considers appropriate design methods for cladding panels and discusses some of the problems which have been encountered in practice, such as manufacturing techniques, fixing details, waterproofing seals, and thermal insulation.

PANEL DESIGN

The design of GRC proceeds from a knowledge of its basic properties under tensile, compressive, bending and shear forces, coupled with estimates of behaviour under secondary loading effects such as creep, and thermal and moisture movement. There are many unknowns, particularly with regard to the nature of long-term effects, and the properties of prototype units can vary enormously from those of laboratory sepcimens due to difficulties with production control.

Nevertheless a sensible set of properties can be decided upon and could normally be as follows—

limit of proportionality	9.0 N/mm^2
ultimate tensile strength	16.0 N/mm^2
modulus of rupture	40.0 N/mm^2
ultimate shear stress (interlaminar)	1.7 N/mm^2
ultimate shear stress (in plane)	8.2 N/mm^2
density	2300 kg/m^3

These values are for a typical mix of ordinary Portland cement with 5% of Cem-FIL glass by weight and a water/cement ratio of 0.3 and have been obtained after curing the composite for 28 days.

Two possibilities are then open, either to use a simple factor of safety or to adopt a limit state philosophy. Early designs used a factor of 3 so that the working stresses for the panels became—

working stress in bending	±3.0 N/mm^2
working stress in shear	0.6 N/mm^2
working stress in bond	0.7 N/mm^2

GRC can also be designed by assessing limit states for failure and for serviceability and by applying appropriate factors to loads. A more rational approach then becomes apparent. The factors given in CP110 for structural concrete are being used for loads in GRC at present but rather conservative values are taken by the authors for ultimate stresses.

Sophisticated methods of analysis are also worth consideration because of the complex structural nature of stiffened plates and shells. The authors have on occasion used finite element programmes with provision for plate and beam elements in 2 and 3 dimensions and generally work with around 100 elements per panel. Such an approach tends to give economy of material when compared with simpler approaches, provided that the results are interpreted with care.

Dead weights are calculated on the basis of a composite density of 2300 kg/m^3 while wind loads are computed in the normal manner. With unusually shaped or lightweight panels there is the possibility of wind excited oscillation and this may require care on the part of the designer who is not accustomed to such phenomena.

The coefficient of linear expansion and contraction is taken at present as 14×10^{-6} per °C which is the same as that for neat cement. Shrinkage and creep are both exhibited by GRC but calculations as to their effects are approximate and to allow for the unexpected, provision must be made in the fixings for contraction or expansion both vertically and horizontally. Deflection under load should also be calculated and a span/deflection ratio that does not exceed 500/1 can be expected to prevent any cracking of finishes or surfaces.

As a final precaution and as a check on the manufacturing process, a prototype unit for each design should be load-tested. The requirements for such conservatism will disappear in time but prudence is recommended because the two factors which most commonly lead to structural failure are the use of a new material or a new design method. Worse still could be a combination of the two.

SINGLE SKIN AND SANDWICH PANELS

Cladding units can be fabricated using either single skin or sandwich panel composite construction. In the first approach the panel acts as a fascia which must be capable of withstanding stresses applied during erection and handling as well as self-weight and wind loadings. It may also have to provide the required integrity against fire. If the second approach is adopted the panel can be used as a complete wall system which not only keeps out the weather but provides the insulation to the building and also forms the interior wall finish. In both cases the spray-up technique is used during manufacture.

The most economical way of utilising single skin construction is to make the skin itself fairly thin and stiffen the panel face with ribs. In practice a minimum skin thickness of 6 mm can be sprayed, but it is often better to use a mean thickness of 10 mm to allow for surface variations. On panels incorporating an exposed aggregate finish the thickness must obviously be increased. Ribs may be solid, but it has been found that box-sections can be formed by moulding a 'green' GRC sheet round an infill material placed on the skin, and then over-lapping the edges onto, and intermeshing them into, the 'green' GRC skin.

In sandwich panel construction skin thicknesses of 10 mm minimum have again been used. It is obviously essential in this approach that the two skins are monolithically connected. The authors recommend linking the skins with thin GRC ribs and using the infill purely to provide the required insulation. Some authorities have proposed bonding the skins to an infill block after the GRC has cured and using this to carry shear forces between the skins. Even if a suitable incombustible inorganic material is available for such purposes it is doubtful at present whether the high degree of workmanship necessary to provide structural interaction can be achieved economically. Difficulties arise both in the preparation of the surfaces and in the selection and application of suitable bonding agents.

If the inner face of a sandwich panel is to provide the final surface finish, very tight tolerances must be attained. In one instance for example, slight 'bowing' of the panels occurred due to insufficient propping of the mould during spraying. This distortion would be unnoticeable on the exterior surface, but it became a problem on the inside face where heating ducts ran the full length of the wall flush with the panels.

Further difficulties of 'bowing' occurred with channel-section single skin panels on another project, and in this case the source of the problem was inadequate curing facilities. Three fixings were incorporated down the height of the panel and cracking of the bowed panels ensued when the third bolt was tightened. With folded plate panels structural redundancy of fixings should be avoided.

FIXINGS AND JOINTS

Fixings into GRC present different problems from those encountered in the detailing of traditional reinforced concrete panels. In reinforced concrete it is common practice to incorporate a cross dowel through the end of cast-in sockets and tie the dowel to the reinforcing bars.

Experiments with GRC indicate that the dowel has little effect on the pull-out load taken by the socket but may serve a purpose in that it prevents the socket turning when the bolts are being tightened. The dowel may however impair the performance of the fixing by destroying the flow of fibre round the socket during spraying operations.

Blocks of GRC are generally incorporated in the panels within which the fixings are set. From test results and theoretical information available, the authors have used a minimum block size of eight socket diameters. As a matter of good building practice stainless steel fixings and bolts are always specified and care is taken to avoid dissimilar metals which could lead to electrolytic corrosion.

Bond strength to fixings is dependent on the ability to spray around the socket. In one case it was found that the use of a common type of window fixing block resulted in a congestion of fibres without any cement matrix. Fibre bridges across the corner of the

Figure 1   Kingston Fairfield West — skin-construction panel.

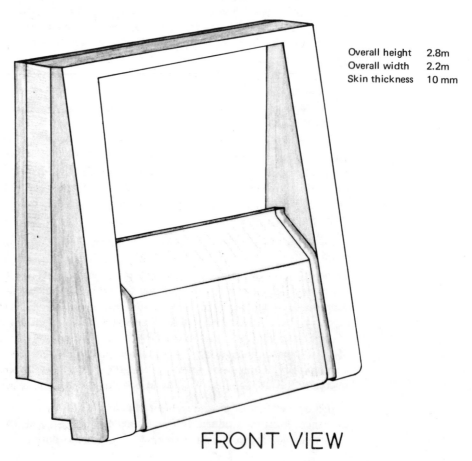

Overall height   2.8m
Overall width   2.2m
Skin thickness   10 mm

FRONT VIEW

Figure 2   **Stourbridge car park – skin-construction panel.**

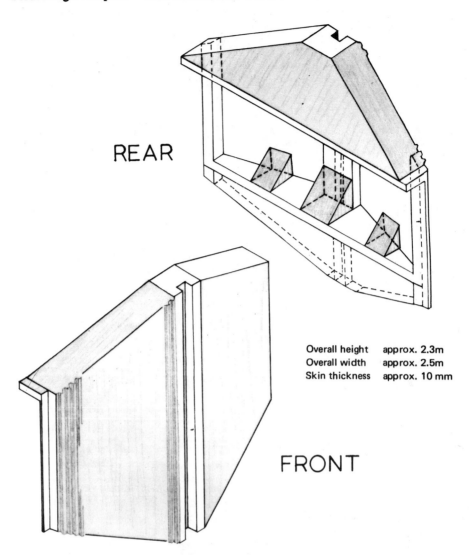

25 mm cube block and the outer face of the window opening thus prohibited the flow of cement. For this reason the working bond stress used by the authors is conservative. A value of 0.7 N/mm^2 has been adopted which is far lower than results obtained under laboratory conditions where an even flow of cement and fibre has been ensured.

Because of fibre bridging it has been found preferable to removable sharp angles and fine projections whenever possible and incorporate radiused curves. Intricate surface details on panels initially envisaged in reinforced concrete require re-thinking before GRC is adopted as a substitute material.

Grooves and slots also present problems in GRC components because of the bridging action. This has influenced the treatment of joints between precast cladding panels. A conventional open-drain joint, favoured under most circumstances in reinforced concrete construction, requires a deep slot to take the baffle and a saw-cut effect in front of the baffle on the panel to drain the water outwards. Difficulties in spraying these features has led GRC manufacturers to prefer butt joints sealed with polysulphides. Further work must be done on joint details because surface to surface bonding with mastics as the sole means of keeping out water may create long-term problems.

Figures 1-5 show typical panels which have either been made or have been tendered for by Glass Reinforced Concrete Limited of Northwich and with which the authors

Figure 3   Felling Security Centre – sandwich-construction panel.

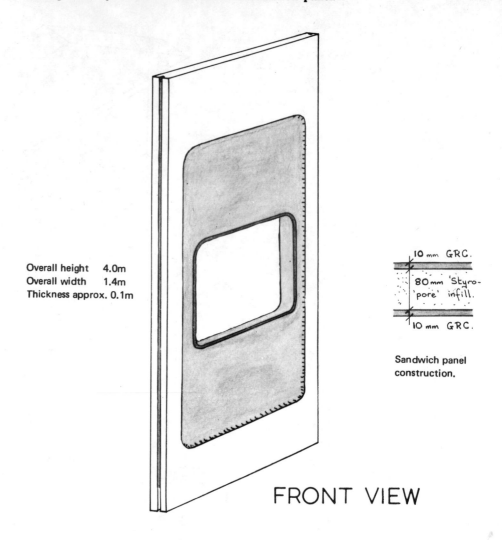

have been involved. Figure 1 shows a single skin panel which was used for an office development while Figure 2 indicates a heavily modelled facade unit for a car park.

A typical sandwich panel is drawn in Figure 3 and the central core is made from 'Styropore', a lightweight material. As a combination of single skin and sandwich construction, Figure 4 illustrates a facade panel. Typical fixing details are shown in Figure 5 which indicates the need to allow for expansion and for tolerances in the three directions to ease erection, to enable the panels to be lined up, and to allow for future movement.

FIRE RESISTANCE

Certain fire tests on GRC panels have shown that with composites based on ordinary Portland cement and Cem-FIL there can be variations in the fire resistant properties. With the incorporation of suitable inorganic fillers to allow entrapped moisture to escape from the composites the resistance to penetration by flame becomes impressive. Insulation against thermal transmission is also improved. In one recent test on a single skin panel the outer face was still at a comfortable hand temperature whilst the inner face had been at the furnace temperature of approximately 900°C for one and a half hours. Initial spalling occurred during the first quarter of an hour but subsequently no damage to the panel was observed. Work on suitable fire resistant mixes is being carried out by various bodies and it is anticipated that the performance of GRC in fires will become an important property.

Figure 4   **Police Traffic Headquarters—Liverpool – skin-construction panel.**

Overall height   approx. 1.3m
Overall width   approx. 4.8m
Rib size approx. 100 mm x 100 mm

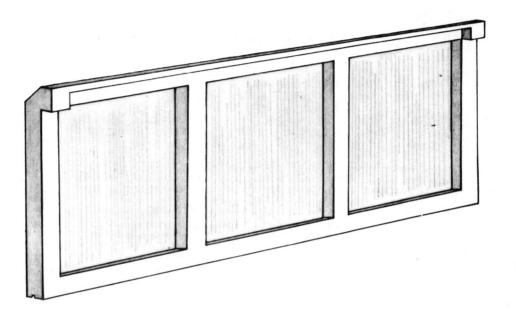

REAR VIEW

Section through a typical rib.

CONCLUSIONS

In any comparison between the relative advantages of GRC and alternative materials it must always be borne in mind that information on the long-term effects of weather and atmospheric corrosion is limited.

Caution must also be exercised when applying design data obtained from laboratory tests to practical problems. The behaviour and rigidity of a panel of complex shape is dependent to a large extent on the spraying technique. Wads of unbonded fibres at a critical position on the structure could be disastrous.

GRC composites display remarkable properties in respect of impact strength, tensile strength and fire resistance and have obvious advantages over reinforced concrete in the correct applications. However the material would appear to be less adaptable to complex facial patterns and requires care during manufacture, handling and erection if it is to be used to its full potential.

With the growth of knowledge and greater confidence in its use, GRC can be expected to provide an exciting addition to the materials available to the construction industry.

Figure 5   **Kingston Fairfield West – typical cladding panel fixing detail.**

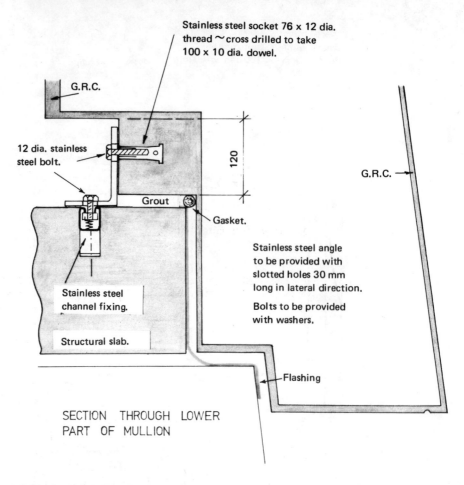

ACKNOWLEDGEMENTS

The authors thank Messrs Pilkington Brothers Ltd, manufacturers of Cem-FIL, and Glass Reinforced Concrete Limited, fabricators of the panels illustrated in the text, for permission to present this paper. Thanks are also given to the architects, whose designs are illustrated in the figures.

REFERENCES

1. Allen, H G, "Stiffness and strength of two glass-fibre reinforced cement laminates", Journal of Composite Materials, Vol 5, April 1971, pp 194-207.
2. Majumdar, A J and Ryder, J F, "Glass fibre reinforcement of cement products", Glass technology, Vol 9, 1968, pp 78-84.
3. Soane, A J M, "Glass fibre reinforced cement composites", International Congress of Precast Concrete Industry, BIBM, Barcelona, May 1972.
4. Glass Fibre Reinforced Cement Technical Manual, Pilkington Brothers Limited, St Helens, Merseyside, 1974.

8.9 Fort hood fibre concrete overlay

Gilbert R Williamson

Construction Engineering Research Laboratory, Champaign, Illinois, USA

Summary

The first full scale non-experimental steel fibrous concrete placement was completed in March, 1974, at Fort Hood, Texas, USA. Two thousand four hundred and twenty-six cubic metres of concrete were placed as an overlay of a Tactical Equipment Park that is used for the maintenance and repair of combat tanks and other tracked vehicles. 100 to 130 mm 1.5 volume percent of steel fibrous concrete was placed over an existing asphaltic concrete pavement. A 0.3 × 12.7 mm round fibre was selected for the project on the basis of low bid. No mixing or placing problems were encountered with this fibre. The overlay was designed for an edge load of 59 421 kg, and the flexural strength requirements of the concrete were 7.5 N/mm^2 at 28 days, and 9.0 N/mm^2 at 90 days. Early evaluation indicates that the overlay is performing well with only one working crack, approximately 4 m long, in the entire pavement.

Résumé

La première installation de béton acier fibreux non expérimentale, à grande échelle, fut réalisée en mars 1974 à Fort Hood, Texas, USA. Deux mille quatre cent vingt-cinq mètres cubes de béton furent installés pour la hausse d'un parc d'equipement tactique utilisé pour l'entretien et les réparations de tanks de combat et autres véhicules de pistes. De dix à treize centimètres de béton acier fibreux à 1.5% de volume furent installés par-dessus le revêtement de béton asphalte existant. Une fibre de 0.3 × 12.7 mm fut sélectionnée pour le project, sur base de sous-enchère. Aucun problème de mélange ou d'installation ne s'est posé pour cette fibre. La hausse fut désignée pour une charge limite de 59 421 kgs, et les exigences de puissance de flexibilité du béton furent de 7.5 N/mm^2 à 28 jours, et 9.0 $N \cdot mm^2$ à 90 jours. Une évaluation précoce indique que la hausse réagit bien, avec seulement une crevasse s'étalant sur 4 m de long pour tout le revêtement.

INTRODUCTION

Definition

Background. The first scale non-experimental steel fibre concrete placement was completed in March, 1974 at Fort Hood, Texas, USA. Two thousand four hundred and twenty-six cubic metres of concrete were placed as an overlay of a Tactical Equipment Park. The existing pavement consisted of 120 to 180 mm of asphaltic concrete on a stabilized limestone base. The area is used for maintenance of M-60 tanks, M-88 tank retrievers, and other tracked vehicles. Because of the severe wear engendered by the action of the tracks, the pavement has required replacement every 3 to 4 years. Fibrous concrete was selected as a more lasting solution to the problem.

Fibre concrete is conventional concrete with randomly dispersed fibres of short length and small cross-sectional area. Materials commonly used as fibres include steel, glass, asbestos, nylon, polypropylene and other synthetic and natural materials. Steel fibres are the most frequently used because of their wider application and overall superior properties. The use of glass fibres has been increasing, but their use has been primarily in the area of non-pavement type applications; ie surface bonding of blocks, exterior building panels and extruded shapes. However, several experimental overlays of slab-on-grade projects have proven the feasibility of using glass fibres in pavements.

Properties and Design Considerations

The use of steel fibres in concrete or mortar will result in a material of improved physical characteristics. The percent of improvement varies with the amount of fibres in the mix. Table 1 shows the percent of improvement that can be expected over plain concrete for several basic properties, for a 1.5 volume percent of steel fibre.

The values of Table 1 are based upon hundreds of tests. Several theories for predicting the strength characteristics of fibrous concrete have been advanced, but all are in a very early stage of development. Ordinary pavement design procedures can be used to determine thickness requirements of pavements and overlays, when appropriate modifications have been made to account for the improved characteristics of the material.

Table 1   Properties of fibrous concrete

Property	Percent increase over plain concrete
First crack flexural strength	30 - 50
Ultimate modulus of rupture	70 - 100
Ultimate compressive strength	10 - 25
Flexural fatigue endurance limit	150 - 200
Impact resistance	200 - 300
Freeze/thaw durability index	100 - 200

SITE PREPARATION

The pavement overlayed at Fort Hood consisted of 120 to 180 mm of asphaltic concrete on a 180 mm lime stabilized base. The surface showed considerable deterioration in several areas (Figure 1), and required complete rebuilding. The following field tests were conducted: 15 plate-bearing tests, 1 CBR on subgrade, and 1 in-place density test on the base course material.

The subgrade modulus, k, as determined from the plate-bearing tests varied from 25 MN/m^3 to 189 MN/m^3, with only three values falling below 90. The design value for k was selected as 94.4 MN/m^3. This value is 1.19 standard deviations below the

Figure 1 A partial view of the site at Fort Hood.

mean value of the remaining data. This provides 90 percent confidence that not more than 25 percent of the foundation area will exhibit k values less than the design value. The 3 areas with low k values were upgraded to this standard. Upgrading consisted of removing all surface, base and sub-base material and replacing to the approved standard. Once all distressed areas had been upgraded, the only additional site preparation consisted of brooming prior to placement of the overlay.

DESIGN CONSIDERATIONS

Overlay thicknesses can be determined by any approved method, such as that given in US Army Manuals TM5-822-6, and TM5-824-3. As stated in these manuals, the slab thickness necessary to provide the desired load-carrying capacity is a function of five principal variables: (a) Vehicle wheel load or axle load, (b) Configuration of the wheels or tracks, (c) Volume of traffic during the design life of the pavement, (d) Modulus of rupture (flexural strength) of the concrete, (e) Modulus of subgrade reaction.

The higher flexural strengths and improved fatigue characteristics of fibrous concrete require a modification in the mathematical expressions and design aids currently used in the pavement and overlay design. These modifications are primarily a judgement of the designer.

Thickness Design The design vehicle for the Fort Hood project was the M-60 tank. The gross load of this vehicle is 47 628 kg. A 25 percent impact factor brought the total design load to 59 421 kg. This load was divided equally between six bogies, and each bogie load was distributed to four grousers. The vehicle was then oriented perpendicular to the edge of the pavement, or the pavement joint. Because fibrous overlays are very thin (up to 50 percent thinner than plain concrete), it was considered more economical to design the slab adjacent to joints as if they were simple edges. This eliminated the use of load transfer mechanisms. The design index was 8, based upon 10 vehicle operations per day. This resulted in 91 250 load repetitions for the 25 year design life. The subgrade modulus design value was 94.4 MN/m^3. The flexural strength requirements of the concrete were 7.5 N/mm^2 at 28 days and 9.0 N/mm^2 at 90 days. A computerized edge-loading Westergaard static analysis was conducted to determine the flexural stress, slab deflection, and subgrade stress. Thin fibrous concrete overlays on non-rigid pavements or slabs-on-grade can result in excessive deflections that may produce pumping. Careful considerations must be given to determining the deflection of the system.

Although load transfer devices were considered unnecessary, dowels were used to tie all exterior lanes to the adjacent lane. This was done to prevent movement of the outside lane slab when tanks and other vehicles were stopped quickly when approaching the edge of the overlay perpendicularly. Joints were sawed every fifty feet to a depth of one-half the slab thickness.

MIX DESIGN

The mix design used on the Fort Hood Project is given in Table 2. It was developed at CERL using aggregates from the job site. The use of fly ash permitted the specifying of two flexural strength requirements, 7.5 N/mm² at 28 days and 9.0 N/mm² at 90 days. This is due to the high strength gain of the concrete between 28 and 90 days when fly ash is present. Both values were used in the thickness design.

Table 2   **Fibrous concrete mix design, cubic metre**

Constituent	Amount, kg
Cement	235
Fly ash	105
Sand	612
0.95 cm Aggregate	612
Fibers	91
Water/Cement Ratio	0.56
Air	4-6 percent
Water reducer, per mfg. rec.	

CONSTRUCTION PROCEDURES

The 0.3 × 12.7 mm round fibre was selected for the project on the basis of low bid. The final mix design was prepared by the Corps of Engineers Southwestern Division Laboratory. The general mixing and placing procedures were as follows: three cu. m of concrete per batch were mixed in a 3.44 cu. m stationary mixer. Conventional concrete mixer trucks then delivered 6.1 cu. m loads to the pavement site.

Fibre Incorporation

The arrangement for incorporating the fibres into the mix is shown in Figure 2. Each 3 cu.m batch required 363 kg of fibres. The fibres were dumped from the packing boxes onto a table with a grizzly at one end. The fibres were raked through the grizzly (to break up lumps) onto a conveyor belt that deposited the fibres on the main charging belt at a point just outside the mixer. The fibres were charged at the same time as the aggregate, and were all in the revolving mixer before the cement, fly ash and water were beginning to be charged. Fibre balling was essentially non-existent on this project. This was attributed to the following: good mixing characteristics of the fibre; placement of the conveyor and the charging belt so as to minimize the direction change of the fibre; charging the fibres with the aggregate; and the use of a high-speed (107 m per min.) charging belt that literally threw the material back into the mixer, rather than just dumping the material at the opening where it would then have to be wound down into the mixer producing a good condition for balling of the fibres. Once the batching sequence was established, satisfactory results were achieved throughout the project.

Placing and Finishing

The overlay was placed in 6.1 m wide lanes with the use of a Cleary paver riding on rails, see Figure 3. Areas adjacent to the building and in the vicinity of the grease racks were placed and finished by hand. No difficulties were experienced with either method. A 100–130 mm slump was maintained, however, a lower slump is recommended for

Figure 2 Fibres being raked through a grizzly onto a conveyor belt.

slip-form type pavers. Finishing consisted of several passes with a bull float followed by brooming. A burlap drag cannot be used with fibrous concrete because the fibres are pulled out by the burlap. Texturing by rollers will also produce satisfactory surfaces. Because of the contractor's preference, wet curing was used for this project; however, curing compounds are usually specified, and are recommended. Joints were cut every 15.2 m as soon as practical.

Figure 3 The overlay was placed with the use of a Cleary paver.

Testing

Standard Corps of Engineers flexural, split tensile and compression test specimens were made at specified intervals for both field and laboratory testing. Air content and slump measurements were taken at the same time.

Table 3 shows the results of several laboratory tests and indicates clearly that a 20 percent strength gain from 28 to 90 days is attainable with fly ash.

Table 3   **Typical flexural and compression test results of Fort Hood Overlay Project**

Test set no	Air percent	Slump mm	Flexural strength, N/mm^2			Compressive strength N/mm^2
			7-day	28-day	90-day	28-day
1	5.3	135	6.84	6.88	8.96	50.87
2	5.8	147	5.87	7.26	7.53	47.40
3	6.5	165	5.42	6.70	7.47	45.90
4	4.5	114	5.42	6.77	8.51	43.06
5	4.4	89	5.49	7.19	7.99	48.99
6	4.2	102	6.63	8.09	8.82	50.07
7	4.4	89	6.46	8.65	8.58	49.20
8	2.5	95	5.49	6.67	8.78	45.31
9	3.7	89	6.35	7.12	8.72	50.73
		Mean	5.92	7.26	8.37	47.92
		Std. Dev.	71	98	81	390
		CO Variation	.0835	.0935	.0676	.0565

NOTE: All values are the average of 3 tests.

The specimens for this set of tests were taken from areas of low subgrade modulus, and will be used for long range evaluation of the overlay. The field tested specimens had an average flexural strength of 7.6 N/mm^2 at 28 days and 9.1 N/mm^2 at 90 days.

DISCUSSION AND CONCLUSIONS

Discussions with both the batch plant operator and the paving foreman have reinforced the claim that the mixing, placing and finishing of fibrous concrete is very similar to that of plain concrete. Once the proper mixing sequence has been established to eliminate fibre balling, most other methods and techniques are applicable.

Early Evaluation

Only a minimum amount of cracking has appeared in the overlay, and this all occurred within a few days after placing, before being opened to traffic. Of the several thousand square metres of overlay placed by hand, only one crack, 2.4–3.0 m long, had appeared. Several transverse cracks appeared in the machine placed areas, but the most common crack was a longitudinal one in the centre of the paving lanes. Sixty to seventy metres of this type of cracking occurred. All of the cracks could be classified as hairline, and none appeared to be working cracks once the overlay was opened to traffic.

A survey of the overlay after being opened for traffic for 9 months has revealed only one working crack. It is across a 3.6 m hand-placed lane, and has opened up about 6.4 mm. The hairline cracks that developed shortly after construction remain tightly closed. The overall performance of the slab is very satisfactory.

Costs

The actual labour and material costs for the 2 426 cu. m of fibre concrete placed was $93.00 per cu. m. This does not include the contractor's profit. The fibres were purchased for approximately 42 cents per kg, and the cost of incorporating them into the mix was approximately 3.3 cents per kg. If a 15 percent profit is added to this cost, the total price would be approximately $107.00 per cu. m. This cost compares favourably with the $101.00 per cu. m cost of other overlays of 200 to 250 mm thickness at Fort Hood. (These thicknesses were designed for loadings similar to the fibre concrete overlay.)

The actual cost of the fibrous concrete at Fort Hood confirms the cost achieved on other projects. However, it is still difficult to get contractors to bid this material at a reasonable price because of the fear of the balling problem, and ignorance of the placing and finishing characteristics. Once contractors become more familiar with fibrous concrete, the competitive position of the material will improve.

Conclusions

a. Fibre concrete is a practical and cost effective overlay material. It can be mixed, placed and finished with conventional concrete equipment.

b. Trial mixes must be conducted with the equipment at the job site to determine the proper charging techniques in order to eliminate fibre balling.

c. The lowest cost fibre that is consistent with mixing and flexural strength requirements should be used.

ACKNOWLEDGEMENT

The overlay design and the original investigative work was done by the US Army Construction Engineering Research Laboratory (CERL), Champaign, Illinois, USA, from which the data for this paper were obtained.

1. Williamson, GR and Gray, BH, Technical Information Pamphlet on Use of Fibrous Concrete, Preliminary Report M-44, Construction Engineering Research Laboratory (CERL), Champaign, IL, May 1973.
2. Gray, BH, Williamson, GR and Batson, GB, Fibrous Concrete, Construction Materials of the Seventies, Conference Proceedings M-28 (CERL), December 1972.
3. Schwarz, AW, Steel Fibre Reinforced Concrete, United States Steel Corporation, Pittsburgh, PA, 1973.
4. Rigid Pavements for Airfields other than Army, US Army Technical Manual TM-5-824-3.
5. Rigid Pavements for Roads, Streets, Walks and Other Open Storage Areas, US Army Technical Manual TM-5-822-6.